Florida's Sandy Beaches

An Access Guide

Beach sunset

Florida's Sandy Beaches

An Access Guide

Published with the assistance of the Division of
Beaches and Shores, Department of Natural Resources,
and the Division of Coastal Management, State of Florida
Department of Environmental Regulation. Additional funding
was provided by the National Oceanic and Atmospheric
Administration.

University Presses of Florida
University of West Florida Press/Pensacola

UNIVERSITY PRESSES OF FLORIDA is the central agency for scholarly publishing of the State of Florida's university system, producing books selected for publication by the faculty editorial committees of Florida's nine public universities: Florida A&M University (Tallahassee), Florida Atlantic University (Boca Raton), Florida International University (Miami), Florida State University (Tallahassee), University of Central Florida (Orlando), University of Florida (Gainesville), University of North Florida (Jacksonville), University of South Florida (Tampa), University of West Florida (Pensacola).

ORDERS for books published by all member presses of University Presses of Florida should be addressed to University Presses of Florida, 15 NW 15th Street, Gainesville, FL 32603.

Copyright © 1985 by the Board of Regents of the State of Florida

This publication was funded by the Department of Environmental Regulation, Office of Coastal Management, 2600 Blairstone Road, Tallahassee, FL 32301, through a grant from the United States Office of Coastal Resource Management, National Oceanic and Atmospheric Administration, under the Coastal Zone Management Act of 1972, as amended.

Printed in the U.S.A. on acid-free paper

Library of Congress Cataloging in Publication Data
Main entry under title:

Florida's sandy beaches.

"Beach Access Project—Office of Coastal Studies, University of West Florida:—p.
  Includes index.
  1. Florida—Description and travel—1981- —Guidebooks.
  2. Beaches—Florida—Guidebooks. 3. Recreation areas—Florida—Guidebooks. 4. Parks—Florida—Guidebooks.
  I. University of West Florida. Office of Coastal Studies.
  F309.3.F53 1985   917.59′0463   84–29922
  ISBN 0–8130–0820–4 (alk. paper)

Beach Access Project
Office of Coastal Studies
University of West Florida

David W. Fischer
Principal Investigator

Jerome F. Coling
Chief Cartographer

Donald E. Henningsen
Project Director, Field Work

Deborah Joy
Chief Writer, Layout

Staff

Cartography
Ernest Barnett
Robert M. Brown

Craig Shipley
Michael Brown

Coordination, Research
Mary Morris

Illustrations
Peggy Riedell
J. Seagle

Administration
Donald Whitman

Writing
David Sengenberger (April–August 1983)
Mary Morris

Secretaries
Maureen Waldron
Sandra Newell

Indian Rocks Beach, upper Pinellas County

Acknowledgments

We gratefully acknowledge the use of the path-finding publication the *California Coastal Access Guide*, prepared by the California Coastal Commission and published by the University of California Press in 1981. This work both helped to cement our idea of the need for an access guide and aided us in designing our own effort. Madge Caughman from the California Coastal Commission acted as a consultant for our project by helping us in arranging the format.

In Florida we acknowledge the early support of State Senator W. D. Childers and his staff for their interest in helping to fund our project. David Worley, director, Office of Coastal Zone Management, Department of Environmental Regulation, funded the project through a grant from the Federal Office of Coastal Resource Management, National Oceanic and Atmospheric Administration, under the Coastal Zone Management Act of 1972, as amended.

We especially appreciate the devotion and enthusiastic support of Deborah Flack, director, Division of Beaches and Shores, Department of Natural Resources, through whom this project was funded. Her office and staff, especially James Balsillie, the contract administrator, helped to secure timely information and aided us in many small but significant ways to ensure completion and publication.

The many county and city directors of parks and recreation units throughout Florida deserve our special thanks for furnishing us with information to complete this guide. We especially thank Walt Rothenbach, director, Sarasota County Parks and Recreation, and his staff, including Bob Cann. Sarasota County was chosen as our pilot study area, and these persons helped us to become acquainted with coastal park and access issues as well as to design our questionnaire.

The artwork was contributed by Peggy Riedell and J. Seagle, illustrators in the Division of Beaches and Shores, and the Marine Research Laboratory, Department of Natural Resources, respectively.

Finally, we appreciate the information and photos received from the Division of Tourism, Department of Commerce, and both the Bureau of Education and Information and the Marine Research Laboratory, Department of Natural Resources.

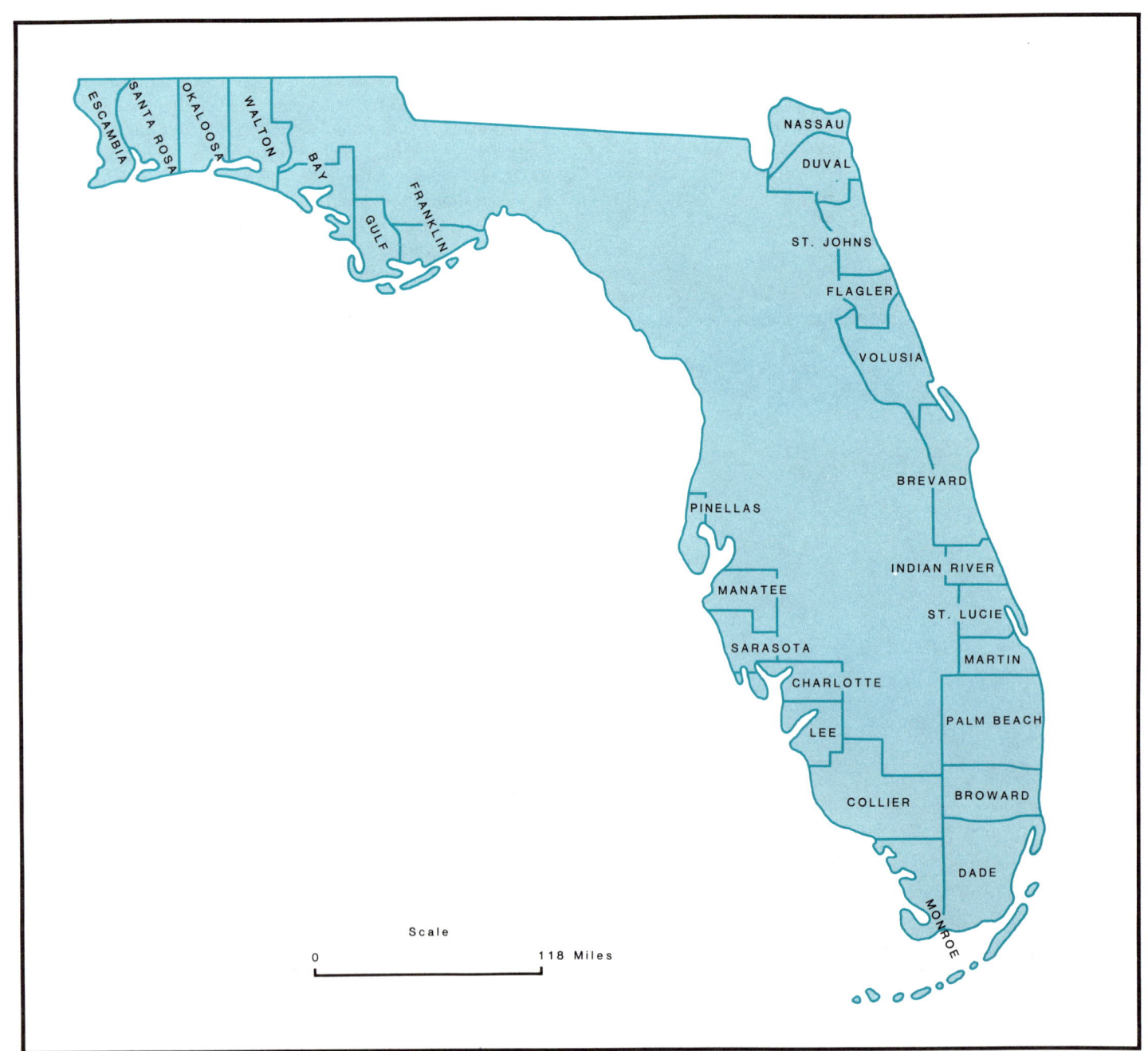

Florida's counties with sandy beaches

Contents

A Note on Sources	x
Foreword by Governor Robert Graham	xi
Introduction	xii
How to Use the Guide	xiii
Public Access	xv
Coastal State Parks	xvii
State Park and Recreation Areas	xviii
State Park Camping	xx
National Beach Areas	xxi

East Coast

Nassau County	3
Atlantic Intracoastal Waterway	6
Duval County	7
Drive-on Beaches	13
St. Johns County	15
Dolphins	20
Flagler County	23
Water Safety	28
Volusia County	31
Sea Turtles	39
Saltwater Fishing Piers	40
Brevard County	43
Indian River County	53
Coastal Management Program	58
St. Lucie County	61
Florida Seafood	66
Forest Zone	68
Martin County	69
Underwater Archaeology	73
Palm Beach County	75
Endangered Species Act of 1973	81
Broward County	83
Beach Erosion and Restoration	89
Dade County	91
Miami Beach Beachfront Park and Promenade	101
Monroe County	103
Coral Reef Parks	112

Southwest Coast

Pinellas County	115
Manatee	126
Manatee County	127
Marine Mammal Protection Act of 1972	133
Artificial Reefs	134
Sarasota County	135
Brown Pelican	142
Charlotte County	143
Florida's Bicycle Laws	146
West Coast Intracoastal Waterway	147
Lee County	149
Southwest Shell Collecting	157
Collier County	159
Mangrove	165

Northwest Coast

Escambia/Santa Rosa Counties	169
Okaloosa County	177
Florida's Saltwater Laws	183
Walton County	187
Bay County	191
Dune Grasses	201
Gulf County	203
Save Our Coast	207
Franklin County	209
Index	215

A Note on Sources

This guide was compiled from a variety of sources. The county descriptions were based on selected pamphlets, brochures, and documents sent to us by the 25 coastal counties, supplemented with library materials. The selections were based on interesting historical features involving the coast and key coastal features and activities that could fit on one page. No attempt was made to provide complete historical summaries. The maps were drawn from U.S. Geological Survey quadrangles. The directions to each beach were taken from the maps used to compile the guide maps, Florida Department of Transportation maps, supplemented with Department of Natural Resource aerial photos (1″ = 200′) and site visit notes. The beach names, public accesses, facilities available, and environmental descriptions were taken from survey forms filled out by county and city park and recreation officials. Each coastal county and the majority of the cities have published a parks and recreation section in their comprehensive plans; these were collected and used as a further source of information. In addition, each county and city was visited to understand the nature of the beaches, accesses, and special coastal features. The feature articles were based on information supplied by state agencies supplemented with library materials.

A draft copy was mailed to each county for verification of written material, maps, and grid information for that county. Twenty-one counties responded with recommendations; the four that did not respond indicated no changes.

Where we agreed with the corrections to the draft material, changes were made. As we went to press, changes in beach names, access points, and acquisition of new public beaches were taking place. Revisions will occur, updating information in the rapidly changing beachfront areas to assist those needing information about Florida's sandy beaches.

Foreword

The beautiful sandy beaches of Florida continue to attract Americans from all fifty states and tourists from other nations. This resource is being preserved so that future generations can also enjoy our white sand and warm seas and generally pursue outdoor recreation on our beaches.

In 1983 we began the acquisition of beaches for public use through the Save Our Coast program. People should be able to have access to the beaches while we work to preserve their fragile dunes and shoreline. This access guide should help in finding the major beach areas along our extensive coastline. The University of West Florida has produced a functional guide to assist in locating state, federal, and local coastal parks and recreation areas.

The feature articles elaborate on many coastal-related subjects. There are interesting summaries of what we have been doing to ensure that all visitors can enjoy our sandy beaches.

I would like to convey my gratitude to the Division of Beaches and Shores, DNR, as well as the Office of Coastal Management, DER, and the other agencies who have participated in the preparation of *Florida's Sandy Beaches, An Access Guide*.

Robert Graham, Governor

Introduction

Florida derives much of its income from tourism, which depends heavily on an understanding of and access to its beaches. Florida's beaches have different settings and character that influence how they are perceived and used. Some are well used, others less so; some are isolated, while others lie in the urban core. But all the beaches are designed for coastal recreation pursuits, although many are not used as intended or are overused.

This guide will assist in matching interests with beaches by showing where they are, how to get there, and what to expect on arrival. It is designed to provide most of the information needed for finding and using Florida's beaches. The authors and publisher are not responsible for changes in beach access and operations as their data came from the cities and counties involved and were checked by them in the spring of 1984.

The guide is divided into three major geographic areas—east coast, southwest coast, and northwest coast (Panhandle). The county descriptions begin at the northernmost county, Nassau, and continue down to Key West, the southernmost city in Florida. The only access to the keys is via Dade County on the east coast. For southwest coast visitors the guide begins at the northernmost county, Pinellas, and continues down to the Everglades. For those arriving from the west the guide begins with the westernmost county, Escambia, and continues to Franklin, the easternmost Panhandle county. To the east of Franklin County and around the bend of coastline to Pinellas County there are eight counties that have only limited sandy beaches, public access, and parking. This coastline is generally characterized by vegetation meeting the Gulf of Mexico rather than by sand and dunes. Because this guide is designed for sandy beach access and use, these counties are not included.

The coastal counties covered in the guide have been highlighted with selected historical episodes and interesting local features. Three maps are normally used for each county: a county map for orientation, district maps for general routing, and inset maps for urban areas needing more detailed routing. Accompanying the district maps is a grid denoting for each beach name the available parking and other transportation, facilities, and the character of the beach site. Supplementing the grid are entries for each beach, giving precise directions and information, such as hours of operation. The terms "developed" and "undeveloped" appear in some beach descriptions. Developed means that there are facilities available on the beach or close by for tourist use; undeveloped means facilities are generally lacking, even though the beach may be located in a highly urbanized area. Conversely, developed beaches are found in remote locations that have parking, showers, drinking water, toilets, picnic tables, etc.

Articles of general interest are also included to provide more information about beaches in Florida. Some discuss the beaches directly, others concentrate on beach activities. Photos and illustrations supplement the features.

All of the information used to compile this guide was accurate in the spring of 1984. As changes are made in the kinds of facilities offered and new public beaches and accessways are created, this guide will be updated. In addition, storms and hurricanes, such as the Thanksgiving storm in 1984, can change the beach contour, access, and facilities available.

How to Use the Guide

The guide is simple to use to locate a county, municipality, or specific area where there is a sandy beach. It begins its coverage in Nassau County at the Georgia border, on the Atlantic coast, and progresses southward down the Atlantic coast to Key West. On the southern Gulf coast, coverage begins in Pinellas County and moves southward to the Everglades. Northwest Florida's coverage begins at the Alabama border in Escambia County and progresses eastward along the Panhandle to Franklin County.

A grid has been included that provides a general description of the beach facilities and environments found in each county. Changes may have occurred after this information was compiled. The authors and publisher are not responsible for these changes. The example below illustrates how this information is displayed, and the glossary on the next page defines the terms used.

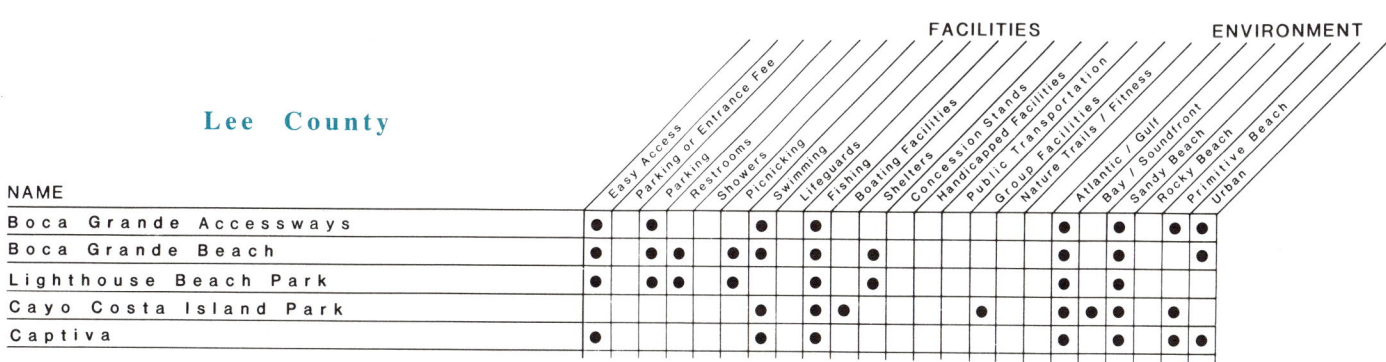

xiii

Facilities

Easy access. Indicates an entrance that is easy to locate and to use. In most cases there will be signs or markers identifying the access point as an entrance to the public beach area.

Parking or entrance fee. Identifies the public parking area as one requiring payment. In some areas it also means that an entrance fee must be paid for all beach users.

Parking. Identifies the availability of parking in the vicinity of the listed beach area.

Restrooms. Facilities provided for beach users close to the beach/swimming area.

Showers. Cold or hot showers provided.

Picnicking. Facilities provided for eating, such as picnic tables, benches, and in some areas fireplaces.

Swimming. Permitted in areas marked, with suitable sandy areas and water.

Lifeguards. On duty during posted hours and within the area indicated. Usually on duty only during the active swimming season.

Fishing. Areas where fishing is permitted if state and local laws are followed. Pier or surf casting permitted where noted.

Boating facilities. Indicates established facilities for boating ramps, piers, mooring, and docks.

Shelters. Indicates covered structures that can be used for eating and for protection in inclement weather.

Concession stands. Small- to medium-sized concession stands selling a limited variety of food.

Handicapped facilities. The area has some facilities to accommodate the handicapped, such as ramps, boardwalks, and special railings.

Public transportation. A public transportation system is available that services that beach area.

Group facilities. Larger covered structures holding more than ten persons that may be used for group activities.

Nature trails/fitness. Trails and fitness areas are associated with the beach area and facilities and are usually marked with signs and suggestions for use by the public.

Environment

Atlantic/Gulf. Indicates beach fronts the Atlantic Ocean or the Gulf of Mexico.

Bay/sound front. Indicates beach fronts a bay or sound.

Sandy beach. Fine-grained sand that is easy to walk on with bare feet.

Rocky beach. Small to large pebbles, or coral pieces.

Primitive beach. No facilities available; few or no homes in vicinity.

Urban. Highly developed areas with numerous motels, hotels, and restaurants adjacent to the beach area.

Public Access

The right of the public to use the shoreline has its beginning in ancient civilizations that enacted laws and codes to protect this use. Justinian (483–565 A.D.), a Roman emperor, first recorded the public trust doctrine, which said that the seashore and sea were "common to all" and could not be privately owned. This understanding of Roman law carried into laws of European nations. England's Magna Carta (1215) mentioned the public's right to fishing and navigation. There followed several English interpretations and clarifications of this premise that gave Queen Elizabeth I title to the tidal lands.

The early American colonists brought with them the English common law concept of keeping coastal access open for all new settlers. As the original American colonies matured, control of the tidelands was assumed by the states as a public trust. Public trust means the people of that state have placed the legal title of the public trust land with the government to protect their rights to its use.

Florida has legally separated private and public portions of the beach at the mean high-water line (average high tide) averaged over a nineteen-year period. The state's constitution says that the beaches below mean high water are held in trust for all the people. This stretch is often called the wet sand area.

State agencies are actively attempting to open up more of the coastline (the dry sand area) for public use through many continuing programs; one of the most recent is the Save Our Coast beach acquisition program. Local governments are funding programs to improve, increase, and post public accesses for the public's benefit. These accesses include not only the dry sand but crossings from public streets to the dry and wet sandy beaches.

The Department of Natural Resources is also funding major coastal projects that require local governments to provide beach access at half-mile intervals. Governments continue to add easements across private property to the beach for the flourishing tourist business and for the use of local citizens.

The legal precedents regarding access rights that go back to pre-colonial days still have effects on our beach access today. Four of the legal concepts affecting access rights are discussed briefly for further understanding of the right of access to public property held in trust.

"Implied dedication" means that in the case of a privately owned beach used by the public whose owner has not attempted to restrict the public's use, the owner is implying that he is giving the land to the public. The public is accepting the land by using it. Two requirements must be satisfied: the private owner's intent to dedicate the beach or accessway, and the public's acknowledgment and acceptance of the dedicated use.

"Express dedication" of a beach or accessway generally means that the public

XV

has been given the right to use the property as access over an extended period, and future governmental actions cannot restrict that use.

"Customary rights" regarding beaches are embedded in English common law. Once there has been public use of a beach for a long time, that use legally establishes the defined area as public, if these seven points are satisfied:

the use must have been without interruption;
the use must be so ancient that nobody remembers otherwise;
the use must have been reasonable and in keeping with the character of the land;
the use must have been peaceable and free from dispute;
there must be certainty as to just what land was being used;
the use must not be repugnant or inconsistent with public policy and other laws;
the use must have been obligatory for the upland owners—the public's use not being subject to the option of each individual upland owner.

"Prescriptive easement" means that in the case of the public using a private area without eviction by the owner over a period of time, it has a right to use it, and the owner cannot restrict the public right to use the easement.

Beaches are major attractions in Florida, and access will continue to receive attention by both state and local governments so that residents and tourists alike can share these magnificent natural resources. This access guide has been prepared to assist in locating Florida's public beaches and their access points. For continuing enjoyment of these beaches the public should obey posted access rules and should be careful not to infringe upon the rights of nearby property owners. The authors and publisher are not responsible for changes in access.

Coastal State Parks

Florida's natural environment lures many to escape the confines of a routine, urban life for the relaxation of the natural world. A mild climate, the nearness and expanse of the sea, and a spectacular assortment of plant and animal life have created a state ideal for outdoor recreation. Surprisingly, the establishment of an agency to regulate state parks in Florida was not actively promoted until after World War II. Until then the state was sparsely settled and relatively undeveloped, and the use of much private land was left unquestioned by the public. With time, however, it was realized that efforts needed to be made to preserve Florida against the overburdening development that could accompany an increasing population and growth in the state's largest industry—tourism.

The purpose of the Florida state parks system is to provide the public with natural outdoor resources that are representative examples of the "original natural Florida" seen by the first Europeans. In an effort to achieve this goal the following state park regulations have been established by the administrative offices of the Department of Natural Resources, Division of Recreation and Parks.

> There is an entrance fee to all parks; there is no additional charge for fishing, swimming, picnicking, or boating.
> Parks are open daily, opening at 8 A.M. and closing at sundown.
> Intoxicants may not be consumed in the park.
> Pets must be on a six-foot, hand-held leash and be well behaved at all times. They are not permitted in campgrounds, on swimming beaches, and at concessions.
> All plant and animal life is protected in state parks.

Cape Florida State Park, Key Biscayne

State Park and Recreation Areas

Anastasia State Recreation Area
St. Johns Co. (904) 471-3033
This is the "coquina" quarry site mined by the Spanish during the settlement of St. Augustine. The park provides access to a wide, drive-on beach and lagoon popular for fishing and skiing. It is reached off S.R. A1A in St. Augustine Beach. Located on the Atlantic Coast migration route, the park is also noted for its diversity of bird life.

Anclote Key State Park
Pinellas Co.
This isolated and primitive park can be reached only by private boat.

Bahia-Honda State Recreation Area
Monroe Co. (305) 872-2353
Noted for tarpon fishing, this is Florida's southernmost state recreation area. The park also promotes swimming and diving in both the Florida Bay and the Atlantic Ocean.

Big Lagoon State Recreation Area
Escambia Co. (904) 492-1595
From an observation tower the expanse of over 700 acres of coastal park and the view toward the Gulf across the lagoon and barrier island can be appreciated. Nature trails, swimming, and camping are park activities.

Bill Baggs Cape Florida State Recreation Area
Dade Co. (305) 361-5811
At the south end of Key Biscayne sits Cape Florida Light, one of the oldest structures in South Florida. This area was also the historic site of a confrontation with the Seminole Indians in 1836.

Caladesi Island State Park
Pinellas Co.
Having 607 acres of developed yet isolated land this park can be reached only by boat. A ferry service operates from the pier at Honeymoon Island State Park to the north.

Flagler Beach State Recreation Area
Flagler Co. (904) 439-2474
Guided nature walks and camping are popular activities at this 145-acre park. It is reached off S.R. A1A in Flagler Beach.

Fort Clinch State Park
Nassau Co. (904) 261-4212
Situated on Amelia Island, 3 miles north of Fernandina Beach on S.R. A1A, Fort Clinch has both historic and natural appeal. Its 1,086 acres have huge sand dunes and pristine beaches. Although built in 1847, the fort remains in excellent condition. Park rangers reenact history as Union garrison soldiers of 1864.

Fort Pierce Inlet State Recreation Area
St. Lucie Co. (305) 461-1570
This 340-acre park is located on the barrier island between the Indian River and the Atlantic Ocean. A museum at Pepper Beach interprets the local history of the Ais Indians, the 1715 Spanish Fleet disaster, the Seminole Wars (1816–58), and the lost American gold payroll.

Grayton Beach State Recreation Area
Walton Co. (904) 231-4210
The sculptured results of salt spray and wind can be seen in park barrier dunes and vegetation. Situated between the Gulf and scenic Western Lake this park is reached off the Miracle Strip (U.S. 98) between Ft. Walton Beach and Panama City. Surf fishing, swimming, and camping are popular activities.

Honeymoon Island State Park
Pinellas Co.
This undeveloped state park is located on the northern tip of Honeymoon Island. Having over 10,000 feet of beach with dune walkovers, it is reached by S.R. 586 west over the Dunedin Causeway.

Hugh Taylor Birch State Recreation Area
Broward Co. (305) 564-4521
Donated to the state in 1942, this 180-acre park was intended to preserve the solitude it had in the late 1800s. A miniature scenic railroad carries visitors through the isolated coastal hammocks and along the waterway. Rental canoes and paddleboats are available at the freshwater lagoon. Buildings are available for group camping.

John Pennekamp Coral Reef State Park
Monroe Co. (305) 451-1202
This is the first underwater state park and part of the largest living coral reef area in the United States. The park spans 8.5 miles into the Atlantic and 21 miles along the shore of Key Largo. A special feature of the park is a nine-foot bronze statue, "Christ of the Deep," symbolizing peace for mankind, located beneath 20 feet of Atlantic water.

John U. Lloyd Beach State Recreation Area
Broward Co. (305) 923-2833
This 244-acre developed park is named after the attorney who aided its acquisition. It offers excellent fishing at the Port Everglades jetties and scuba diving at an offshore reef.

Little Talbot Island State Park
Duval Co. (904) 251-3231
An isolated oceanfront park, Little Talbot Island covers 2,500 acres of beaches, dunes, tidal creeks, and salt marshes. Picturesque coastal hardwood lies in the northwest portion of the island. The park is about 17 miles northeast of Jacksonville, off S.R. A1A.

Long Key State Recreation Area
Monroe Co. (305) 664-4815
Small and isolated, this park is known for the wading birds that are a big attraction during the winter months.

Perdido Key State Preserve
Escambia Co. (904) 492-1595
Consisting of 250 acres of white sand, sea oats and rosemary-covered dunes, and pine flatwoods, Perdido Key State Preserve is located on the barrier island between the Gulf of Mexico and Old River. Activities in this undeveloped area are picnicking, swimming, fishing, and nature study.

Sebastian Inlet State Recreation Area
Indian River and Brevard Co. (305) 727-1752
Fantastic fishing and surfing are reasons this area is one of Florida's most popular recreation parks. It is located on the north and south sides of Sebastian Inlet.

St. Andrews State Recreation Area
Bay Co. (904) 234-2522
This park is known for its crystal white beaches, towering dunes, and clear blue-green Gulf waters. Fishing is allowed in the Gulf, Ship Channel, and Grand Lagoon. Access to this 1,063-acre state park is on S.R. 392, three miles east of Panama City Beach.

St. George Island State Park/Dr. Julian G. Bruce
Franklin Co.
This 1,883-acre park is located 10 miles southeast of Eastpoint off U.S. 98. It has 9 miles of beach access on the Gulf of Mexico, 16 miles on St. George Sound. Four miles inland at Gap Point is a primitive camp area.

St. Joseph Peninsula State Park/T. H. Stone Memorial
Gulf Co. (904) 227-1327
Almost completely surrounded by water, this state park has 10 miles of the most scenic stretch of unspoiled beaches and dunes on the Gulf coast. There is a 1,650-acre wilderness preserve where hiking and primitive camping is allowed.

St. Lucie Inlet State Recreation Area
Martin Co.
This 808-acre park will be open in early 1985.

Washington Oaks Gardens/"The Rocks"
Flagler Co. (904) 445-3161
Formal landscaped gardens and a unique beach studded with huge coquina boulders lie two miles south of Marineland. The beach, locally known as "The Rocks," is the only east coast beach of its kind and is the home of the endangered scrub jay.

Wiggins Pass State Recreation Area
Collier Co. (813) 597-6196
The beaches around Wiggins Pass are a popular nesting area for sea turtles during summer nights. Cast-netting for mullet is a popular fishing activity in the area.

State Park Camping

Camping is a popular outdoor recreational pursuit in Florida. In a sophisticated camping vehicle or in a sleeping bag under the stars, the traveler can enjoy an extended outdoor experience or a convenient overnight stop.

The majority of coastal state parks that offer the camping experience have specific camping sites situated near such developed facilities as electricity, water, bathhouse, tables, grills, and a park store. All these campsites are available year-round. Only half of all campsites may be reserved; the remaining half are available on a first-come, first-served basis. Calls for reservations must be made during park hours. Collect calls and mail requests will not be accepted.

Reservations will not be accepted more than 60 days in advance of check-in date.

Camping period may not exceed 14 days. Reservations will not be held after 5 P.M., unless officials are notified by phone of a later arrival. Check-out time is 2 P.M.

Camping fees are $6 per night per campsite at interior state parks. In coastal parks, $7 per night per campsite is charged; in the Florida Keys, the cost is $8 per night per campsite. Senior citizen rates are $3.

Four persons are allowed per campsite with an additional $1 per person or car; maximum of 8 campers and/or 2 cars per site. An additional $2 per night per campsite is charged if electricity is used.

Pets are not allowed in camping areas or on beaches.

An annual family camping permit is available to Florida residents for $100 per year. It does not include the use of electricity, museum fees, concessions, or tours.

Backpackers are required to register with the park office where they can also obtain hiking information. The camping fee is $1 per person per night; maximum of 12 persons per site. Camping is allowed only within designated areas.

State Recreational Areas (S.R.A.) and State Parks (S.P.)

Anastasia S.R.A.	(904) 471-3033
Bahia-Honda S.R.A.	(305) 872-2353
Flagler Beach S.R.A.	(904) 439-2474
Fort Clinch S.P.	(904) 261-4212
John Pennekamp Coral Reef S.P.	(305) 451-1202
Little Talbot Island S.P.	(904) 251-3231
Long Key S.R.A.	(305) 664-4815
St. Andrews S.R.A.	(904) 234-2522
Sebastian Inlet S.R.A.	(305) 727-1752

Summer Season (June 1–Labor Day)

Big Lagoon S.R.A.	(904) 492-1595
Grayton Beach S.R.A.	(904) 231-4210
St. Joseph Peninsula/T. H. Stone Memorial S.P.	(904) 227-1327

Two coastal state parks have the vast expanse of open country necessary for the isolation that makes primitive/backpack camping desirable to many enthusiasts:

St. George Island S.P./Dr. Julian G. Bruce, Eastpoint, Franklin Co. (904) 670-2111

St. Joseph Peninsula S.P./T. H. Stone Memorial, Port St. Joe, Gulf Co. (904) 227-1327

National Beach Areas

The National Park Service, U.S. Department of the Interior, has under its supervision in Florida six coastal areas that are readily accessible to the public. These include parks, seashores, monuments, and a memorial that are very distinctive and can be readily enjoyed by most tourists.

Canaveral National Seashore

The Canaveral National Seashore occupies parts of two counties, Volusia and Brevard, and is a neighbor to the John F. Kennedy Space Center. It has 26 miles of sandy beach separated into three areas: Playalinda, 5 miles; Apollo, 5 miles; and Klondike, 16 miles. This region is overlapped by the Merritt Island National Wildlife Refuge. Playalinda is the primary beach area that has some facilities and services for beach users. There is a snack bar, and picnicking is permitted on the beach. Drinking water is available at the headquarters building, but there are no beach showers. A specific area on Playalinda Beach has lifeguards during summer and weekends in the spring and fall. The beach area and road are often closed during space launches. Playalinda has 750 parking spaces. At the north end of the National Seashore, Apollo Beach has 175 parking spaces and dune crossovers. The area between Apollo and Playalinda is about 16 miles of beach where there are no facilities and few people. For additional information contact Superintendent, Canaveral National Seashore, P.O. Box 2583, Titusville, FL 32780, (305) 867-4675.

DeSoto National Memorial

On the southwest coast of Florida in Manatee County is DeSoto National Memorial, commemorating Hernando de Soto's landing in Florida in May 1539. How the sixteenth-century Spaniards lived—their weapons, food, and dress—is demonstrated from December through April. The DeSoto Memorial is located on the Manatee River 5 miles west of Bradenton. For additional information contact Superintendent, DeSoto National Memorial, 75th Street, NW, Bradenton, FL 33505, (813) 792-0458.

Gulf Islands National Seashore

One of the best known National Park Service beaches is the Gulf Islands National Seashore complex with beaches located in Escambia, Santa Rosa, and Okaloosa counties. This entire area is characterized by its sugar-white sand, dunes covered with sea oats, and the conservation measures that have been taken by the Park Service. Perdido Key has 14.2 miles of beach with unrestricted access and 250 parking spaces. Fort Pickens has 17 miles of beach and a parking area for 600 vehicles. On Santa Rosa Island there are 18 miles of beach with dunes and vegetation that extend to the bay. There are numerous picnic facilities, freshwater showers, restrooms, and open shelters. Lifeguards are on duty in the marked areas during the months of active summer use. There is no entrance fee. Further to the east in Fort Walton Beach there is a small beach and hobie cat launching area on Choctawhatchee Bay. The area is approximately 220 feet along the shore with room to park 50 vehicles. There are limited facilities and no lifeguards. An entrance fee is charged only at the Fort Pickens area. For information contact Superintendent, Gulf Islands National Seashore, P.O. Box 100, Gulf Breeze, FL 32561, (904) 932-5302.

Fort Matanzas National Monument

Interesting areas in the east coast county of St. Johns are the Fort Matanzas National Monument and the Castillo de San Marcos National Monument, a massive fort dating back to 1672 located in St. Augustine. Close by there is beach access for cars. Information regarding the forts and surroundings can be obtained from Superintendent, Fort Matanzas National Monument, 1 Castillo Drive, St. Augustine, FL 32084, (904) 829-6506.

Biscayne National Park

Biscayne National Park, located in the southern portion of Biscayne Bay in Dade County, consists of 175,000 acres of water, reefs, and barrier islands. Within this Park Service jurisdiction are three locations that provide information, guidance, and rules. Activities such as snorkeling, scuba diving, boating, fishing, and wildlife viewing have Park Service guidelines to be followed. Nine miles east of Homestead, on the mainland, is the Park Service headquarters at Convoy Point. Eight miles to the east, on the barrier island, is Elliot Key Harbor with a marina and campground. University Dock is approximately two miles to the north. The Elliot Key Park Ranger headquarters is at Elliot Key Harbor, and there is a ranger station on Adams Key, at the southern end of Elliot Key. The Biscayne National Park is famous for its living corals, tropical fish, shipwrecks, birds, and fascinating islands and shoals. For additional information contact Superintendent, Biscayne National Park, P.O. Box 1369, Homestead, FL 33030, (305) 247-2044.

Everglades National Park

Everglades National Park, on the southern tip of Florida, contains about 1,400,533 acres of land and water that also include part of the Ten Thousand Islands, where the average elevation is 5 feet or less above sea level. During the 1930s concern for the preservation of the Everglades finally led to the formation of Everglades National Park in 1947.

The earliest inhabitants of the Everglades were the pre-Columbian Calusa Indians. This is also the area where those remaining members of the Seminole Indian tribes retreated as the Europeans settled. Originally known as the "River of Glades," the name was changed to "Everglades" in 1823. While there is only wilderness within the park, the Everglades represents a unique tropical ecosystem supporting a wide variety of plants and animals. It is an important coastal area that has many island beaches accessible only by boat. Activities include fishing, picnicking, swimming, boating, shelling, and camping. No lifeguards are available. Contact Superintendent, Everglades National Park, P.O. Box 279, Homestead, FL 33030, (305) 247-6211.

East Coast

Historic Fernandina

Nassau County

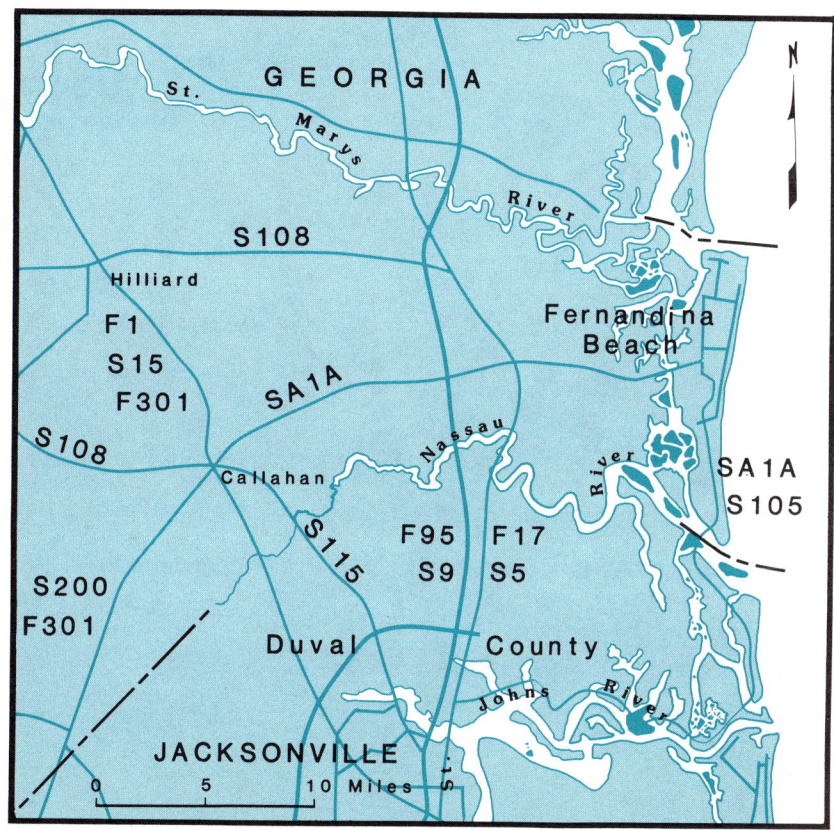

Nassau County lies in the northeastern corner of the state. The St. Marys River is its northern boundary with Georgia; the Nassau River is part of the southern border of Duval County. It is assumed that Bahamian immigrants, who came to this area during the British occupation, named the Nassau River after the principal town of the Bahamas; Nassau Sound and the south county town of Nassauville were also named during that period. When the county was established in 1824 the name followed local tradition.

Between the Intracoastal Waterway and the Atlantic Ocean lies Amelia Island. The first recorded visitor to the island was the French admiral and explorer Jean Ribault. He described the island as a land of great riches with wondrous harbors, fine soil, and a good climate. When he arrived in May 1562, the island was populated with Timucuan Indians. Ribault named the island "Isle de Mai" (Island of May) and claimed it as French territory, initiating a turbulent history of occupation and conquest. Spanish explorer Pedro Menéndez de Avilés arrived three years later to protest the French claim. The island was captured and renamed after the Santa Maria Mission. In 1702 the British destroyed the mission and overthrew Spanish Fort Fernando. The island was eventually claimed for England in 1735 by General James Oglethorpe and later renamed in honor of Princess Amelia, sister of King George II of England.

On July 10, 1821, Florida was ceded to the United States after many years of battling among England, Spain, France, and the United States. Three years later Nassau County was created, and Fernandina Beach was chosen as the county seat.

Fort Clinch was built in 1847 to protect the harbor. Its architecture is outstanding and exists in an excellent state of preservation. It became the first fort held by the Confederacy during the Civil War; however, it was evacuated in March 1862 when it was learned that a federal fleet was sailing to Fernandina. Fort Clinch is one of the finest military-nature seaside parks in Florida. It covers 1,086 acres on S.R. A1A, three miles north of Fernandina Beach, and overlooks Cumberland Island, a National Seashore Park in Georgia. The fort is operated under a living history program: park rangers, dressed in Union private uniforms, reenact the daily activities of the garrison soldiers of 1864 the first weekend of every month.

The City of Fernandina Beach has a 30-block historic district, Centre St. Ferdinand, facing the Amelia River. Victorian mansions and gingerbread residences dating from the 1850s, when Florida's first cross-state railroad ran from Fernandina to Cedar Key, are listed in the National Register of Historic Places. At the west end of Ash Street sits a renovated railway station serving as the Chamber of Commerce and the Marine Welcome Station for those traveling the Atlantic Intracoastal Waterway. The picturesque shrimp fleet marina is the birthplace of the modern shrimping industry.

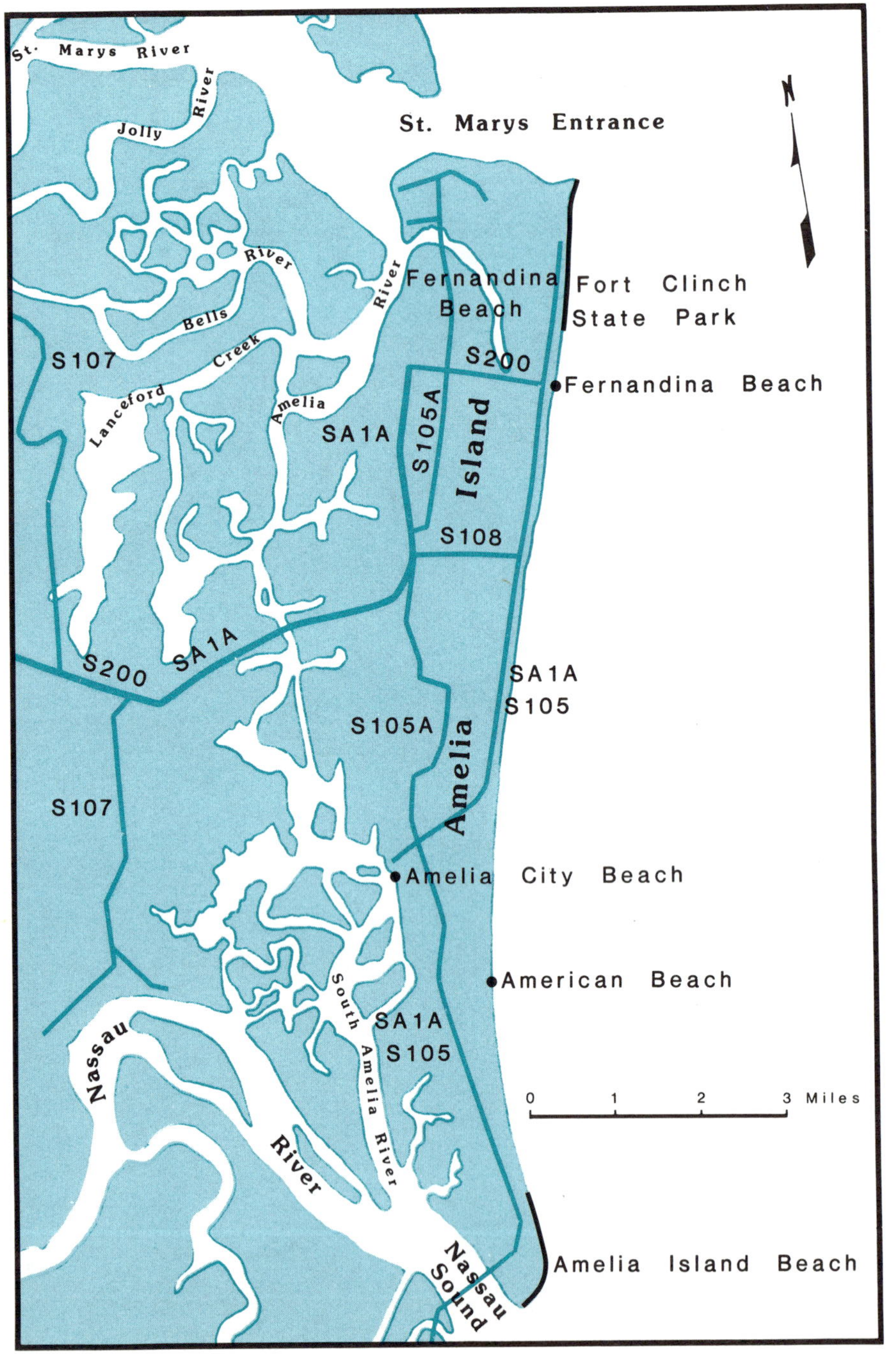

Amelia Island

Nassau County

FACILITIES / ENVIRONMENT

NAME	Easy Access	Parking or Entrance Fee	Parking	Restrooms	Showers	Picnicking	Swimming	Lifeguards	Fishing	Boating Facilities	Shelters	Concession Stands	Handicapped Facilities	Public Transportation	Group Facilities	Nature Trails / Fitness	Atlantic / Gulf	Bay / Soundfront	Sandy Beach	Rocky Beach	Primitive Beach	Urban
Ft. Clinch State Park	●	●	●	●	●	●	●	●	●	●	●	●	●		●	●	●	●	●			●
Fernandina Beach	●		●	●		●	●	●	●		●	●		●	●		●		●			●
Amelia City Beach	●	●	●				●	●	●								●		●			
American Beach	●	●	●	●			●	●	●		●						●		●			
Amelia Island Beach	●	●	●	●		●	●	●	●		●						●		●			

Ft. Clinch State Park covers over 1,000 acres off S.R. A1A on the north end of Amelia Island. Access is at the end of North 14th St., 3 miles north of Fernandina Beach. This is a developed recreation and historic area. No alcohol allowed. Fishing pier and Camping-by-the-Sea.

Fernandina Beach is a developed beach with lifeguards and 325 feet of beach frontage. Access is 3 miles north of Amelia City on S.R. A1A at Atlantic Ave.

Amelia City Beach has 10,000 feet of beach with lifeguards and is situated north of American Beach on S.R. A1A. Parking and driving on the beach with annual permit. Access ramps to beach at Peters Point, Belle Glade, Scott, and Sadler roads.

American Beach has 150 feet of developed beach with lifeguard areas. Access is marked on the Buccaneer Trail (S.R. 105/A1A), 1.5 miles south of Amelia City. Fishing pier. Parking and driving permitted on the beach with annual permit. Overnight camping. Access points at Julia St., Lewis St., and Burney Ave.

Amelia Island Beach lies south of American Beach on S.R. A1A and includes Amelia Island Plantation Resort. This developed beach has 23,500 feet of ocean and sound beaches with dune walkovers. Overnight camping. Boat ramp to Nassau Sound.

Fernandina Beach

Atlantic Intracoastal Waterway

In 1881 the Florida East Coast Canal Company began construction of a channel 5 feet deep and 50 feet wide, extending from Jacksonville to Miami. Its purpose was to aid transportation along the east coast, which had suffered the ravages of ocean storms and currents. The construction of the Intracoastal Waterway connecting a chain of rivers, lakes, and lagoons was completed in 1912. With time came the desire to link the Florida Channel with the Intracoastal Waterway extending from Trenton, New Jersey, to the St. Marys River on the Georgia-Florida border, as well as increasing the depth to 12 feet and width to 125 feet. Sponsored by the Florida Inland Navigation District (FIND) and the U.S. Army Corps of Engineers as a federal project by the Rivers and Harbors Act of 1927, the project has been completed to the authorized depth and width from Jacksonville to Fort Pierce. From that point south to Miami, the project has been completed to a depth of 10 feet and to the full project width of 125 feet. These dimensions will remain until an economic feasibility study shows whether the benefits of the work outweigh the costs.

The Intracoastal Waterway is under the jurisdiction of both the U.S. Coast Guard and the Florida Marine Patrol. Those cruising the channel should refer to the latest official navigational charts. The charts show the location and characteristics of navigational aids and indicate water depths, bridge and overhead cable clearances, distance markers, notes of caution, and tide information. They may be obtained from marine supply houses or the Distribution Division C44, National Ocean Survey, Riverdale, MD 20840, (301) 436-6990.

Most of the waterway from Fernandina Beach to Miami is protected from winds and rough water by barrier islands, with the exception of the open waters of Mosquito Lagoon, Indian River, and Lake Worth. Currents in the Atlantic Intracoastal Waterway in Florida are generally not significant except near inlets and through bridges near inlets during maximum current flows on flood and ebb tides.

All fixed bridges crossing the channel have a clearance of 65 feet above mean high water and a horizontal clearance of 90 feet between fenders, except the Julia Tuttle Causeway in Miami, which has a 56-foot clearance, and the Venetian Causeway, which has a horizontal clearance of 60 feet between fenders.

Between Fernandina and Miami there are numerous drawbridges operating under regulations established by the U.S. Coast Guard. Information on the bridges—location in statute miles south of Fernandina, vertical clearance in closed position at mean high water, horizontal clearance between fenders, and special operating regulations—is available from the Jacksonville District, U.S. Army Corps of Engineers, P.O. Box 4970, Jacksonville, FL 32201. A summary of the data is also available from the Florida Inland Navigation District, 2725 Avenue E, Riviera Beach, FL 33404, (305) 848-1217.

Note: The signal to request opening of a drawbridge is one prolonged blast followed by one short blast.

Boating on the Intracoastal Waterway

Duval County

Duval County, named for William P. Duval, first territorial governor of Florida, is the northeastern gateway for air, rail, water, and highway traffic entering Florida. Jacksonville, the state's largest city area and the county seat, has one of the largest ports in Florida. It is located on the beautiful, broad St. Johns River, 20 miles from where the river empties into the Atlantic Ocean. The River of Currents and the River of Lakes are early Spanish and Indian names for the St. Johns River that describe the spectacular way its currents meet the ocean surf. The source of the St. Johns River had been a matter of speculation since Menéndez' time. In the 1870s it was said that the St. Johns went "all the way to Hell and Blazes!" (Lending credence to this claim was the existence of a Lake Helen Blazes west of Melbourne.) Recently, however, the U.S. Geological Survey has identified Blue Cypress Lake as the river's source.

Over four and one-half centuries ago Ponce de Leon discovered this area, claiming the land for Spain and establishing the Catholic religion. Spanish Florida soon attracted the interest of France. In 1562, Jean Ribault, a French Huguenot, landed in Indian wilderness near the mouth of the St. Johns River, where he held Florida's first Protestant service. France desired to establish a base from which French ships could raid Spanish vessels, to establish a French point of entry and seat of commerce, and to found a French colony where Protestants could worship freely and safely. The sod and timber construction of Fort Caroline on Alimacani (present-day Fort George Island) in 1564 was a serious attempt by the French to claim a colonial empire in this part of the New World. The attempt was short lived as the French were soon evicted by the Spanish in a manner that gave the name Matanzas, Spanish for "slaughters," to the area.

Two hundred fifty years later Fort George Island came into the possession of Zephaniah Kingsley, slaver, planter, and great-uncle to the painter James McNeil Whistler. Kingsley Plantation became a training center for slaves who were then sold by this notorious trafficker. The Kingsley slave empire endured the depredations of both white and red outlaws, assaults of pirates, corrupt practices of Spanish officials, and annoying interference from the U.S. Army. The Kingsley Plantation House has been restored to illustrate its unusual story.

Visitors to the 16 miles of Duval County coast, almost half of which is available for public access, enjoy hard-packed sandy shores from Nassau Sound south to Ponte Vedra Beach. Red snapper, amberjack, Spanish mackerel, and flounder make the area popular with deep-sea fishermen.

Today, Jacksonville continues its role as a leading southeastern commercial, financial, and industrial center. Wholesaling, insurance, banking, shipbuilding, and forest products are among the city's principal economic activities.

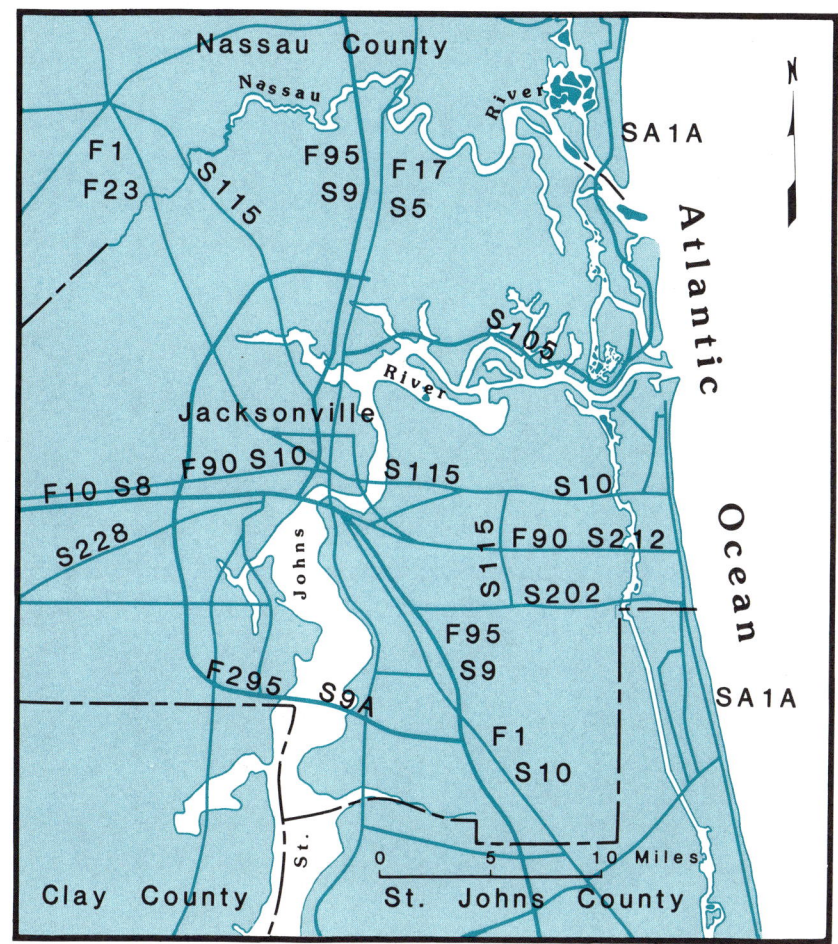

Twenty-six miles north of Jacksonville at the mouth of the St. Johns River lies Mayport, a picturesque little fishing village. The name Mayport is a reminder that the French called the St. Johns "Rivière de Mai" (May River).

Mayport Village, south of the St. Johns River mouth below Huguenot Memorial Park

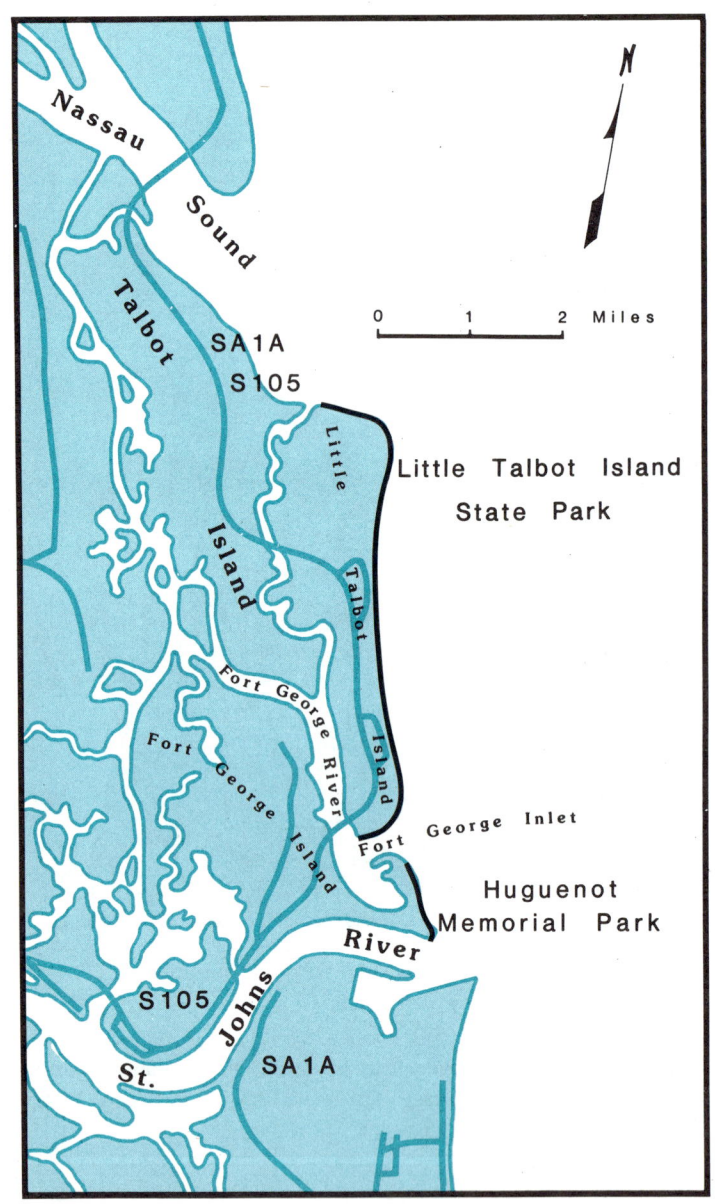

Northern district

Duval County

NAME	Easy Access	Parking or Entrance Fee	Parking	Restrooms	Showers	Picnicking	Swimming	Lifeguards	Fishing	Boating Facilities	Shelters	Concession Stands	Handicapped Facilities	Public Transportation	Group Facilities	Nature Trails / Fitness	Atlantic / Gulf	Bay / Soundfront	Sandy Beach	Rocky Beach	Primitive Beach	Urban
NORTHERN DISTRICT																						
Little Talbot Island State Park	●	●	●	●	●	●	●	●		●		●	●	●	●		●		●			
Huguenot Memorial Park	●		●		●	●		●	●						●		●		●			
SOUTHERN DISTRICT																						
Hannah Park	●	●	●	●	●	●	●	●	●		●	●	●		●	●	●		●			●
Atlantic Beach	●					●											●		●	●		
Neptune Beach	●				●	●	●	●				●					●		●	●		●
Jacksonville Beach	●		●	●	●		●	●	●			●	●	●	●		●		●			●

Little Talbot Island State Park lies between Fort George River and the Atlantic Ocean. There are two access points, one and two miles north of Fort George Island on S.R. A1A/S.R. 105. This developed state park has 5.5 miles of beach. No alcohol allowed.

Huguenot Memorial Park has two miles of beach with boating facilities. The park lies at the southern tip of Fort George Island and has marked access off S.R. A1A.

Hannah Park, formerly Seminole Beach, is located on S.R. A1A, about 3 miles north of Atlantic Blvd. (S.R. 10) and just south of Mayport Naval Station. One of the largest oceanfront parks in Florida, it is a developed beach (1.6 miles) with lifeguards. No alcohol allowed.

Atlantic Beach is one of the most urbanized of the Jacksonville beaches, accessible from 12 street-end points off Beach Ave. Approximately 480 feet of this strand is undeveloped oceanfront.

Neptune Beach lies south of Atlantic Beach. This undeveloped, urban beach has 1,250 feet of oceanfront. Twenty-three street ends off N. First St. provide access to this beach.

Jacksonville Beach is accessible from Atlantic Blvd. (S.R. 10), Beach Blvd. (U.S. 90), or J. Turner Butler Expressway. The beach has 2,400 feet of frontage reached by 64 street ends off N. First St. No alcohol allowed. Fishing pier at beach.

There is a city-owned public marina on the Intracoastal Waterway on 2d Ave. N. off Beach Blvd. under the McCormick Bridge.

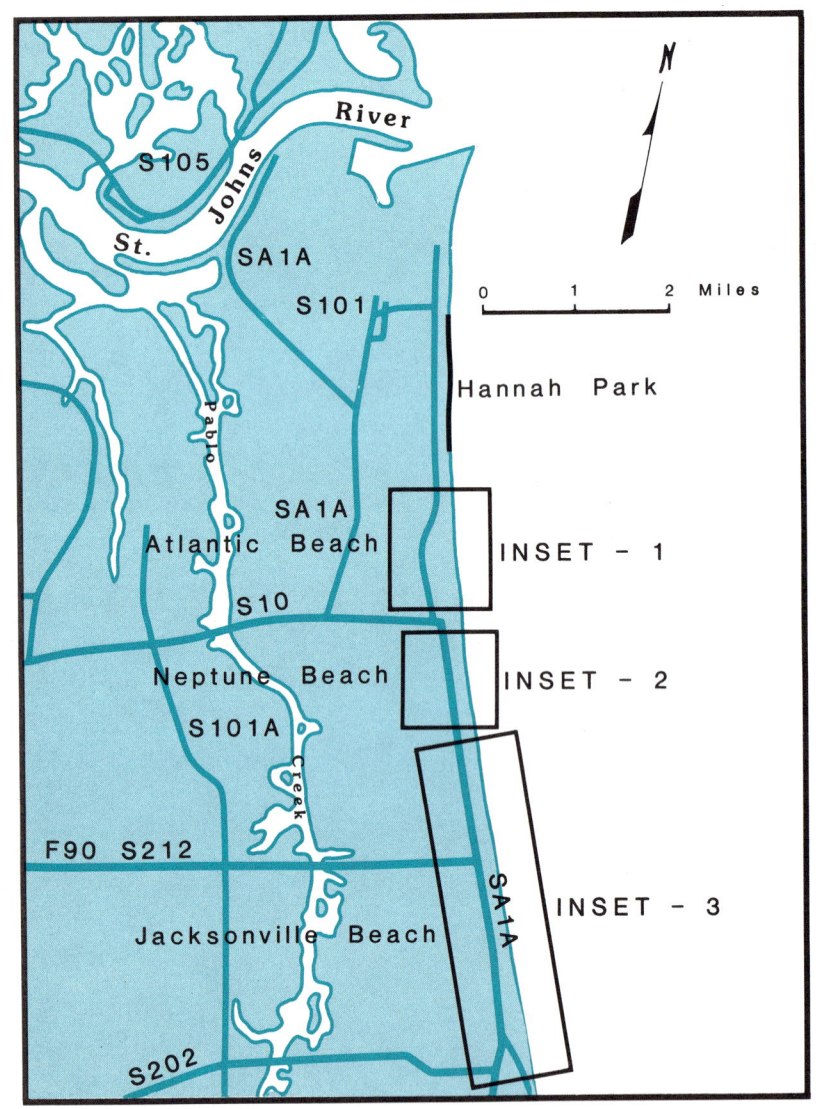

Southern district

Jacksonville Beach

10

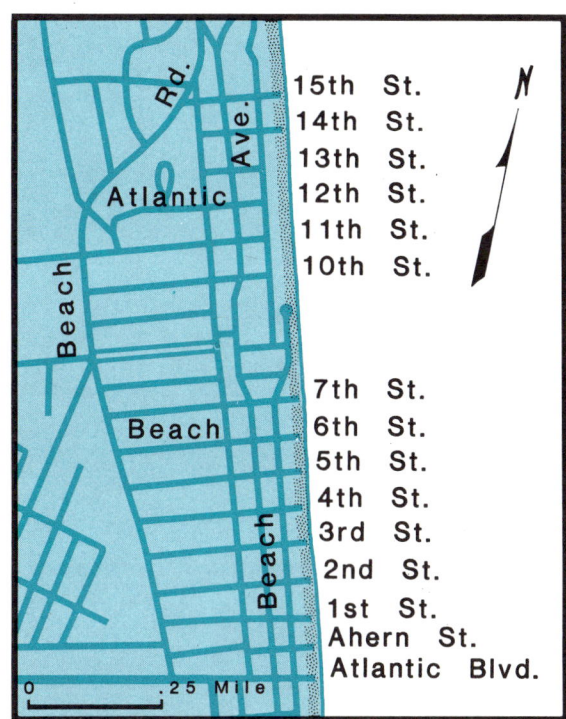

Atlantic Beach (inset 1)

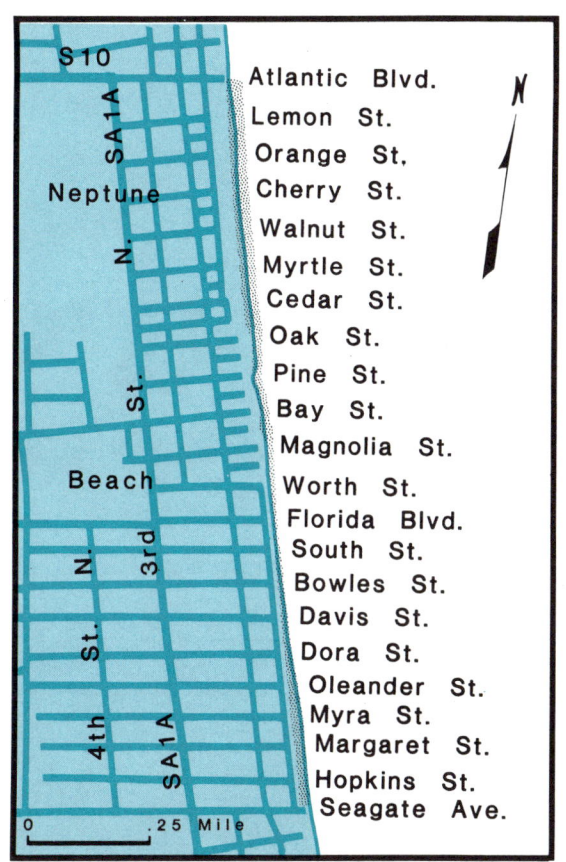

Neptune Beach (inset 2)

Jacksonville Beach (inset 3)

Little Talbot Island

Drive-on Beaches

Few visitors leave Florida without experiencing a drive to the beach. After all, this activity is a primary attraction in a state having more extensive ocean beaches than any other state. Aided by a map and perhaps a vague sense of direction, the determined visitor will follow signs with traffic instructions. After locating the beach access road, a visitor may find the road continues to the water's edge.

A few counties in Florida have drive-on beaches. Drive-on beaches are legally considered county roads and, therefore, are subject to the traffic regulations of the county. A beach must be smooth, wide, and hard to be suitable for driving. These qualities are found in the northern part of the east coast, primarily Amelia Island, Ponte Vedra Beach, St. Augustine Beach, Ormond Beach, and Daytona Beach. These smooth beaches have a fine quartz sand with few pebbles or shell fragments. The strip of hard-packed sand is as wide as the difference between high and low tide. The fine sand and large tidal fluctuations also form a slight beach slope which eases driving parallel to the shore. From 1903 to 1935 the wide strip of smooth, hard-packed wet sand on Daytona Beach was used for auto racing, and Sir Malcolm Campbell set a land speed record of 276.82 mph there. Stock-car racing followed until 1959, when the speedway was built.

The desirability of drive-on beaches appeals to the many who have had to park and walk far to enjoy the freedom of surf, sand, and sun, each step a reminder of an exhausting return trip to come. Drive-on beaches offer the convenience of almost unlimited access to the beach and the liberty of claiming a private domain by merely parking your car. It becomes an oasis in the sand, nearby and secure, offering a quick escape from overexposure and exhaustion.

Regardless of the allure, drive-on beaches do have potential hazards to both the environment and those seeking the pleasures of the beach. The intertidal ocean beach *appears* to be resistant to long-term vehicle use, but this is primarily due to the continuous change from sand transport during tidal changes. As long as vehicle traffic does not interfere with the more stable areas of the beach, specifically the dunes, damage from vehicle stress appears minimal. Environmental studies have yet to show the effects on intertidal beach creatures such as the mole crab, coquina clam or periwinkle, ghost crabs, and mole shrimps. However, sea turtles, and particularly their nesting areas, are seriously endangered by the stress of vehicles. Fortunately, most known nesting areas are protected against this hazard. As for the beachgoer, vehicles present the same hazards as anywhere else that cars and people encounter one another. While enjoying the pleasures of the county beach "road" one must not forget the dangers.

Most state beaches are unsuitable for driving, and the unsuspecting visitor may end up paying a high towing charge to have his auto pulled from the soft sands. There may also be a fine to pay. Check carefully to see if beach driving is permitted and possible. Each county that has drive-on beaches has humorous stories of cars trapped by soft, dry sand or by incoming tide.

A final warning note: salt spray and sand can corrode and damage a car.

St. Johns County

On his famous search in 1513 for the mythical source of youth-restoring waters, Don Juan Ponce de Leon landed on the shores of St. Johns County. As his discovery coincided with "the Feast of Flowers," he named St. Augustine "Pascua Florida."

Fort Matanzas National Monument lies 14 miles south of St. Augustine on Rattlesnake Island. In 1565 three hundred French Huguenots were put to death at Fort Matanzas (Spanish for "slaughters"). Ferries cross Matanzas Inlet to Rattlesnake Island daily in the summer and on weekends the rest of the year. Just across the Matanzas River lies a strip of land, approximately 12 miles long and less than one mile wide, known as Anastasia Island. Part of the land fronts the Atlantic Ocean where fishing and swimming are excellent; on the soundside is Salt Run Lagoon, abundant with many species of game fish. The land between is a vast, dense landscape sculpted by wind and blowing sand. It was through these waters in 1565 that Spanish explorer Pedro Menéndez de Avilés waded ashore to become the founder of Florida's first permanent colony, St. Augustine.

Throughout the centuries, St. Augustine has been continuously settled, making it the nation's oldest city. Each year in June, St. Augustine is the host city for the performance of Paul Green's *Cross and Sword*, the official state play depicting the battle for the settlement of Florida.

The central plaza and street plan remain almost as they were laid out in 1598 by the Spanish: 31 houses built in colonial times still stand on those streets. The city gates constructed of impressive coquina pylons that guarded the city in 1738 now lead visitors into the restored colonial district. It was on July 10, 1821, that the Spanish flag was last lowered at the Castillo de San Marcos to be replaced by the Stars and Stripes, ending 300 years of Spanish rule of Florida.

Each year hundreds of thousands of tourists visit St. Johns County, both for its fascinating history and for the pleasures of its beaches.

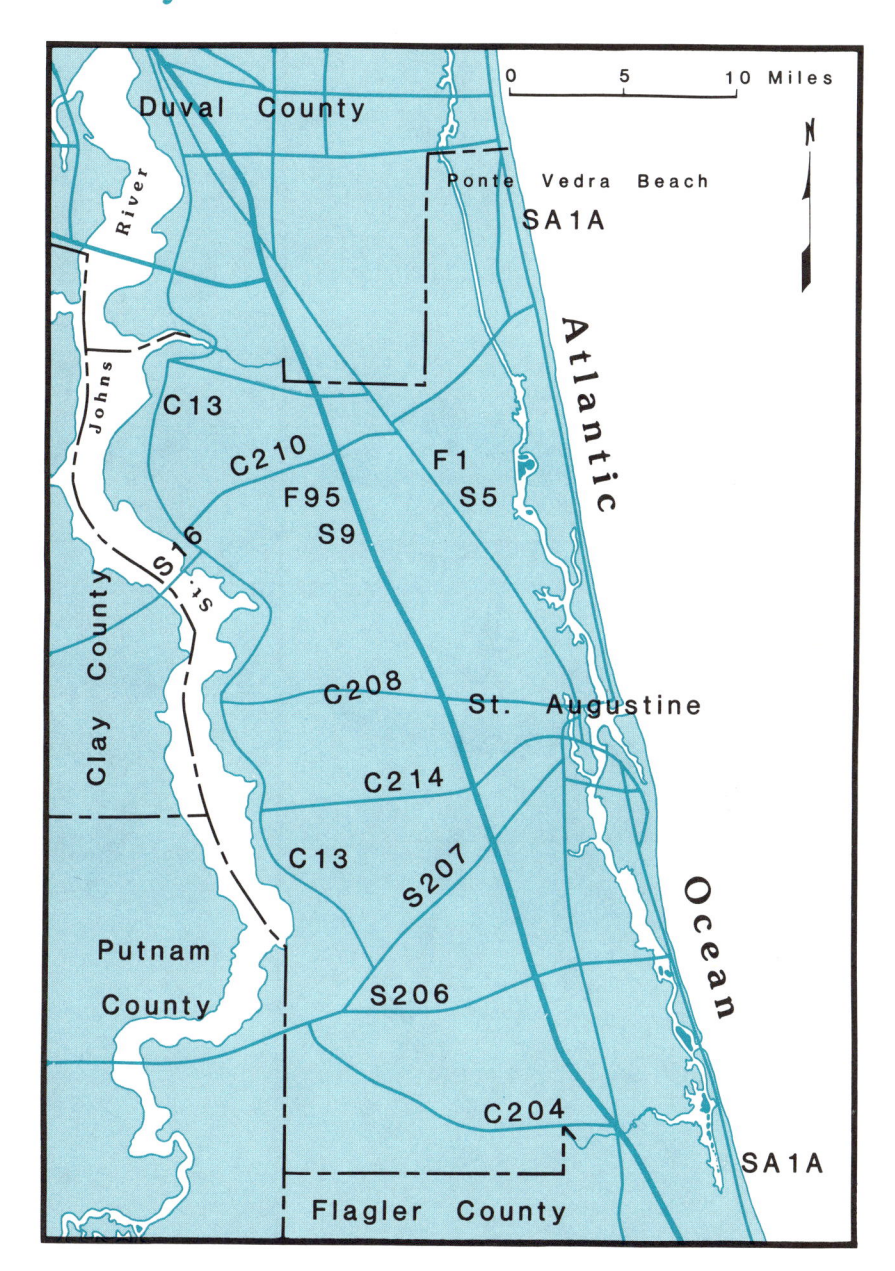

15

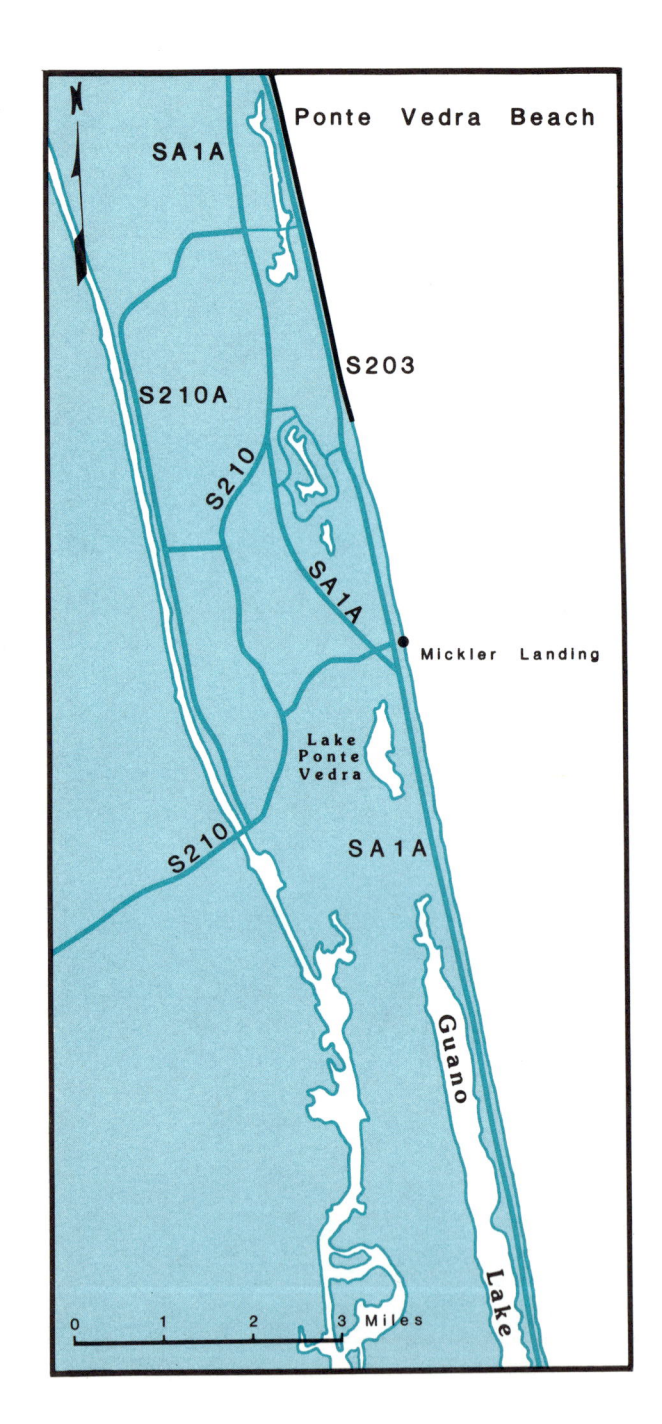

Castillo de San Marcos National Monument

Northern district

St. Johns County

NAME	Easy Access	Parking or Entrance Fee	Parking	Restrooms	Showers	Picnicking	Swimming	Lifeguards	Fishing	Boating Facilities	Shelters	Concession Stands	Handicapped Facilities	Public Transportation	Group Facilities	Nature Trails/Fitness	Atlantic/Gulf	Bay/Soundfront	Sandy Beach	Rocky Beach	Primitive Beach	Urban
NORTHERN DISTRICT																						
Ponte Verda Beach						●	●										●		●		●	●
CENTRAL DISTRICT																						
North Beach						●	●										●		●		●	
South Ponte Verda	●		●			●	●										●		●		●	
Usina's Beach	●	●	●			●	●										●		●		●	
Vilano Beach	●		●			●	●	●									●		●		●	
Anastasia State Recreation Area	●	●	●	●	●	●	●	●	●	●	●	●		●	●		●	●	●			●
St. Augustine Beach	●			●		●	●	●				●		●			●		●			●
SOUTHERN DISTRICT																						
Unnamed Beach	●		●			●	●										●		●			
Butler Beach/Frank B. Butler Park	●	●	●	●	●	●	●		●	●		●		●	●		●	●	●			
Crescent Beach	●		●	●	●	●	●	●			●			●			●		●		●	
Matanzas Beach	●		●			●	●										●		●			

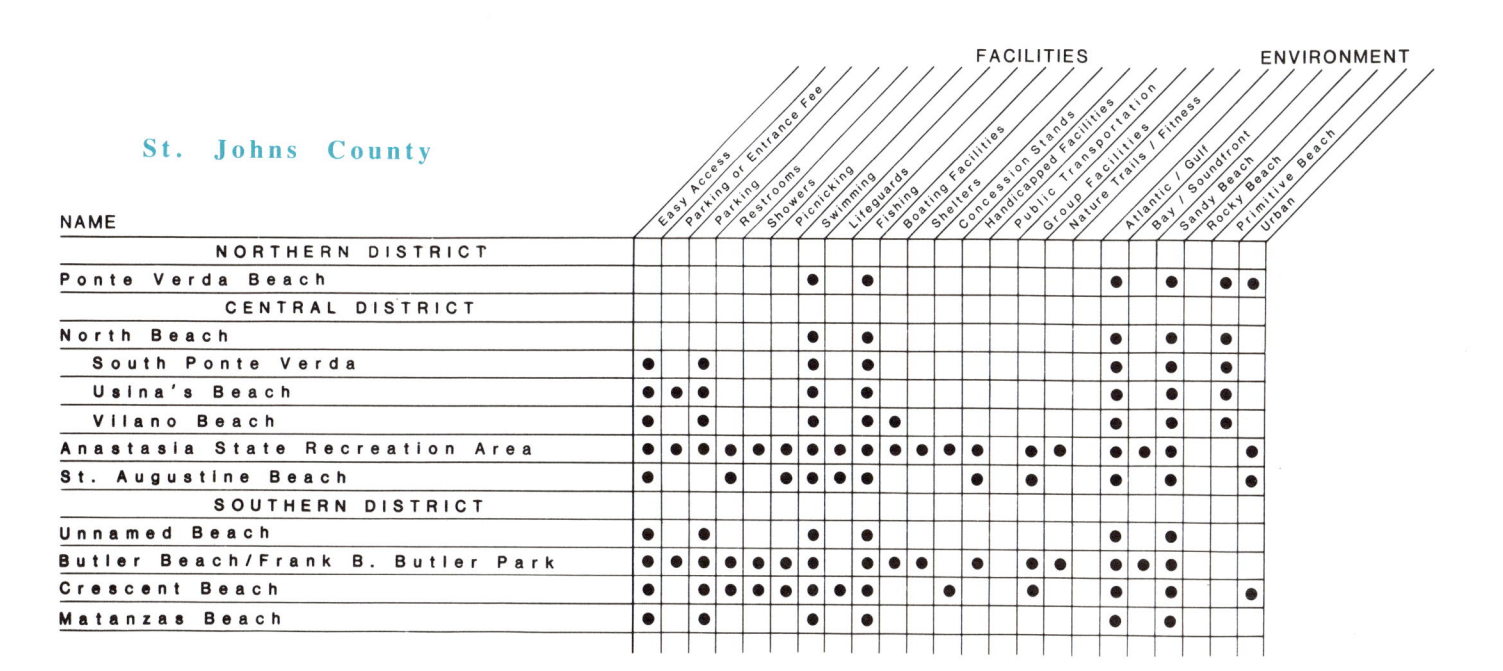

Ponte Vedra Beach extends from the Duval County line a distance of 6.5 miles south. Walkover access is at Mickler Landing on S.R. A1A; there is also parking and access along the shoulder of S.R. A1A.

South Ponte Vedra Beach, an undeveloped urban beach area off S.R. A1A, has several access easements.

Usina's Beach has 2 miles of shoreline accessible via one walkway.

Vilano Beach extends north of St. Augustine Inlet 2 miles. One street end off S.R. A1A provides access for cars. Boating facilities are on the mainland, just left before crossing Vilano Beach Bridge.

Anastasia State Recreation Area fronts 3.5 miles of shoreline from the St. Augustine Inlet southward to the city limits. Access to the beach is 2,000 feet north of Pope Rd. No alcohol is allowed.

St. Augustine Beach extends from Pope Rd. south to St. Augustine-by-the-Sea. There is public access at 32 locations. From St. Augustine Beach south to Matanzas Inlet, the beach has been designated a county road open to traffic and parking. There is a fishing pier.

Unnamed Beach lies between St. Augustine Beach and Butler Beach. About 2 miles of undeveloped frontage is accessible via street ends off S.R. A1A.

Butler Beach/Frank B. Butler Park has 0.75 mile of undeveloped, urban beach off S.R. A1A from Matanzas Ave. south to Frank B. Butler Park and a county-maintained park fronting the Intracoastal Waterway. Five streets provide access to the ocean beach; on the west side of S.R. A1A there is marked access to the park and boat ramp.

Crescent Beach has 5 miles of shoreline from Frank B. Butler Park to Ft. Matanzas National Monument. There are 5 street-end access points off S.R. A1A, as well as a runway to the beach where S.R. 206 and S.R. A1A meet.

Matanzas Beach, 3 miles in length, lies within the boundary of the Ft. Matanzas National Monument. No alcohol is allowed.

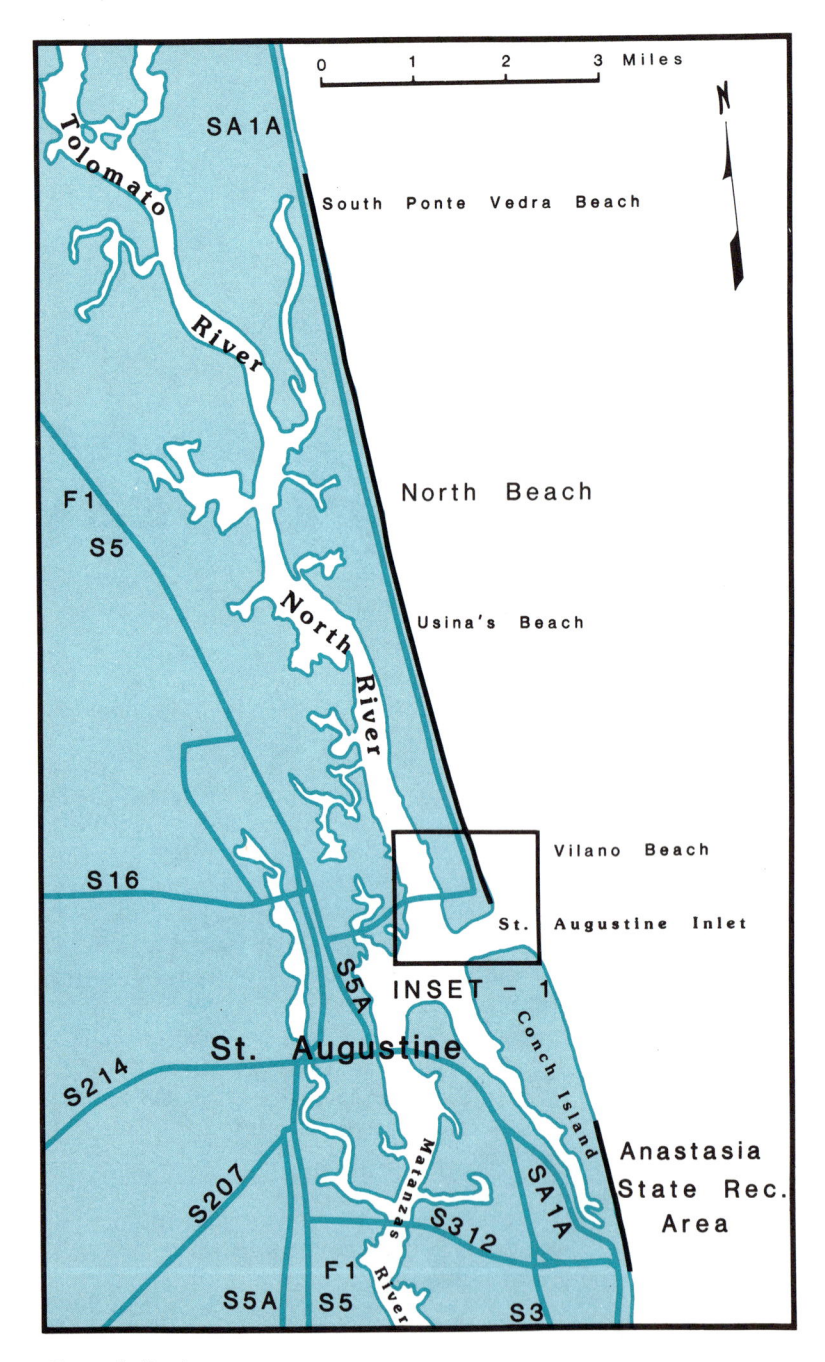

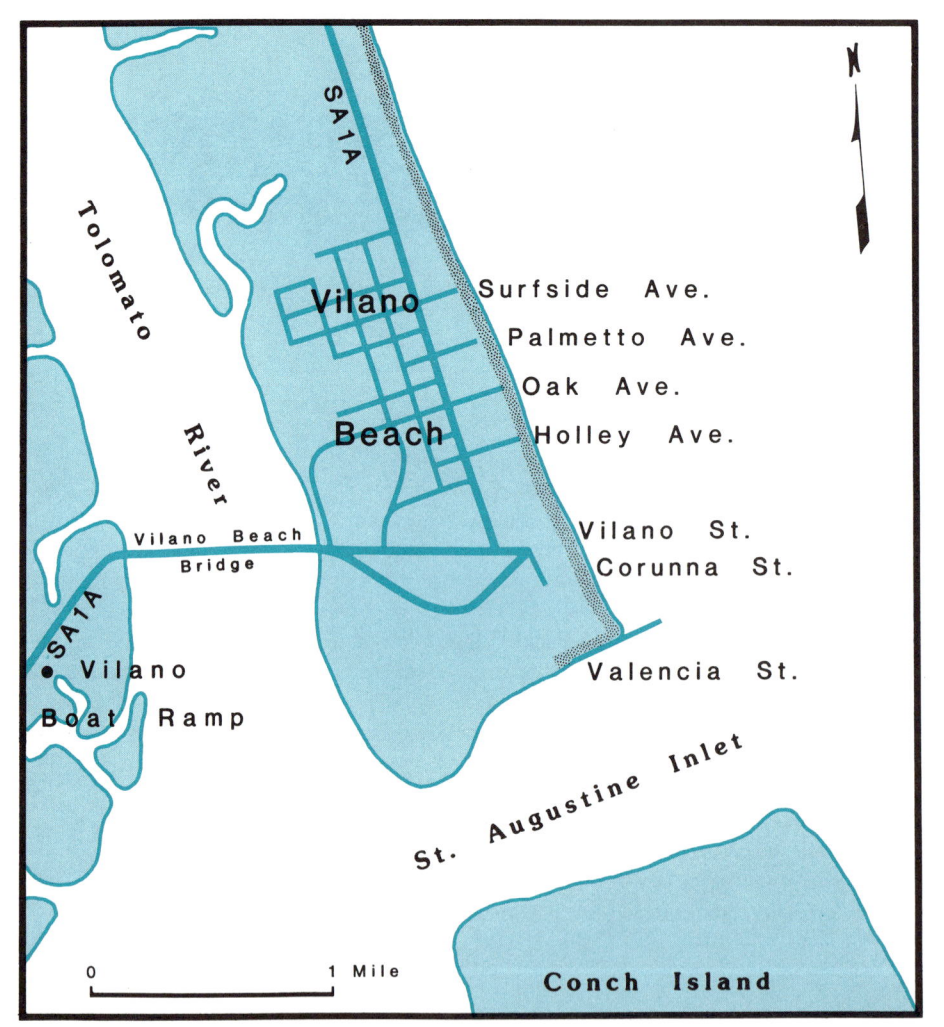

Central district

Vilano Beach (inset 1)

St. Augustine Beach Lighthouse

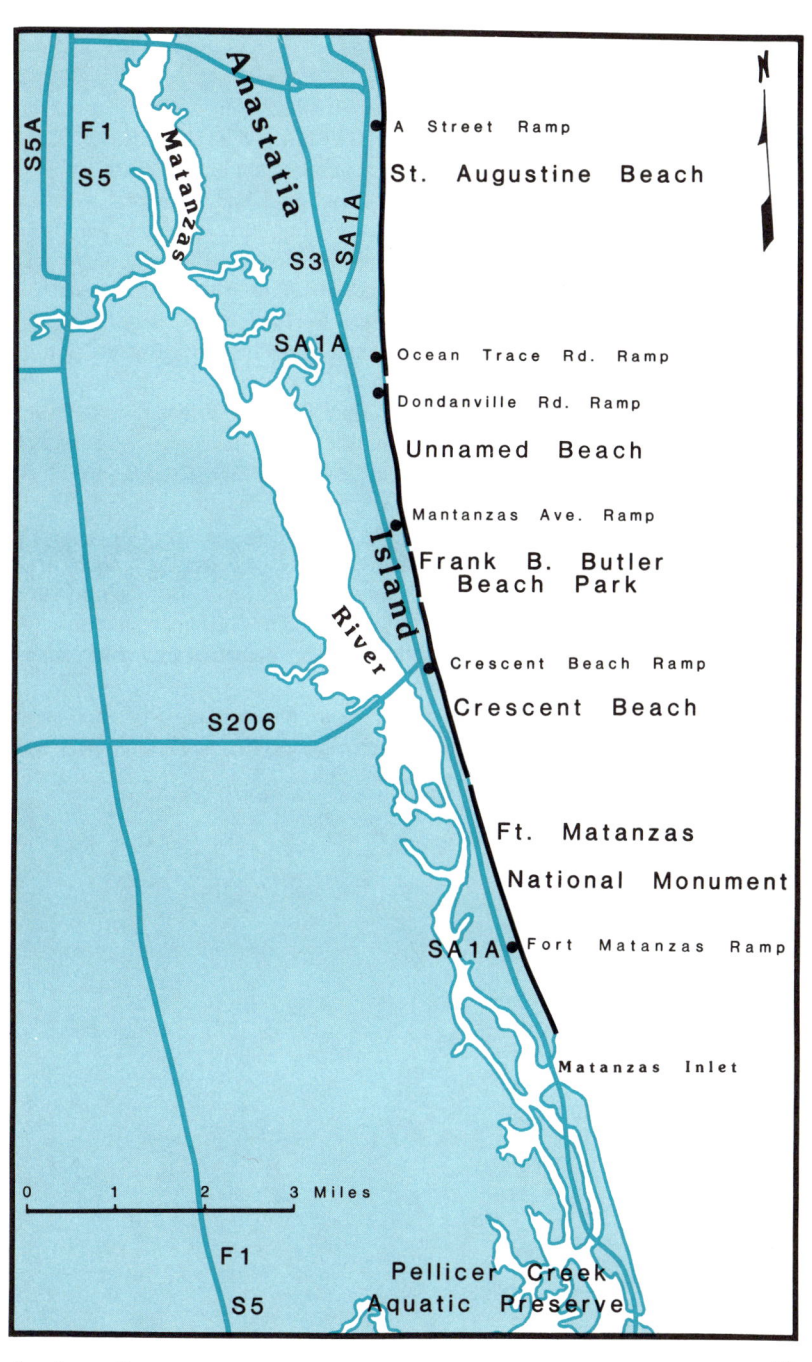

Southern district

19

Dolphins

Dolphins have interested humans throughout recorded history. Once commercially fished for food and oil, they are now protected by law in the United States. Even the capture, handling, and care of dolphins for entertainment is regulated.

The name "dolphin" has been given to both a fin fish and a mammal. Among the many species of mammals, the bottle-nosed dolphin, *Tursiops truncatus*, is commonly spotted in the shallow coastal waters of Florida. They have streamlined bodies sometimes reaching 8 feet in length. Gray on top with a back-curved dorsal fin, and pale on the underside, dolphins appear evenly colored in water. A powerful tail formed by two outward-pointing horizontal flukes is used for propulsion. Side flippers or paddles are steering mechanisms.

Like all mammals, dolphins breathe air. They regularly surface from the water to exchange air through a single blowhole on the top of their heads. Characteristic traits are protruding snouts, puffy-looking foreheads, and extended lower jaws. Their strong jaws are equipped with 88 conical, pointed teeth well adapted for capturing fish.

Friendly animals, dolphins often assist fishers by encircling schools of fish, making easy capture possible. Boaters and swimmers report of playing with dolphins in the surf. Playful antics taught by signals and rewards are well known. As if to reflect its gregarious disposition, the dolphin has a perpetual smile on its face and "speech" that mimics human laughter.

Porpoises, easily confused with dolphins, are also toothed whales. They are, however, heavier bodied and do not exhibit such playful behavior. The porpoise lacks a well-defined snout and has an even jawline.

Dolphins are social animals usually found in schools. Mating is preceded by a courtship ritual of elaborate movements and swimming side-by-side with flippers touching. Dolphins conceive internally and have a year-long pregnancy. A single "pup," weighing 30 pounds and reaching three and a half feet in length, is born tail first. Newborns are fed milk for six months but feed independently on fish within a year. Young dolphins do not leave the mother's side during their first 18 months. Reaching maturity at 12 to 14 years, dolphins live about 25 years.

The dolphin brain is as large and complex as a human's. Their vision is equally good both in or above water. Researchers are trying to communicate with dolphins through the use of specially adapted computers that would allow dolphins to recognize words through sonar echoes. Although dolphins have "pinhole" ears, sounds are detected by sensory perceptors located in the lower jaw. They emit clicking noises that "bounce" off objects. This natural sonar is so refined they can detect a pebble in murky water. The military is training dolphins to protect harbors from enemy penetration.

Dolphins seem to have the human quality of compassion. Countless sea stories tell of dolphins rescuing swimmers, fending off sharks, and dislodging or guiding boats. A New Zealand dolphin in the years between 1888 and 1912 guided motorboats through the rocky channels of the Pelorus Sound and French Pass.

The dolphin's disposition is endearing; few animals are as loyal or gentle. With a blow from its snout, a dolphin could easily kill a person, yet they do not attack unless provoked.

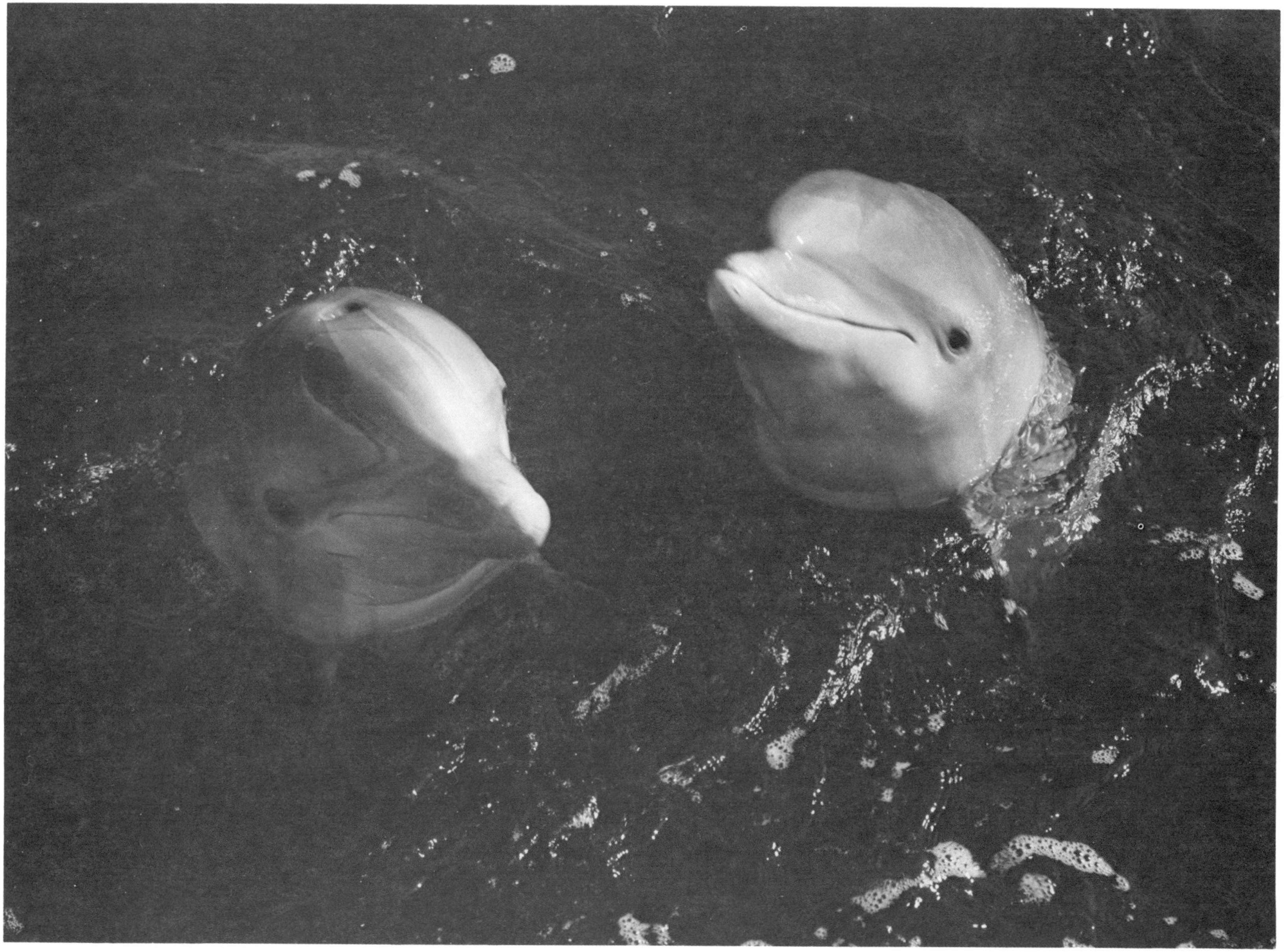

Flagler Beach

Flagler County

French Huguenots and Spanish conquistadors exploring the east part of Florida in search of religious freedom and gold crossed what is now known as Flagler County.

The county was named for Henry Morrison Flagler a northern businessman and associate of John D. Rockefeller, who opened up the east coast of Florida by building railroads and motels, operating steamships, and encouraging land development. Brought up in poverty, Flagler was considered a shrewd and ruthless businessman. However, at the age of 55, when he assumed responsibility for the development of Florida, his attitude changed toward benevolence and accountability in this venture of creating Florida's first resort area.

Just south of the northern county line is Marineland of Florida, one of the first sea research attractions pioneering in dolphin training. Originally an underwater motion picture studio, Marineland has evolved into a unique resort community. It is a first-rate oceanfront resort attraction. In addition to its seclusion and beautiful landscaping, Marineland maintains the University of Florida Graduate School of Marine Biology.

Washington Oaks State Park and its unique oceanfront preserve, "The Rocks," is located just south of Marineland off S.R. A1A. This land was part of Belle Vista Plantation during the Second Seminole War (1835–42). The gardens are cultivated in a lush coastal swale shaded by towering live oaks and magnolias. Along the beachfront, ocean waves have washed away the sand, exposing a picturesque coquina rock beach sculpted by the sea; it attracts many species of shorebirds and captures starfish and crabs in tidal pools.

Flagler Beach State Recreation Area extends across the barrier island 3 miles south of Flagler Beach. The beach, hard-packed and wide at low tide, is a favorite feeding area for many shorebirds. During the summer the area above the high tide line is used for nesting by sea turtles and the rare Florida scrub jay.

Fifteen miles north of Daytona Beach in a dense jungle hardwood grove stand the coquina ruins of Bulow Plantation. In the 1820s this magnificent frontier plantation was one of the most productive and prosperous in East Florida. It consisted of a stone sugar works, sawmill, blacksmith shop, slave quarters, and grand mansion. All were destroyed by the Indians in 1837 during the Second Seminole War. Today the ruins stand as an open-air museum, capturing the plantation aura of the past.

Located on the upper east coast of Florida, Flagler County is the third fastest growing county in the state. Those who love the water and the freedom of open spaces will enjoy its long stretches of beach sand and the unhurried, sun-filled outdoor life.

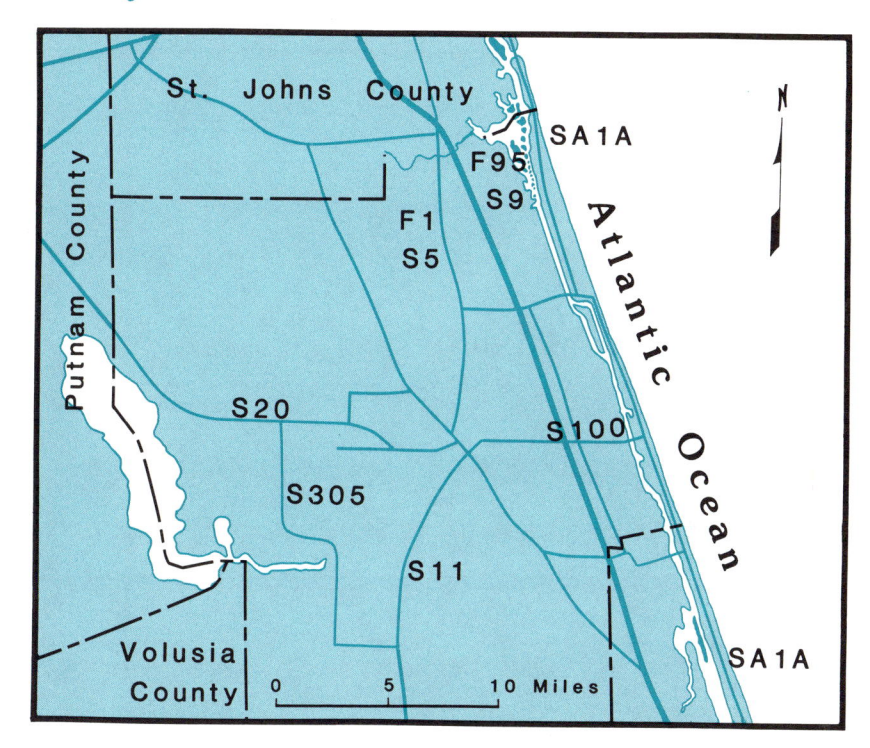

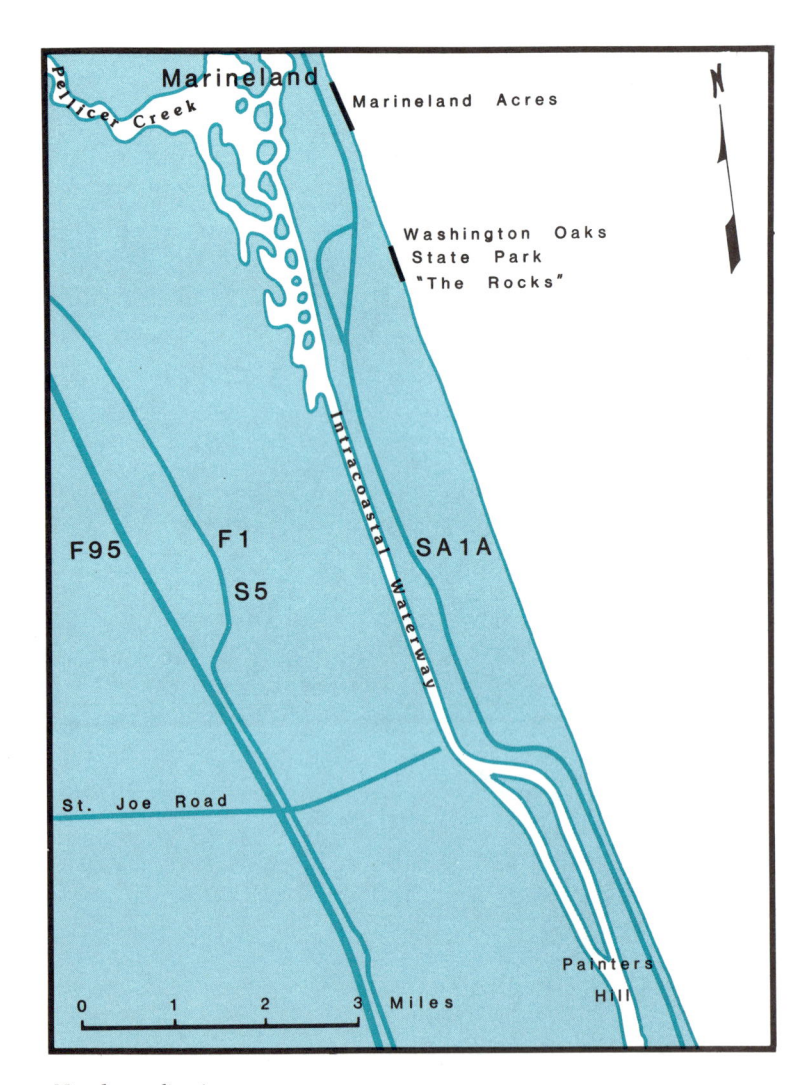

Northern district

"The Rocks"

24

Flagler County

NAME	Easy Access	Parking or Entrance Fee	Parking	Restrooms	Showers	Picnicking	Swimming	Lifeguards	Fishing	Boating Facilities	Shelters	Concession Stands	Handicapped Facilities	Public Transportation	Group Facilities	Nature Trails / Fitness	Atlantic / Gulf	Bay / Soundfront	Sandy Beach	Rocky Beach	Primitive Beach	Urban
NORTHERN DISTRICT																						
Marineland Acres		•		•		•											•		•	•		
Washington Oaks/"The Rocks"	•	•	•	•	•	•			•		•	•		•	•		•	•		•		
SOUTHERN DISTRICT																						
Flagler Beach Municipal Beach	•		•	•	•		•	•	•		•	•				•	•		•			•
Flagler Beach State Recreation Area	•	•	•	•	•	•	•	•	•	•		•		•	•	•	•		•			

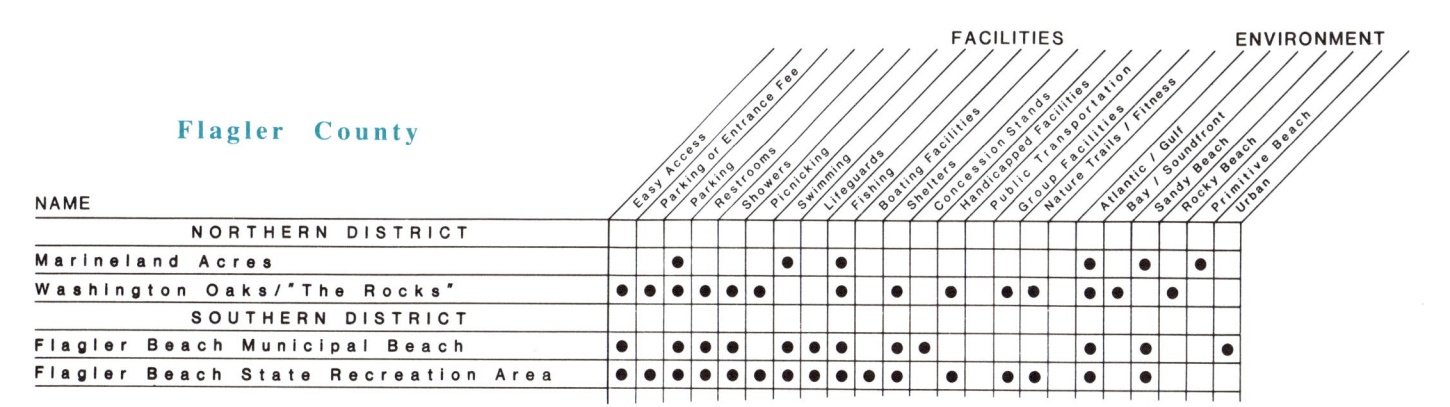

Marineland Acres lies 0.5 mile south of St. Johns Co. off S.R. A1A. The county has 3 streets providing access to this undeveloped coquina beach.

Washington Oaks State Park/"The Rocks" extends from the Atlantic Ocean to the Matanzas River, covering 340 acres. Access to this undeveloped area with 2,200 feet of ocean frontage of coquina rock outcropping is marked on S.R. A1A, about 2 miles south of Marineland.

Flagler Beach Municipal Beach has 12 acres of developed, urban oceanfront park and a public pier. Dune walkovers are off N. Ocean Shore Blvd. (S.R. A1A): N. 22d, 21st, 14th, 12th, 9th, and 7th streets, Moody Blvd. (pier access), S. 2d, 3d, 5th, 13th, 15th, 17th, 19th, 21st, 22d, and 25th streets.

Flagler Beach State Recreation Area extends across the barrier island 3 miles south of the City of Flagler Beach. Access to the undeveloped coquina shell–sand beach is marked on S.R. A1A.

Flagler Beach

25

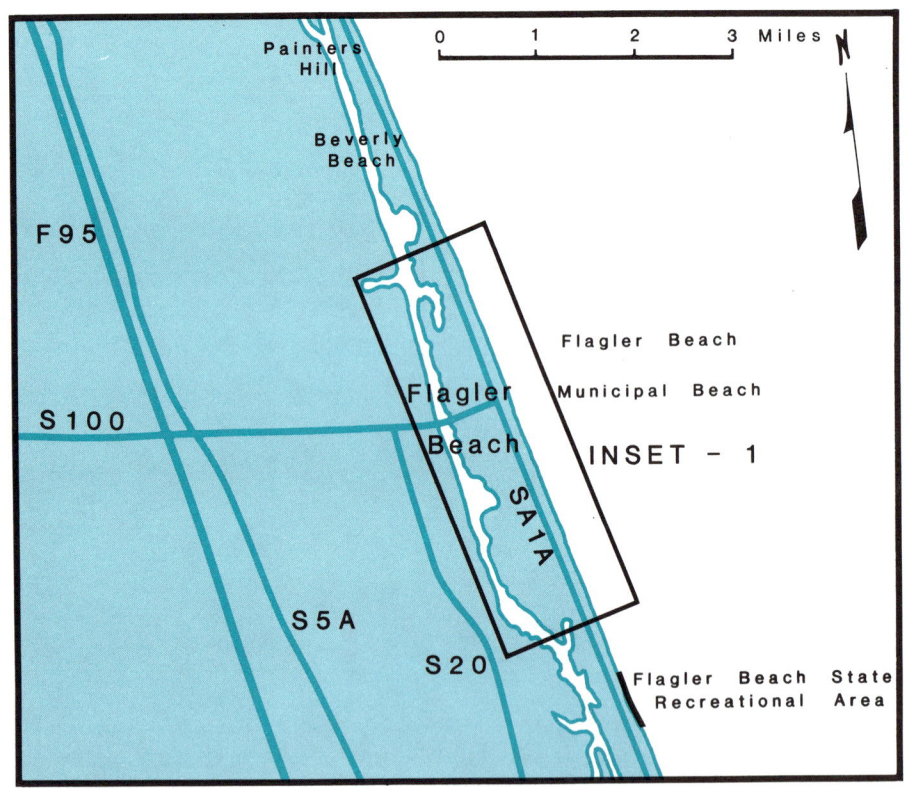

Southern district

Flagler Beach

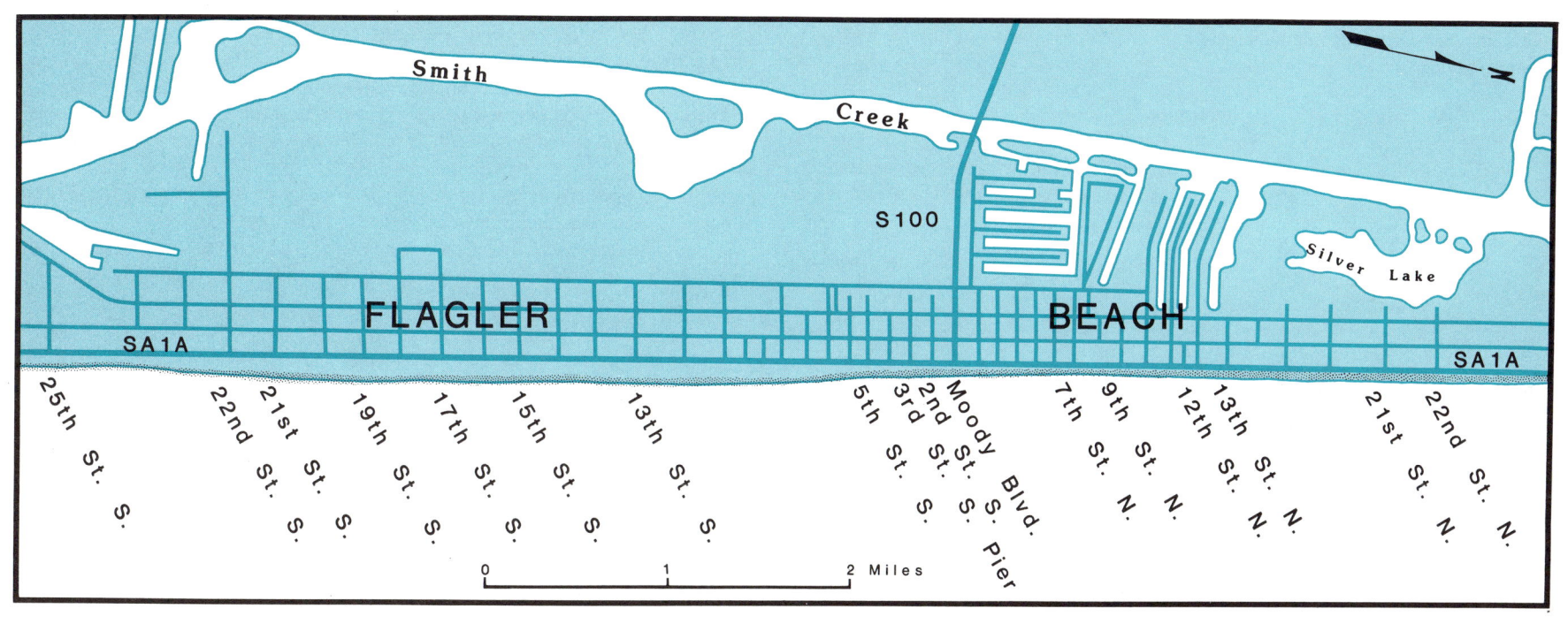

Flagler Beach (inset 1)

Water Safety

Water-based recreation, such as boating and swimming, can involve certain risks. Taking precautions can minimize risks and ensure that a visit to the coast is a pleasant experience.

Swimming Tips

Swimmers should select swimming sites carefully. Do not swim in areas designated for fishing or surfing. When possible, swim in supervised, protected areas. The American Red Cross offers guidelines for safe swimming:

> never swim alone;
> instruct nonswimmers to use personal flotation devices;
> don't depend on a tube or float;
> never dive into strange water;
> supervise children at all times;
> keep basic rescue and lifesaving equipment available;
> use planks, lifesavers, or towels in rescues;
> do not swim when overheated;
> do not swim during an electrical storm.

Dangerous Marine Life

Diving, swimming, and fishing are activities that bring direct contact with marine life. Some sea creatures can be dangerous. Unless you know an animal to be harmless or nonpoisonous, avoid it.

Jellyfish and Portuguese man-of-war are stinging animals adrift in the open ocean. They may be found near shore depending upon prevailing wind and current directions. Stings from jellyfish can cause pain, skin irritation, and localized redness. It is best to avoid contact with any part of a jellyfish, even those that wash onshore, since tentacles can continue to sting for some time.

The stinging cells can be neutralized with alcohol, ammonia, or meat tenderizer. Stings by Portuguese man-of-war should receive prompt medical attention. Extreme pain from these stings can cause a victim to go into shock. Do not attempt to pull off tentacles; it may trigger stinging cells that have not yet discharged. Instead neutralize them and seek medical attention.

Stingrays can be found in seagrass beds or partially buried in open, sandy areas, often in shallow water. They have a venomous spine at the base of the tail; if stepped on, the ray responds instinctively by lashing its tail. Sometimes a spine breaks off in a wound, requiring medical attention to remove it. If toxin is released, weakness and nausea may result. Less severe wounds should be disinfected. The ray's spine is strictly defensive. Avoid handling them and shuffle your feet when wading in shallow water.

Crabs have powerful claws that can pinch or puncture skin. They are found in a variety of habitats ranging from seagrass beds to hard bottoms.

Sharks and other large fish can attack swimmers. It is best to avoid these fish by swimming close to shore in protected areas. Heed all shark warnings by the lifeguards on duty.

Safe Boating Tips

Gasoline vapors are explosive. Close all doors, hatches, and ports when fueling. Smoking is prohibited. Do not operate electronic equipment including radios.

Know the fuel capacity. Ventilate all compartments after fueling.

Wear an approved personal flotation device.

Do not permit persons to ride on areas not designed for use such as the bow, seatback, or gunwale.

Do not overload.

Keep alert for areas where swimmers or divers may be. Divers are marked by a red flag with a white diagonal stripe through it.

Watch the wake.

Don't overpower; it is dangerous.

Do not make high-speed turns. They can cause accidents.

Know and obey the "rules of the road."

Carry all required and recommended safety equipment including an anchor, strong line, flares, flashlight, mirror, and compass.

Know and heed all distress and weather signals.

The American Red Cross offers a boating safety course. Information on title, registration, reporting an accident, or safety equipment requirements can be obtained by contacting a local Marine Patrol Office or Marine Patrol Headquarters at Department of Natural Resources, 3900 Commonwealth Blvd., Tallahassee, FL 32303.

Daytona Beach

Volusia County

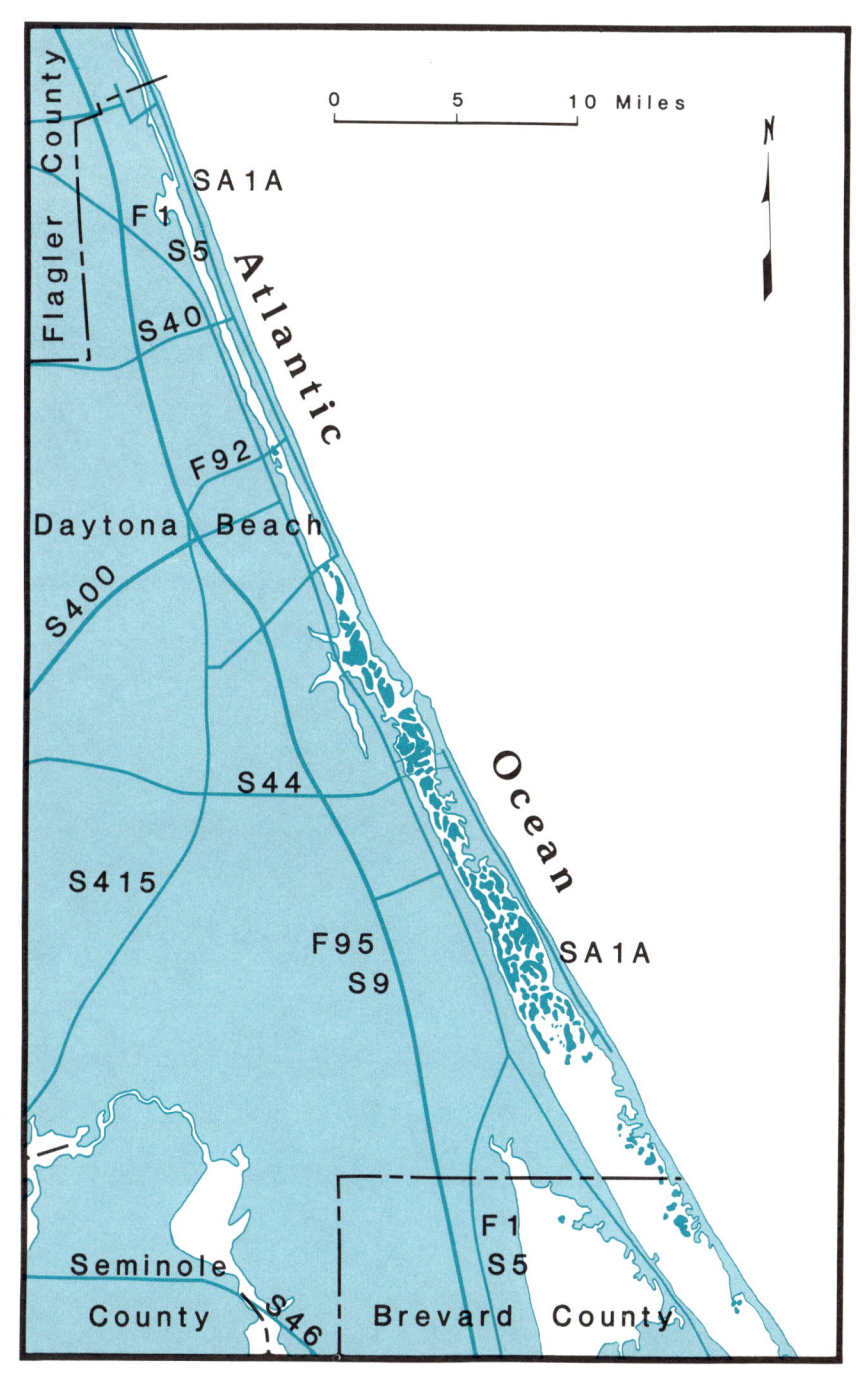

Daytona Beach is visited annually by hundreds of thousands of tourists. The city fronts the Atlantic Ocean; it is separated from the mainland by the saltwater Halifax River; Tomoka River is to the north and Ponce de Leon Inlet is at the south end. Excellent fishing earns this area a reputation as a "fisherman's paradise," a reputation overshadowed by the fame and popularity of the beach and speedway.

One of Volusia County's attractions is Daytona Raceway, 23 miles of wide, firm, white quartz sandy beach, used from the early 1900s until 1961. Today the beach is used for more leisurely recreational activities; driving is still permitted and carefully monitored on the beach road during daylight hours.

Ormond Beach is situated to the north near the confluence of the Tomoka and Halifax rivers. It is speculated that the first settlers were victims of seventeenth-century shipwrecks. In 1804, grants of land were offered by Spain to entice English colonists living in the Bahamas.

Volusia County was first settled by Franciscan friars, who established missions in the area around 1587. The earliest mission was located at which is now Daytona Beach.

The climate has always been a primary attraction in the area's development. In the late 1800s the area became known as the "Millionaire's Colony," frequented in the winter by Vanderbilts, Astors, and Rockefellers.

In addition to the famed raceway and beach, Volusia County has a number of other points of interest. Hontoon Island State Park, located on the St. Johns River west of DeLand, is a preserve of over 1,000 acres of cypress swamp, open savannah, and hammocks of pine, oak, and palm. The island was originally owned by a soldier who fought in the Seminole wars around 1850. The park's primary focus is on boating; however, an Indian ceremonial mound, 300 feet long, 100 feet wide, and 35 feet high, provides evidence of the Timucuan Indians who were the earliest known inhabitants of the island.

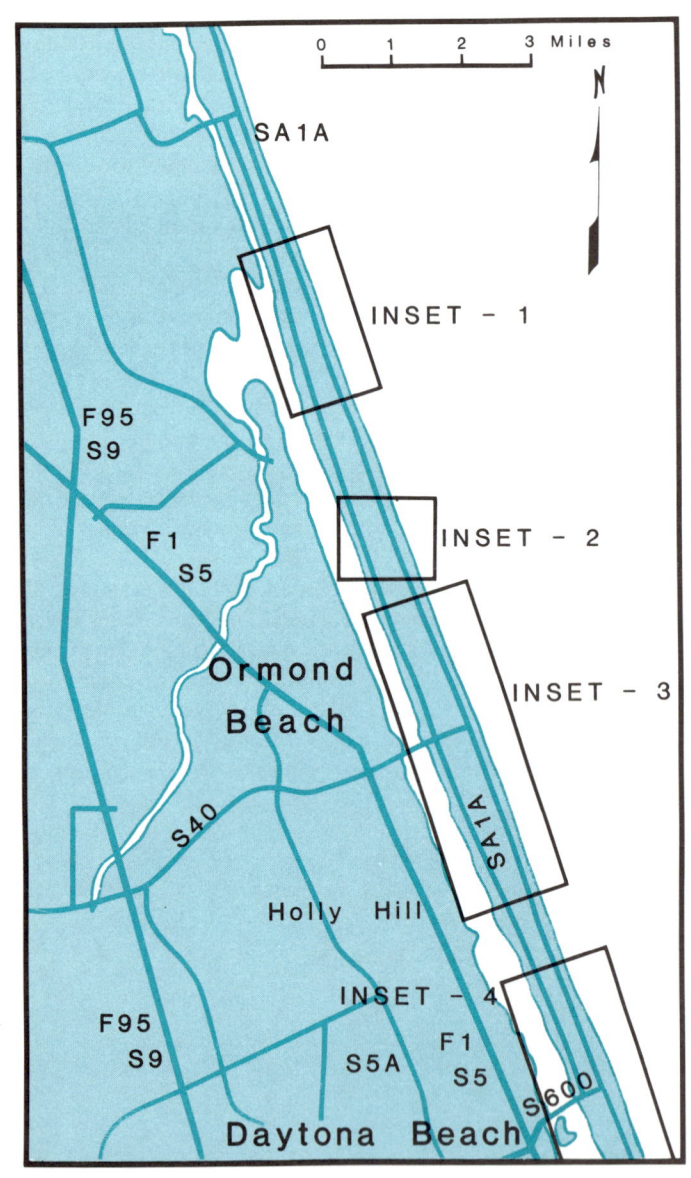

Northern district

32

Volusia County

NAME	Easy Access	Parking or Entrance Fee	Parking	Restrooms	Showers	Picnicking	Swimming	Lifeguards	Fishing	Boating Facilities	Shelters	Concession Stands	Handicapped Facilities	Public Transportation	Group Facilities	Nature Trails / Fitness	Atlantic / Gulf	Bay / Soundfront	Sandy Beach	Rocky Beach	Primitive Beach	Urban
NORTHERN DISTRICT																						
Ormond-by-the-Sea		•		•	•				•			•	•	•	•	•	•					
Ormond Beach	•		•	•	•	•	•	•	•			•	•	•	•	•	•	•				•
CENTRAL DISTRICT																						
Daytona Beach																						
North End	•		•	•		•		•			•	•	•	•	•		•					•
Boardwalk	•		•	•	•	•	•	•	•		•	•	•	•			•					•
South End	•		•	•	•		•		•			•	•	•			•		•	•		•
Wilber-by-the-Sea	•		•			•	•							•		•	•	•				
Ponce Inlet Park	•		•			•	•							•			•	•	•			
New Smyrna Beach	•		•	•	•		•	•	•				•				•	•				•
SOUTHERN DISTRICT																						
Edgewater Beach						•	•									•	•	•				
Canaveral National Seashore/Apollo Beach	•		•	•		•	•	•								•	•				•	

Ormond-by-the-Sea is an undeveloped, urban beach situated on the barrier island between Halifax River and the Atlantic Ocean. The 2.5 miles of beach are reached by street ends off Ocean Shore Blvd. (S.R. A1A): Essex, Lynnhurst, Ocean Edge, Ocean View, Oceanside, Ocean Beach, Ocean Dunes, Ocean, Sea Shore, Breezeway, Bass, Ocean Air, Marlin, Surfside, Imperial Hgts., and Ocean Breeze. No alcohol allowed.

Ormond Beach has 4 miles of developed, urban beach with lifeguard areas. Access points off S.R. A1A are Harvard, Milsap, Cardinal (lifeguard area), River Beach, Rockefeller, Seminole, Granada (S.R. 40), Neptune, Amsden, Standish, Bent Palm Condo, Deer Ln., Arlington Way, and Bosarvy. Granada Ave. (S.R. 40) provides access to parking and driving on the beach up to the pier. No alcohol allowed.

Daytona Beach, north end, boardwalk, and south end has 4.5 miles of beach frontage. This is a highly developed, urban beach area with lifeguards. Street approaches off S.R. A1A to the north end: Plaza, Belair, Boylston, Williams, Hartford, Seaview, Horizon, Zeina, Seabreeze, and University. Access to the boardwalk: Riverview, Glenview, Seabreeze, Oakridge, Ora, Main, Harvey, Kemp, and Broadway (U.S. 92). Street approaches to the south end: Vermont, Goodall, Revilo, Lenox, and Silver Beach. No alcohol allowed.

Daytona Beach Shores has access to 4 miles of undeveloped beachfront off S. Atlantic Ave. (S.R. A1A) at Botefuhr Ave., Van Ave., Dunlawton Blvd., Demott Ave., Emilia Ave., and Phillips Ave. No alcohol allowed.

Wilbur-by-the-Sea has 1.5 miles of undeveloped beachfront reached off S.R. A1A on Heron St., Toronta Ave., and Curlew St. No alcohol allowed.

Ponce Inlet Park undeveloped beach area is reached by a beach drive at the end of Ponce Blvd., 1.5 miles north of the U.S. Coast Guard Reservation on S.R. A1A. No alcohol allowed.

New Smyrna Beach has 7 miles of undeveloped, urban beach. Access is off S. Atlantic Blvd. (S.R. A1A) on Beachway, Crawford, Flagler Ave. (lifeguard area), E. 3d Ave., and 27th Ave. Only that portion of beach area between the average high-water mark and the average low-water mark is open to motorized traffic.

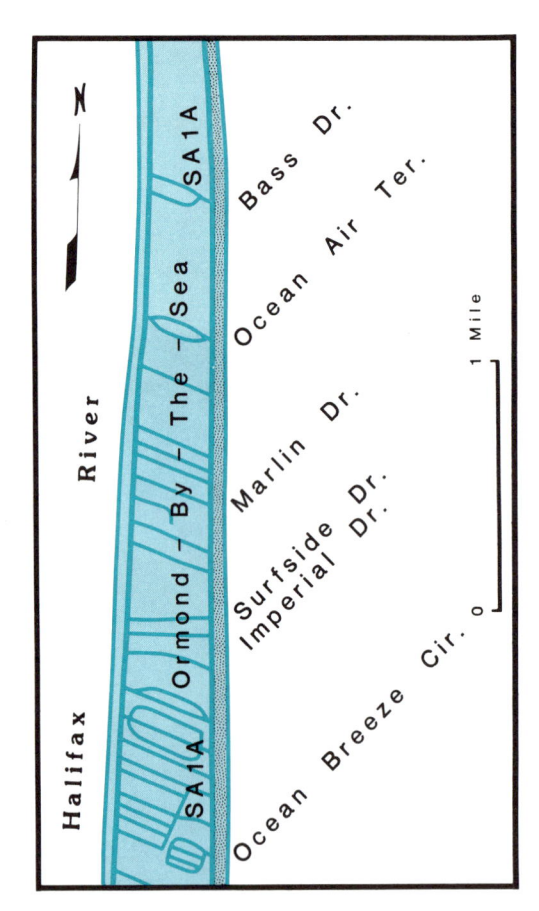

Ormond-by-the-Sea (inset 1)

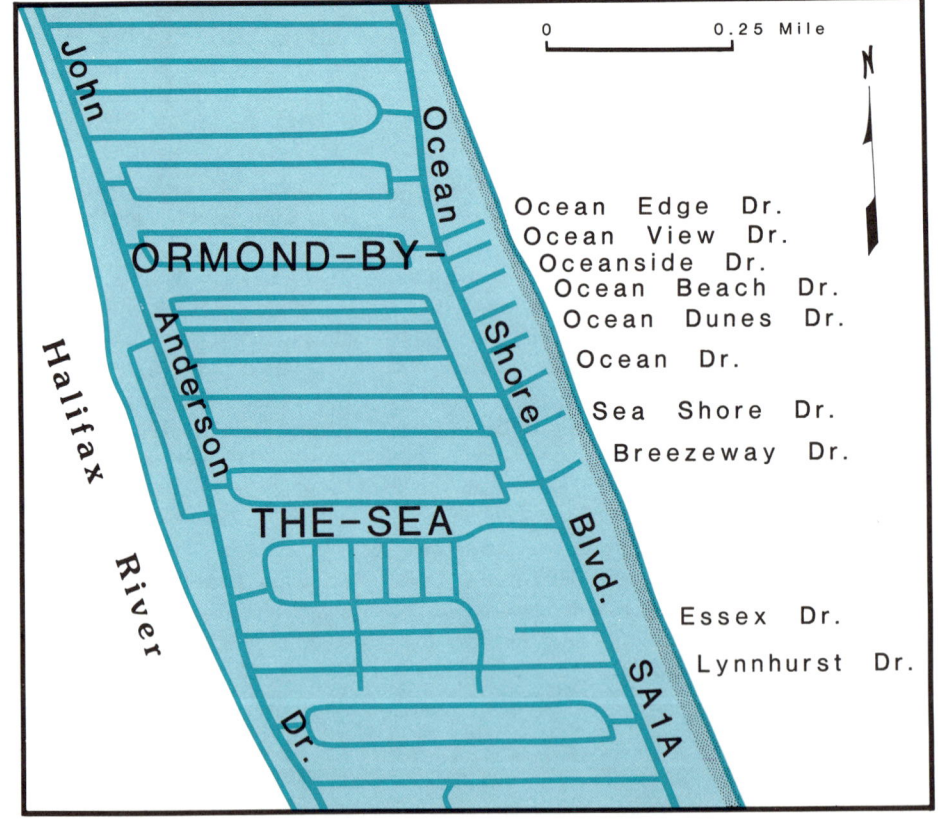

Ormond-by-the-Sea (inset 2)

34

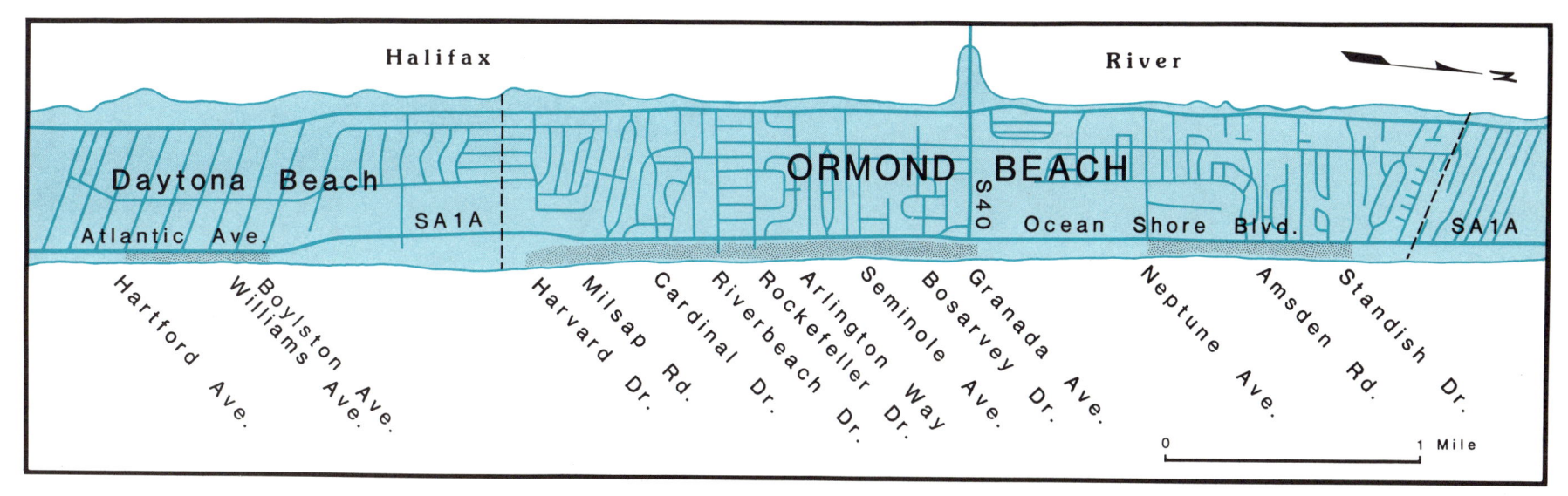

Ormond Beach (inset 3)

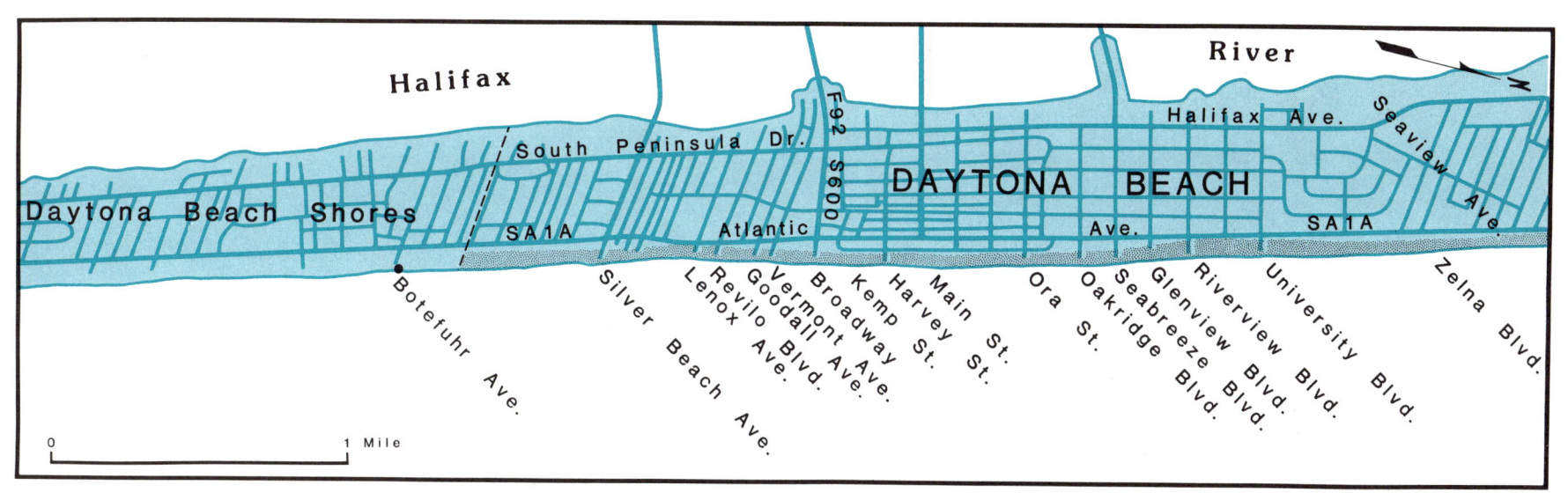

Daytona Beach (inset 4)

35

Surf casting

Central district

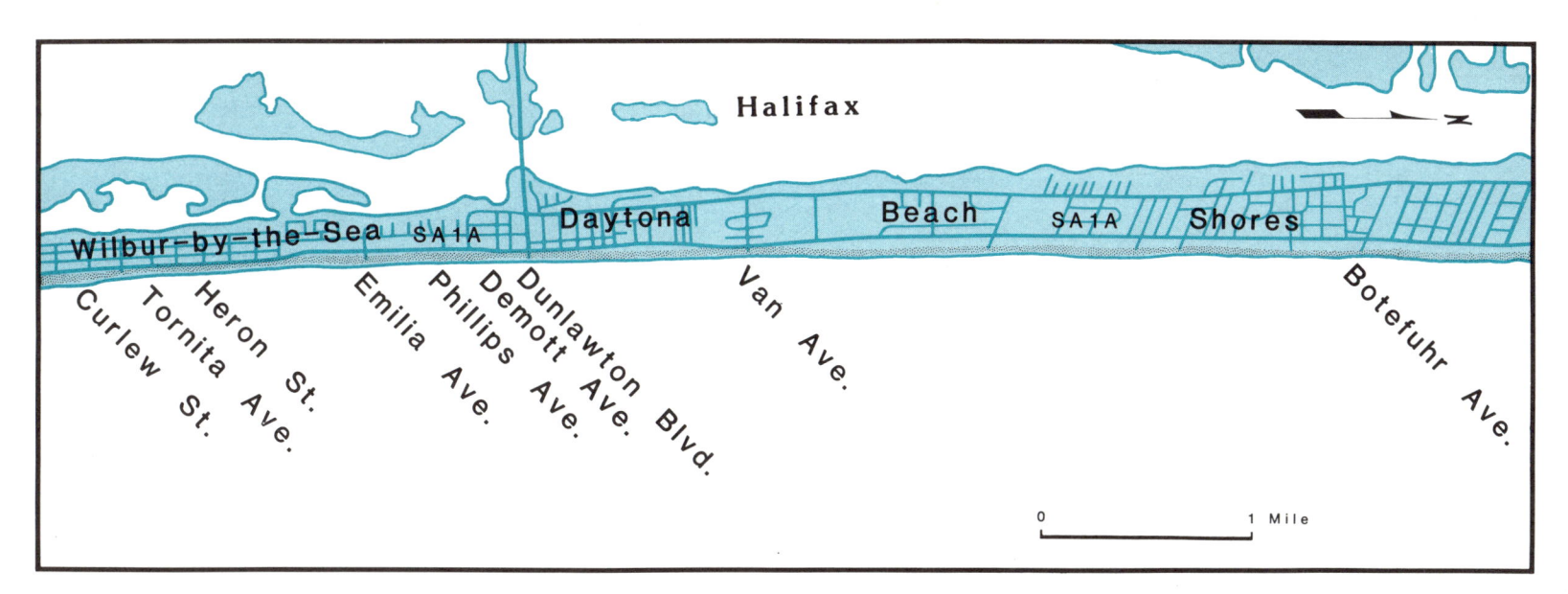

Daytona Beach Shores/Wilbur-by-the-Sea (inset 5)

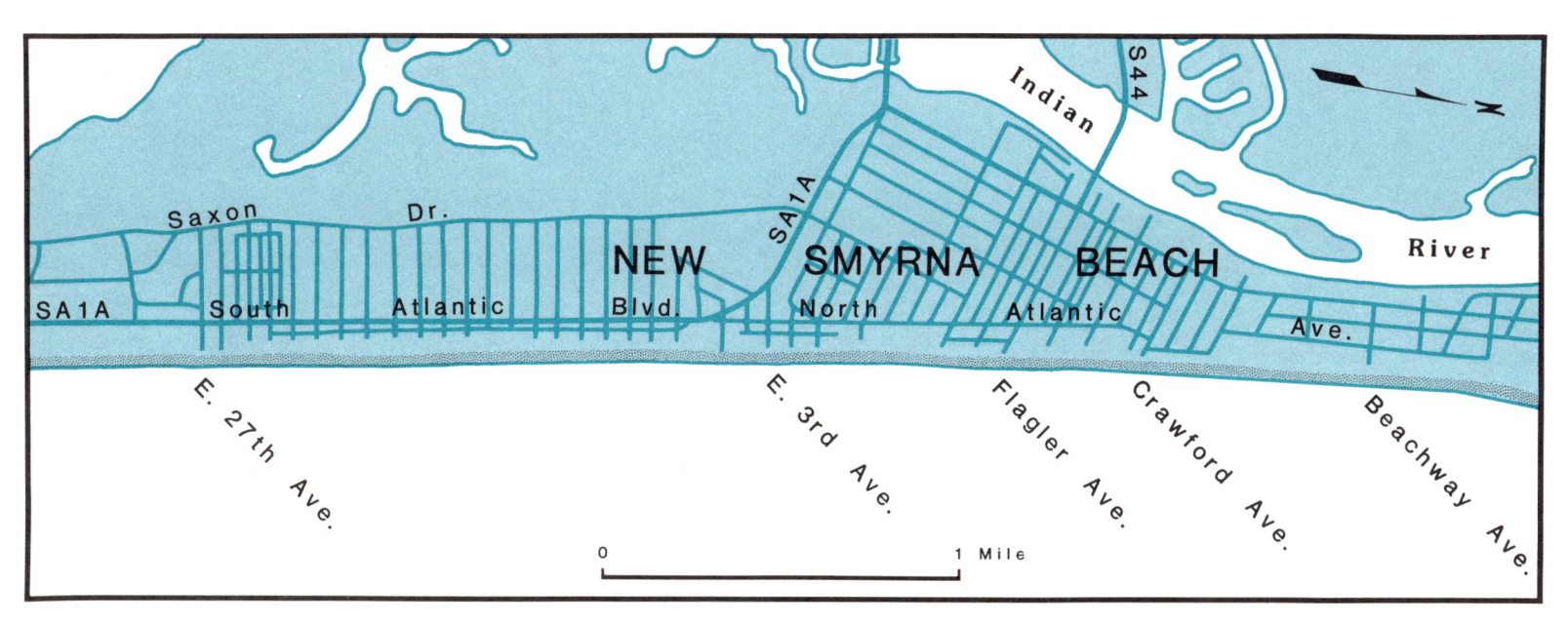

New Smyrna Beach (inset 6)

37

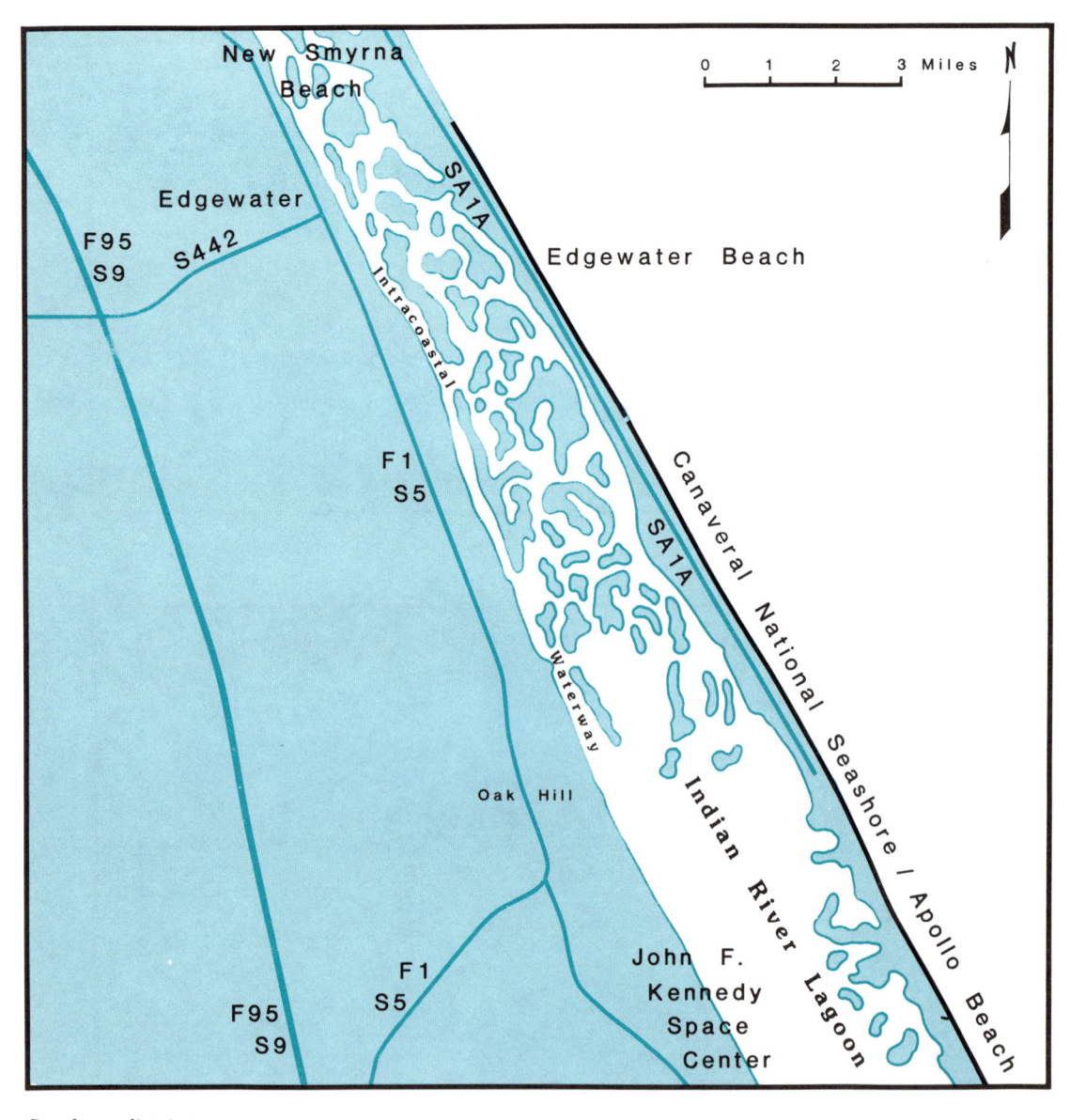

Edgewater, south of New Smyrna Beach, has 31 miles of undeveloped beachfront. Vehicular beach approach is on Hiles Blvd., off Atlantic Ave. (S.R. A1A).

Canaveral National Seashore/Apollo Beach has 5 miles of undeveloped quartz beach. Access is off S.R. A1A with roadside parking and dune walkovers. Fishing pier.

Merritt Island, a barrier island situated between Indian River and Banana River, has 45 dune crossovers off S.R. 3 to the undeveloped county beach that lies south of J.F.K. Space Center. No alcohol is allowed.

Southern district

Sea Turtles

Sea turtles are one of the only ancient reptiles surviving today. They are said to go back 90 million years (Cretaceous Period) and were contemporaries of the dinosaurs. There are two families of sea turtles found in eastern North America, Cheloniidae (four genera) and Dermochelyidae (one genus). The four genera are the loggerhead (*Caretta caretta*), the Atlantic ridley (*Lepidochelys kempii*), the hawksbill (*Eretmochelys imbricata*), and the green turtle (*Chelonia mydas*). The only representative genus of the family Dermochelyidae is the leatherback (*Dermochelys coriacea*). All five of these genera are common to Florida's Atlantic and Gulf waters and beaches.

Turtles are identified by their size and the color and shape of their shells. The green turtle is the main commercial species used for turtle soup. It has an olive-brown shell with "scutes" or plates radiating yellow, green, and black. These gentle creatures are primarily vegetarian and prefer shallow seas where algae and seagrasses grow. The green turtle found in Florida waters is known locally as the black turtle. Its oval shell and long flippers make it an excellent marathon swimmer. Mature specimens weigh 200 to 650 pounds.

Sea turtles have a nesting season from April through August. Females come onshore at night to lay their eggs in the sand. One turtle, the Atlantic Ridley, nests only in Mexico. The beach trek is laborious for the water-adapted turtles, so the female comes onshore at high tide, lessening the distance she needs to travel. She digs a hole with her hind flippers and deposits many white, leathery eggs slightly larger than ping-pong balls. The number of eggs, or clutch size, varies among species from 10 to 300. She attempts to hide the exact site of her nest by covering the eggs and whisking sand in every direction. However, she leaves large tracks on her retreat to the sea, providing poachers with a path to the nest.

If left undisturbed, each egg develops from a mass of cells to a fully formed, tiny sea turtle within 60 days. The baby turtle uses a carnucle, or "egg tooth," to scratch a hole in its shell. The members of the clutch work together to break through the sand, choosing night to lessen their exposure to heat and predators. Once free, the two-ounce hatchlings race toward an ocean they have never seen. Many are eaten by ghost crabs. More fall prey to fish, especially sharks, as they swim across reefbeds to the open sea. Only one in 10,000 eggs survives to maturity.

Tampering with nesting turtles or eggs is a serious violation of the law. Use a toll-free number to report a violation or to seek help for a stranded turtle (800-432-6404), or call the Florida Marine Patrol toll-free for violations (800-432-1829). These protective measures will help ensure a future for the last of the ancient reptiles.

For more information about sea turtles contact Sea Turtle Rescue Fund Center for Environmental Education, Inc., 624 9th St., NW, Washington, DC 20001.

Turtle Watch group (early summer): Chamber of Commerce, 1919 Commercial St., Jensen Beach, FL (315) 334-3444.

First aid station and release program: Mariners Aid to Sea Turtles (MAST), Palm Beach Shores, FL 33404.

Saltwater Fishing Piers

Florida's saltwater fishing piers offer fishermen of all ages and degrees of experience the opportunity to enjoy fishing for the numerous species that appear in the state's coastal waters. The proximity of the Gulf Stream to the Atlantic coast and the broad and shallow continental shelf on the Gulf of Mexico coast provide habitat, food, and water temperatures that bring many of the popular game species within reach of the pier fisherman.

Piers provide a stable platform for fishing that is popular among persons not wanting to fish from a boat. There is a modest fee (in some cases no fee at all), and piers offer such advantages as day and night fishing, service facilities, nearby parking, and immediate evacuation in the event of inclement weather.

Since weather and seasonal variations influence the catch rates and composition at the piers, it is suggested that interested fishermen contact the pier to determine availability of bait and current fishing reports.

Note: This information was compiled as a result of a Florida Sea Grant funded survey. The State University System of Florida Sea Grant College is supported by award of the Office of Sea Grant, National Oceanic and Atmospheric Administration, U.S. Department of Commerce (grant number NA80AA-D-00038) under provisions of the National Sea Grant College and Programs Act of 1966. This information is published by the Marine Advisory Program which functions as a component of the Florida Cooperative Extension Service.

Canaveral Pier

SALTWATER FISHING PIER LOCATIONS

County	City/Pier Name	Daily Fee
Nassau	Fernandina Beach/Ft. Clinch Fishing Pier	$0.50
Duval	Jacksonville Beach/Jacksonville Beach Pier	2.50
St. Johns	St. Augustine/Lighthouse Park Pier	
Flagler	Flagler Beach/Flagler Beach Pier	1.50
Volusia	Ormond Beach/Ormond Beach Pier	2.00
	Daytona Beach/Ocean Pier Casino	1.00
	Sunglow Ocean Pier	2.00
Brevard	Cape Canaveral/Jetty Park Campground	0.50
	Cocoa Beach/Canaveral Pier	1.50
	Melbourne/Eau Gallie Fishing Pier	
	Melbourne Beach Fishing Pier	
	Sebastian Inlet Recreation Area	
	Merritt Island/Barge Canal Tingleys Fishing Camp	
	Titusville/Veterans Memorial Park	
Palm Beach	Juno Beach/Juno Beach Fishing Pier	2.00
	Lake Worth/Lake Worth Pier	1.50
Broward	Dania/City of Dania Fishing Pier	2.00
	Deerfield Beach/Deerfield Fishing Pier	1.25
	Lauderdale By-The-Sea/Anglin's Fishing Pier	2.00
	Pompano Beach/Fisherman's Wharf	1.45
Dade	Miami Beach/Haulover Park Fishing Pier	2.00
	Sunshine Pier	
Collier	Naples/Naples Municipal Pier	
Lee	Bokeellia/Bokeellia Seaport Pier	2.00
	Cape Coral/Municipal Pier	
	Ft. Myers/Ft. Myers Beach Pier	
	Tarpon Street Pier	
	Sanibel/Lighthouse Fishing Pier	
Charlotte	Port Charlotte/Charlotte Harbor Fishing Pier	
	Anglers Pier (Lemon Bay) (2)	
	Port Charlotte Beach Pier	
	Punta Gorda/Punta Gorda Municipal Pier	
Sarasota	Sarasota/Municipal Pier (Hart's Landing)	
	Venice/Venice Municipal Fishing Pier	0.50
Manatee	Anna Maria/Anna Maria City Pier	
	Rod & Reel Pier	
	Bradenton Beach/Bradenton Beach Pier	
Pinellas	Clearwater/Big Pier 60	3.00
	Indian Rocks Beach/Big Indian Rocks Fishing	3.50
	Oldsmar/Oldsmar Pier	
	Redington Shores/Redington Long Pier	3.50
	St. Petersburg/The Pier	
	Safety Harbor/Safety Harbor City Pier	
	Tierra Verde/Andrew Potter Pier	
	Family Pier I	
Franklin	Apalachicola/Battery Park	
	Lafayette Park	
Bay	Panama City/Bay County Public Pier	
	County Pier	
	Panama City Beach/The Dan Russell Pier	2.00
	Mexico Beach/Mexico Beach Pier	
Okaloosa	Ft. Walton Beach/Okaloosa Island Pier	2.50
Santa Rosa	Cantonment/Sound Fishing Pier	
Escambia	Gulf Breeze/Ft. Pickens Fishing Pier	
	Navarre Beach Fishing Pier	4.00
	Pensacola/Municipal Fishing Pier	
	Pensacola Bay Fishing Pier	0.25
	Pensacola Beach/Pensacola Beach Fishing Pier	3.00

41

Cape Canaveral/Cocoa Beach

Brevard County

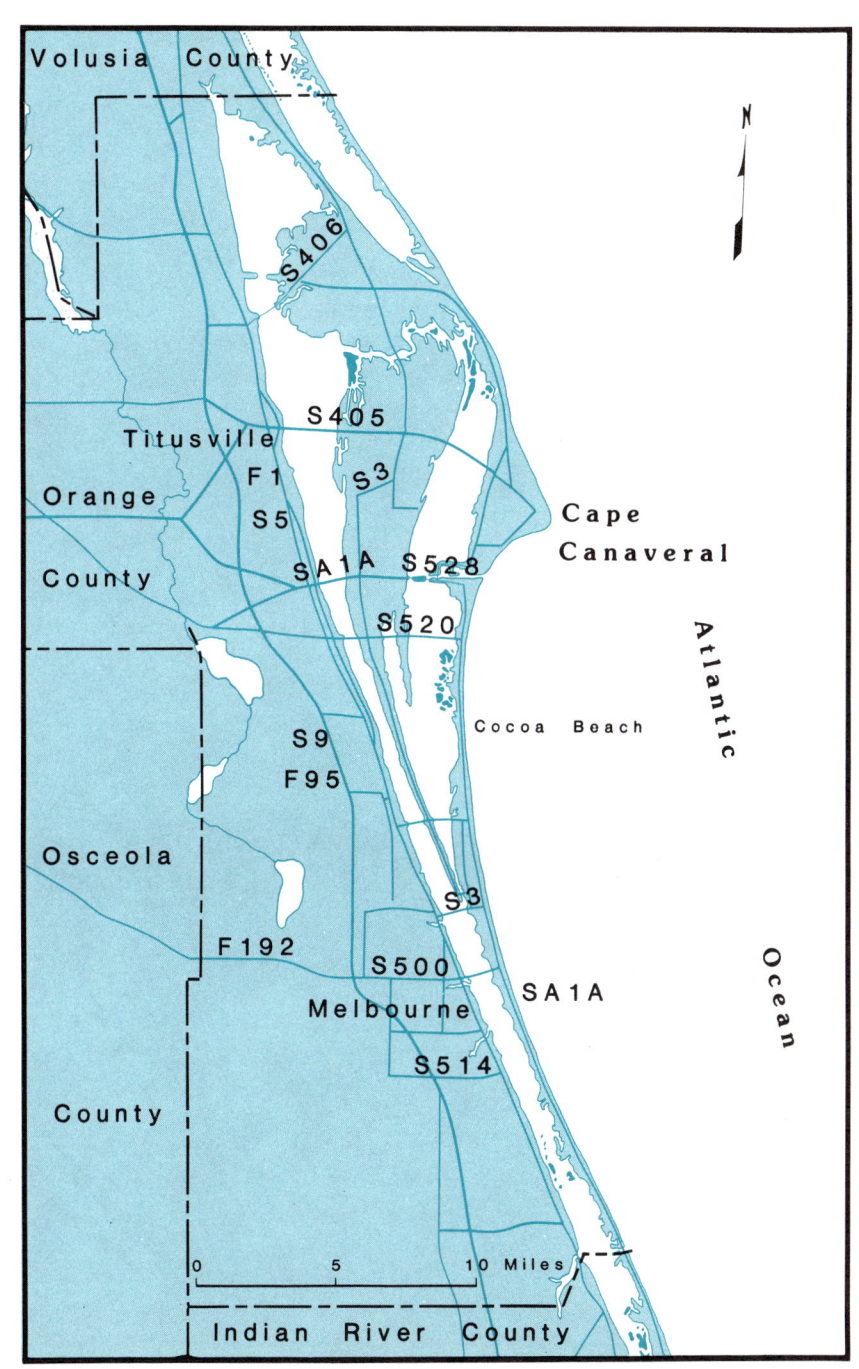

Brevard County is located in central Florida on the Atlantic Ocean. Although only 25 miles wide, it is nearly 75 miles long, making it Florida's longest county. The area has all the pleasures of sand, surf, and sun as well as international attention for its role in the space exploration program.

The north end of Merritt Island, between Indian River and Banana River, is devoted to wildlife management, public recreation, and the national space program. Northernmost is Canaveral National Seashore under the direction of the National Park Service, which has total responsibility for preserving the primitive beach while providing public access to its natural environment.

Just south lies Merritt Island National Wildlife Refuge. These 131,143 acres, established and maintained by the Fish and Wildlife Service, are sanctuary for endangered species, such as the manatee, green sea turtle, brown pelican, and peregrine falcon.

The seashore and refuge lie along the Atlantic flyway, a major migration route for birds. The journey for many birds begins in northern tundra regions of Alaska and Canada and ends in the tropics of Florida and the Caribbean, a journey of 4,000 to 6,000 miles. Merritt Island is a major wintering area; hunting is permitted within the boundaries of the refuge as an important aspect of waterfowl management.

The remaining area of Merritt Island under government control is John F. Kennedy Space Center (NASA). Visitors can tour the rocket launch site and a visitors' center that includes exhibits, museums, and movies about the space program.

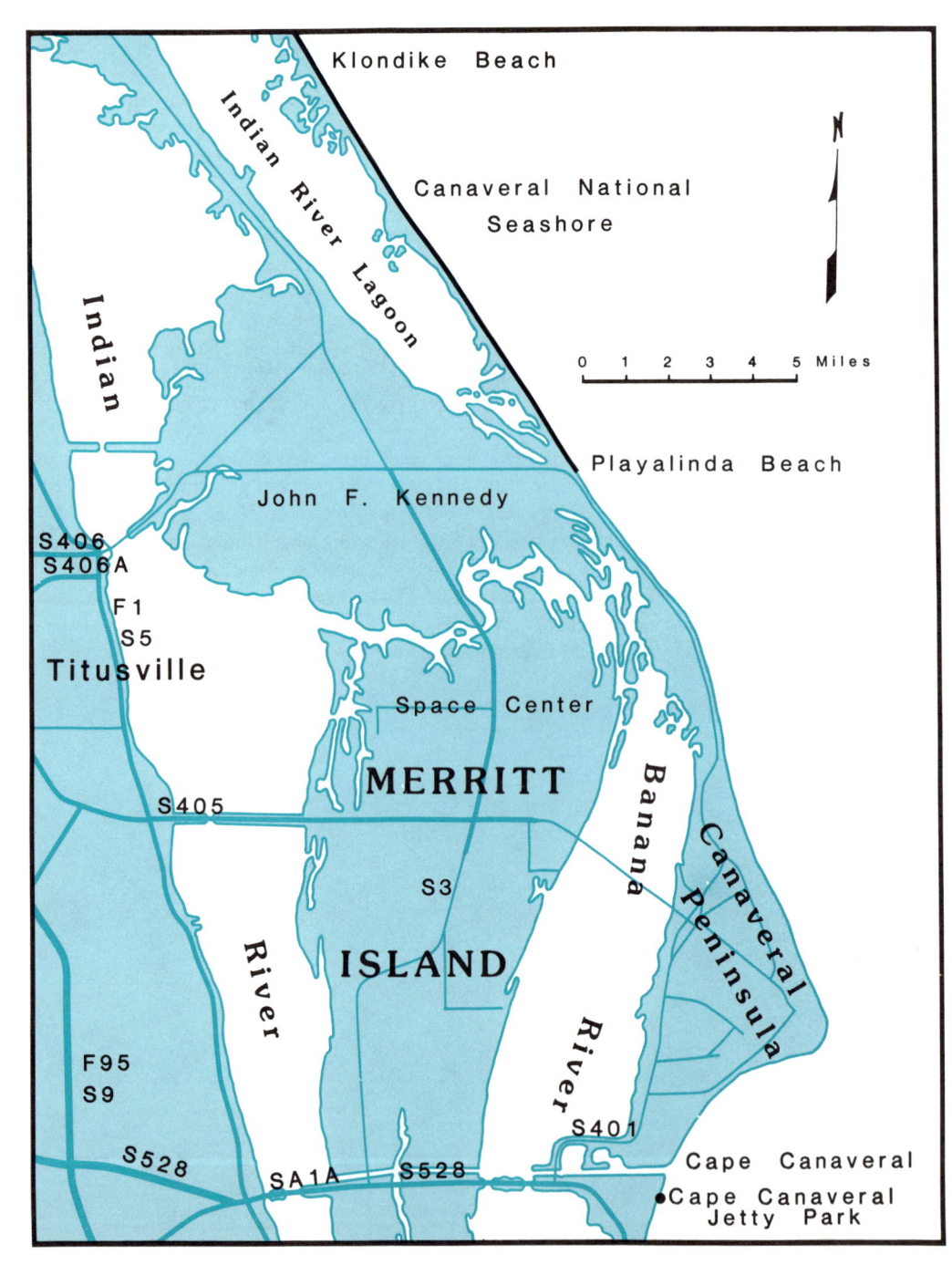

Canaveral National Seashore/Klondike Beach has 16 miles of remote, undeveloped oceanfront. This section of beach between Apollo and Playalinda, at the Volusia-Brevard county line, is accessible by foot only.

Canaveral National Seashore/Playalinda Beach is on the seashore's southern edge within easy viewing distance of the NASA rocket installations and may be closed at times of Kennedy Space Center activities. Access is west on S.R. 402 from Titusville.

Cape Canaveral Jetty Park has 26 developed acres on the barrier island at the east end of S.R. 528 (Bennett Memorial Causeway). There are camping, picnicking, fishing, and beach facilities.

Northern district

Brevard County

NAME	Easy Access	Parking or Entrance Fee	Parking	Restrooms	Showers	Picnicking	Swimming	Lifeguards	Fishing	Boating Facilities	Shelters	Concession Stands	Handicapped Facilities	Public Transportation	Group Facilities	Nature Trails / Fitness	Atlantic / Gulf	Bay / Soundfront	Sandy Beach	Rocky Beach	Primitive Beach	Urban
NORTHERN DISTRICT																						
Canaveral National Seashore																						
Klondike Beach						●	●										●		●		●	
Playalinda Beach	●		●	●		●	●	●	●			●			●		●		●			
Port Canaveral Jetty Park	●	●	●	●	●	●	●		●		●	●	●		●		●		●			●
CENTRAL DISTRICT																						
Sheppard Park	●	●	●	●		●	●				●		●				●		●			
Sidney Fisher Park	●	●	●	●		●	●				●						●		●			
Lori Wilson Park	●		●			●	●	●									●		●			
Cherie Down Park	●		●		●	●	●	●	●								●		●			
Pelican Beach Park	●	●	●	●	●	●	●		●			●	●	●			●		●			
Bicentennial Park/Indian Harbor Beach	●		●			●	●		●				●				●		●			
Paradise Beach Park	●	●	●	●	●	●	●				●	●	●	●	●		●		●			
SOUTHERN DISTRICT																						
Indialantic Boardwalk	●		●			●	●					●					●		●			
Spessard Holland Park	●	●	●	●	●	●	●				●	●					●		●			●
Long Point Park	●	●	●	●	●	●	●	●	●	●			●	●			●	●	●			
Sebastian Inlet State Recreation Area	●	●	●	●	●	●	●	●	●	●	●	●	●		●	●	●	●	●			

FACILITIES — ENVIRONMENT

Port Canaveral Jetty Park has 35 developed acres on the barrier island at the east end of S.R. 528 (Bennett Memorial Causeway). Camping facilities and a fishing pier.

Paradise Beach Park is a developed, urban park about 2.25 miles north of Melbourne Causeway (S.R. 516) on S.R. A1A. Group facilities and 270 parking spaces.

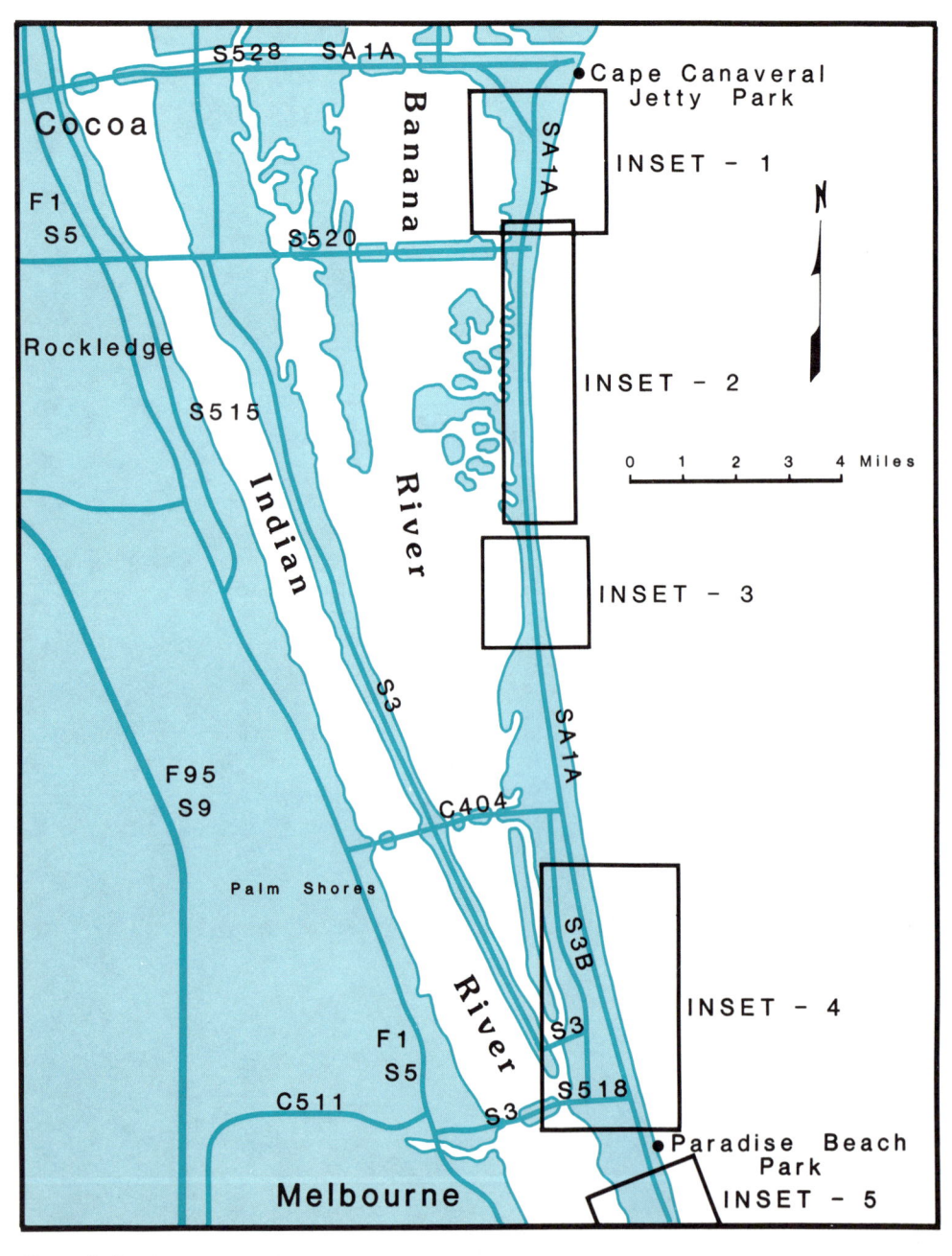

Central district

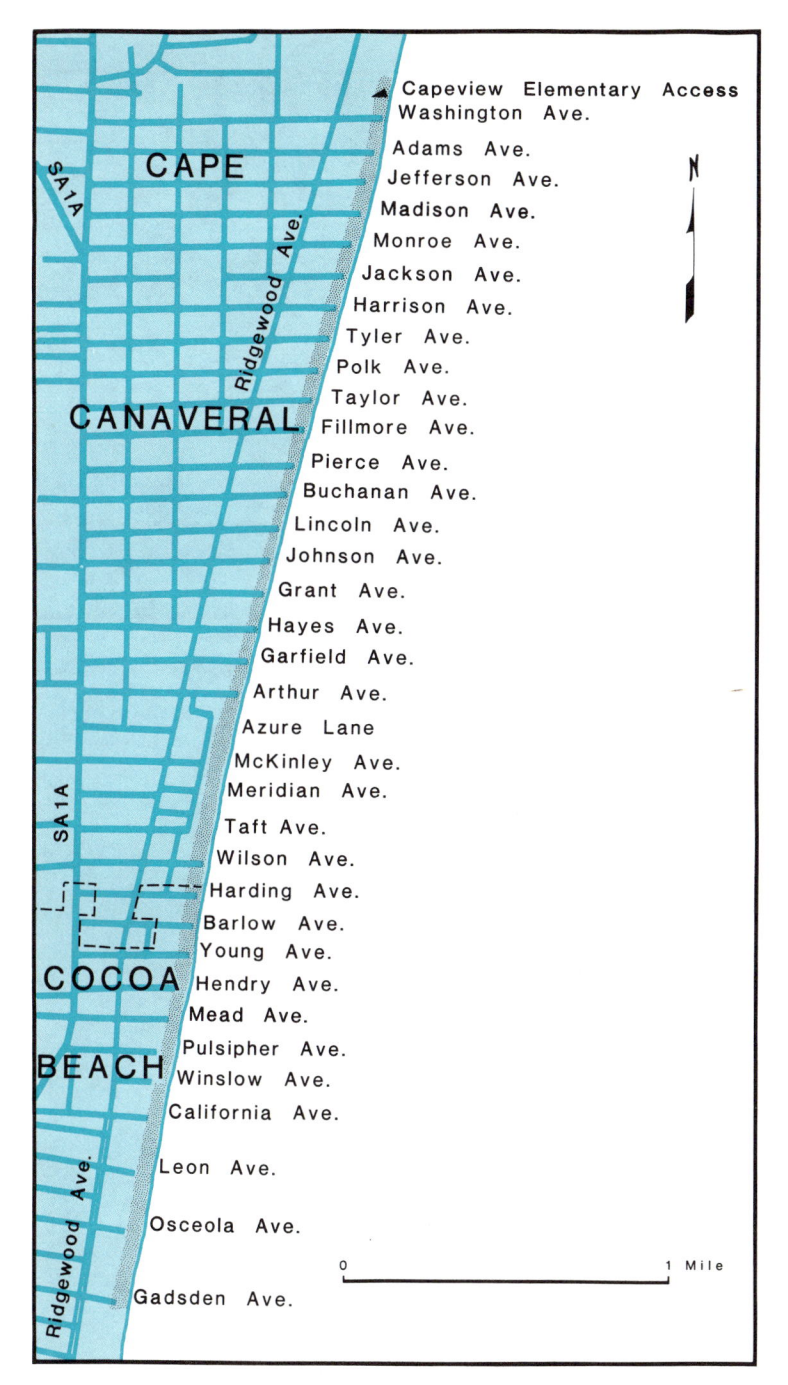

Cape Canaveral access, see map. (Only those accessways having special features are listed.)

    Grant Ave. (crossover with ramp, 15 parking spaces)
    Hayes Ave. (crossover with ramp, 30 parking spaces)
    Garfield Ave. (crossover with ramp and platform, 20 parking spaces)
    Arthur Ave. (crossover with ramp, 10 parking spaces)
    McKinley Ave. (long handicap-accessible boardwalk with platform)
    Wilson Ave. (crossover with ramp, 4 parking spaces)

Cocoa Beach access:

    Harding Ave. (4 parking spaces)
    Barlow Ave. (6 parking spaces)
    Young Ave. (no parking)
    Hendry Ave. (no parking)
    Mead Ave. (21 parking spaces)
    Pulsipher Ave. (40 parking spaces)
    Winslow Ave. (handicap car ramp, 16 parking spaces)
    California Ave. (no parking)

Cocoa Beach access (Cocoa Ocean Beach):

    Leon Ln. (30 parking spaces)
    Osceola Ln. (handicap ramp, 25 parking spaces)
    Gadsden Ln. (14 parking spaces)
    Marion Ln. (14 parking spaces)
    Palm Ln. (handicap ramp; 31 parking spaces)
    Flagler Ln. (handicap ramp, 21 parking spaces)

Cape Canaveral (inset 1)

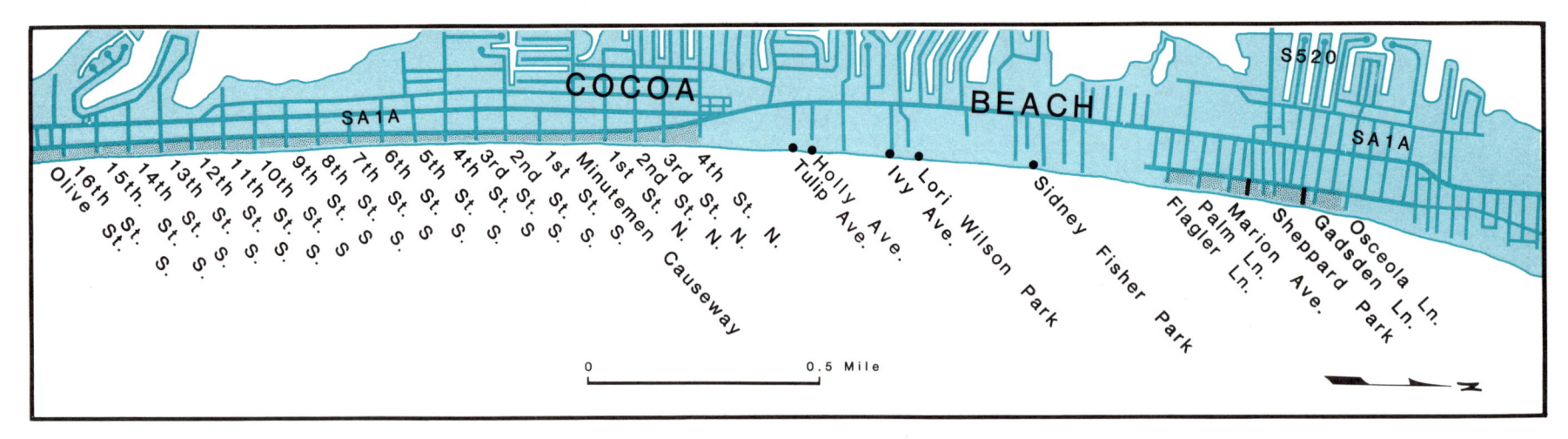

Cocoa Beach (inset 2)

Sheppard Park is located at the ocean end of S.R. 520. These 9 acres of developed beach have a promenade along the ocean and bicycle facilities.

Sidney Fisher Park has 300 feet of beach with dune walkovers. Parking for 200 cars and bicycle facilities.

Lori Wilson Park is located 1.5 miles south of S.R. 520 on S.R. A1A. This 20-acre park has 2 large pavilions and 2 dune crossovers. Parking for 46 cars.

Cocoa Beach access (Seacrest Beach):
Ivy Ave.
Holly Ave.
Tulip Ave. (handicap ramp, 80 parking spaces)

Cocoa Beach access (Cocoa Beach):
4th St.–1st St.
Minuteman Causeway (Cocoa Ave.)
1st St. S.–16th St. S. (handicap ramps at 13th and 10th Ave. S., 1st St. N., Tulip Ave., Flagler Ave., and Palm Ave.)

Brevard Co. access:
Olive St. (crossover with ramp, parking)
Summer St. (large dune crossover, parking)
Fern St. (large dune crossover, 4 parking spaces)
Crescent Beach Dr. (crossover, 10 parking spaces)
24th St. (crossover, no parking)
Sunny Ln. (crossover, some parking)
26th St.–35th St.
S. Patrick Dr.
Patrick Dr.

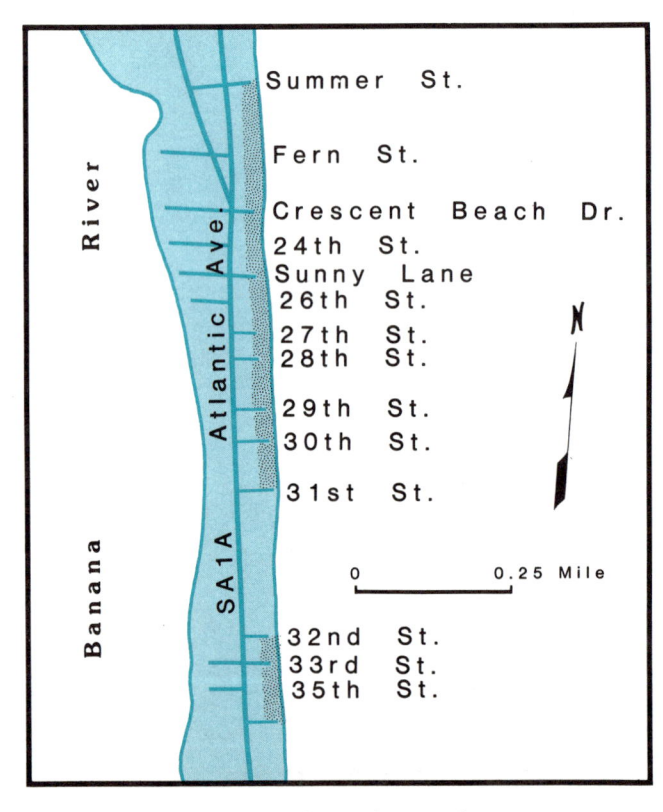

48

South Cocoa/North Patrick Beach (inset 3)

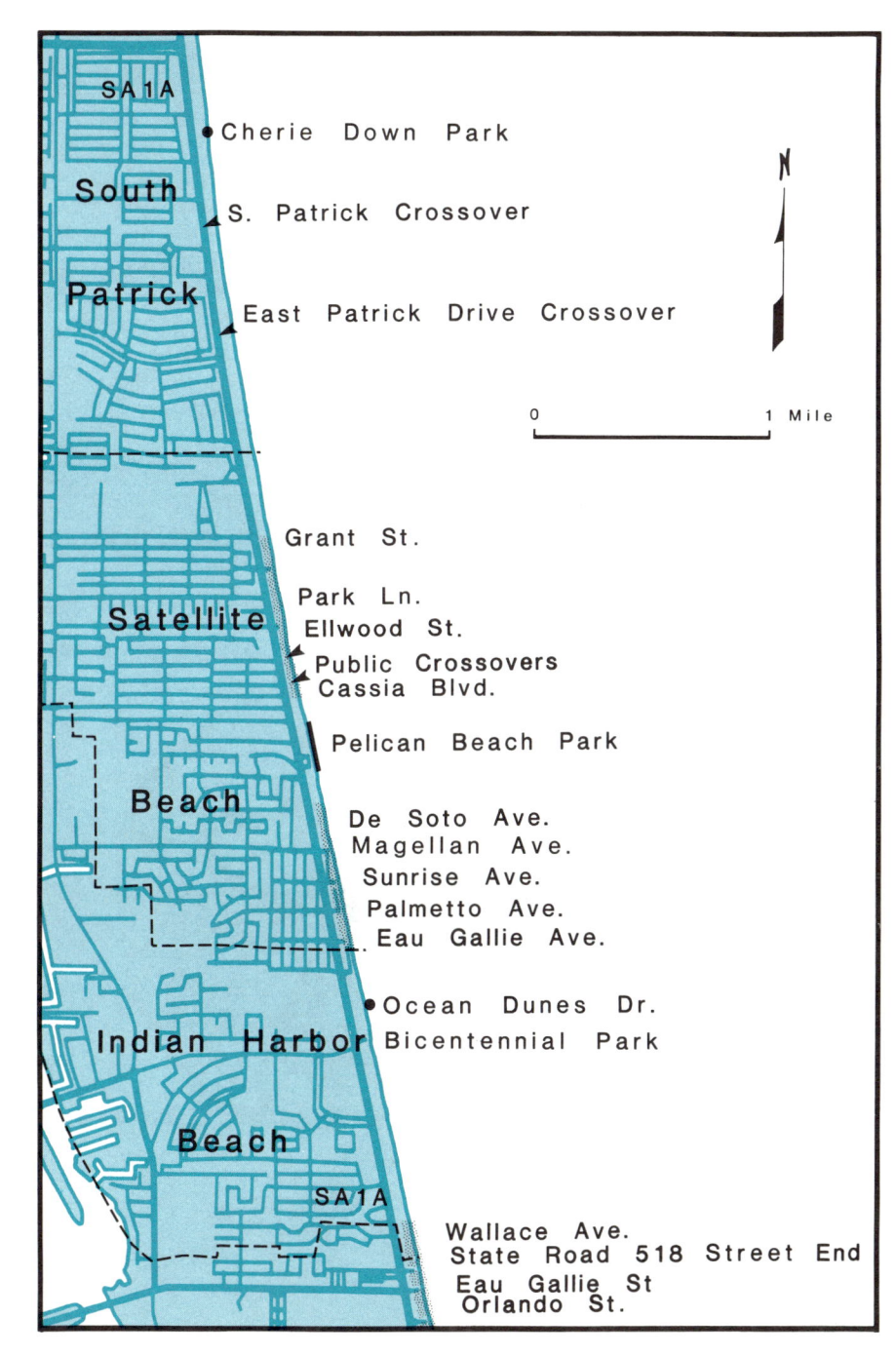

Cherie Down Park has 3 acres of undeveloped beach off S.R. A1A about 0.5 mile south of S.R. 404. Wide handicap-accessible boardwalk.

Pelican Beach Park is located 2.5 miles north of the intersection of S.R. 518 (Royal Palm Blvd.) and S.R. A1A. This large beachfront park has paved parking for 200 autos, 2 beach crossovers, and an observation tower.

Bicentennial Park/Indian Harbor Beach has one acre of developed beach at the end of Ocean Dunes Dr. in Indian Harbor Beach.

Satellite Beach access:
   Grant St.
   Park Ln.
   Ellwood St.
   Cassia Blvd.
   DeSoto Parkway
   Magellan Ave.
   Sunrise Ave.
   Palmetto Ave.

Brevard Co. access (Canova Beach):
   Wallace Ave. (handicap access)
   Eau Gallie St. (handicap access)
   Orlando St. (handicap access)

State of Florida public access:
   S.R. 518 (stone walkway to beach)

South Patrick/Indian Harbor Beach (inset 4)

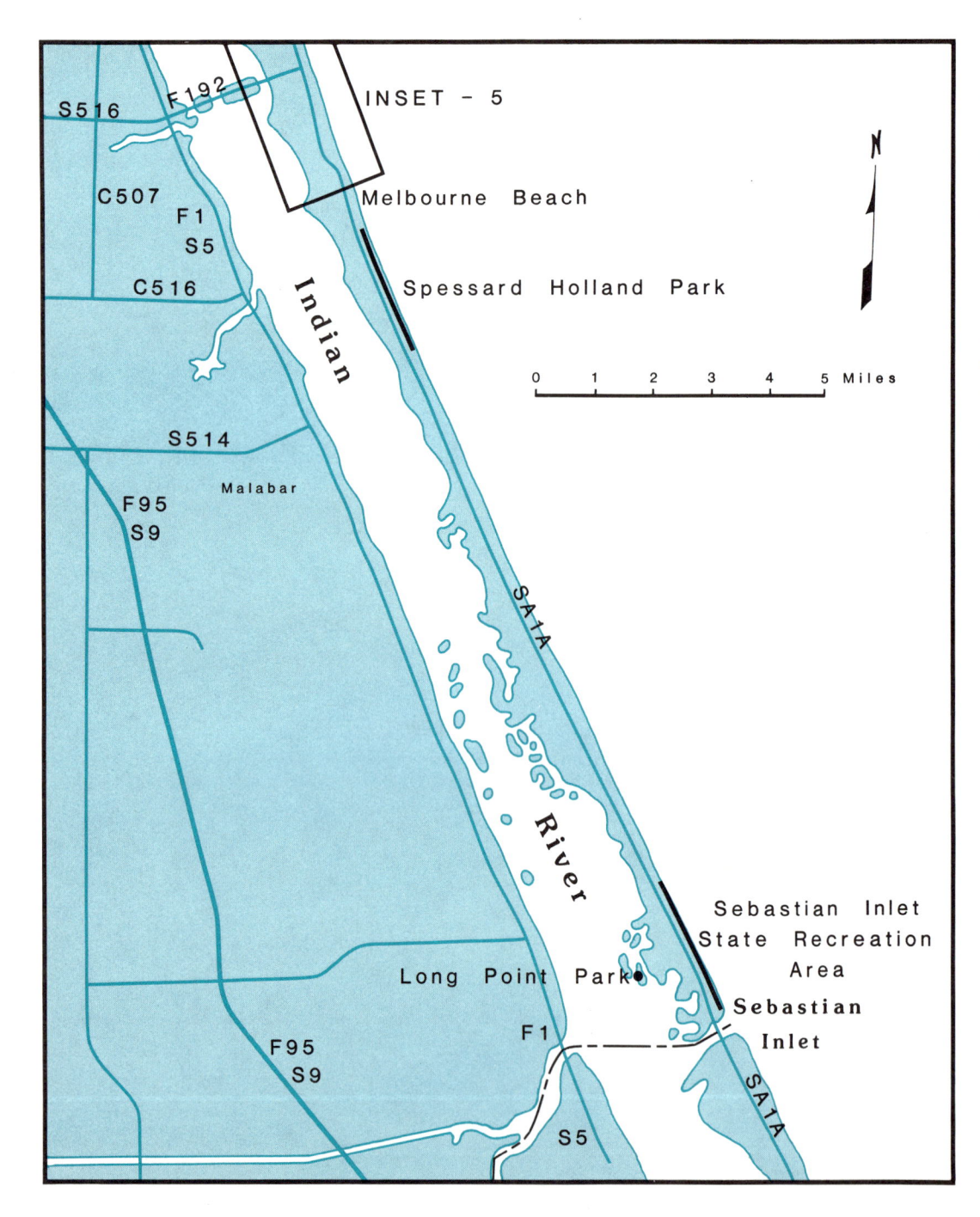

Spessard Holland Park has about 4,000 feet of developed oceanfront situated 0.5 mile south of Melbourne Beach on S.R. A1A. The park is separated by a Patrick A.F.B. tracking station to become eventually a permanent landmark or be dismantled. The park has an 18-hole golf course, lighted ballfields, and 180 parking spaces.

Long Point Park lies 2 miles northwest of the south county line on the west side of the barrier island. Campsites around the island circle a pavilion and concession-picnic area.

Sebastian Inlet State Recreation Area has 4,000 feet of beachfront and extends across the barrier island from the ocean to the Indian River.

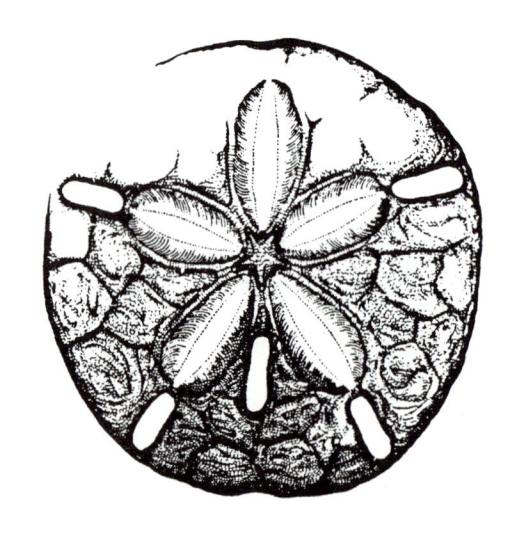

Southern district

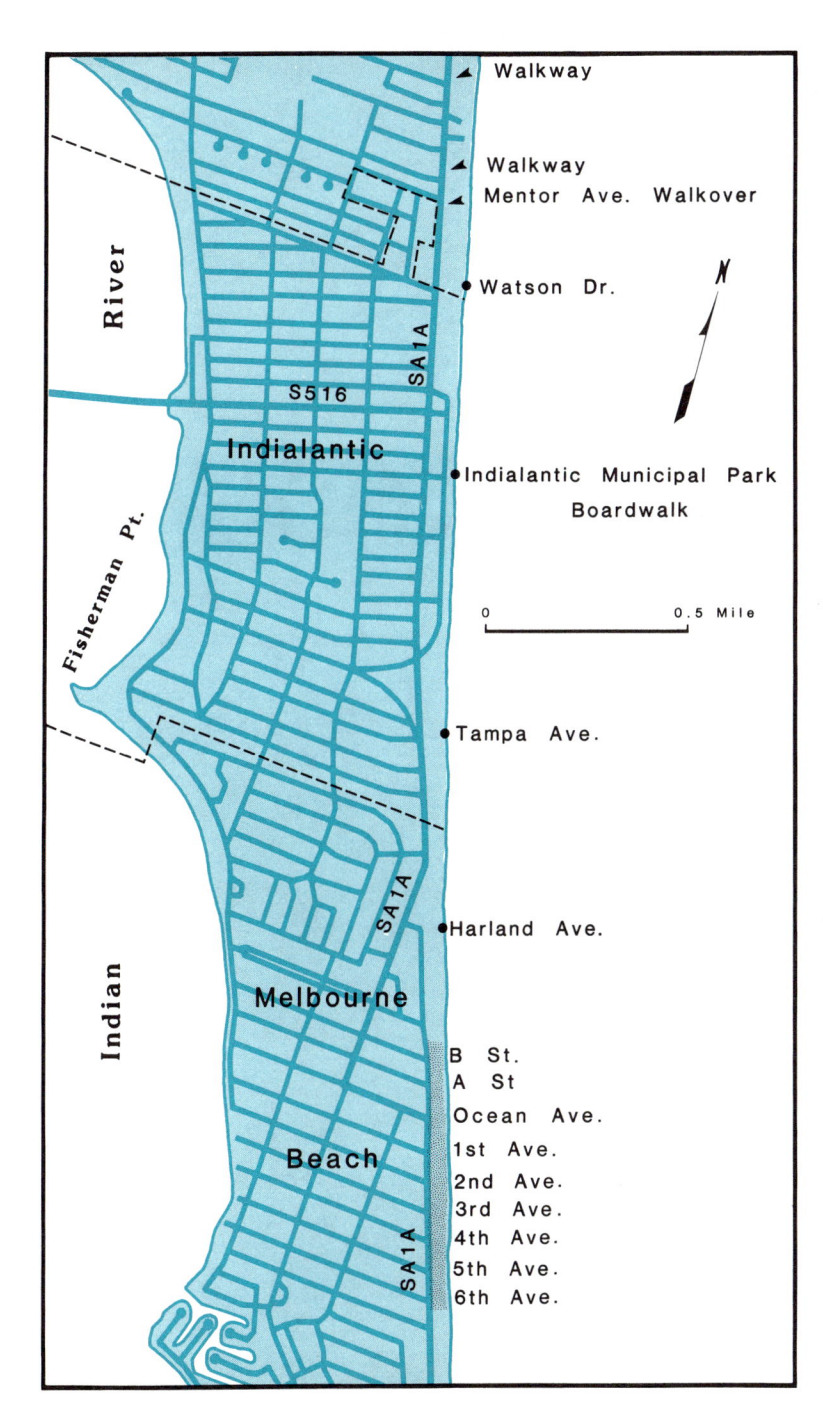

Indialantic Boardwalk is a municipal park with no facilities other than parking (400 spaces) and walkway. Located in Indialantic at the beach end of Memorial Causeway.

Indialantic access:
- Watson Ave. (16 parking spaces)
- Tampa Ave. (4 parking spaces)
- B St.
- A St.
- Ocean Ave. (concrete boardwalk)
- 1st Ave.
- 2d Ave.
- 3d Ave.
- 4th Ave.
- 5th Ave.
- 6th Ave.

Indialantic/Melbourne Beach (inset 5)

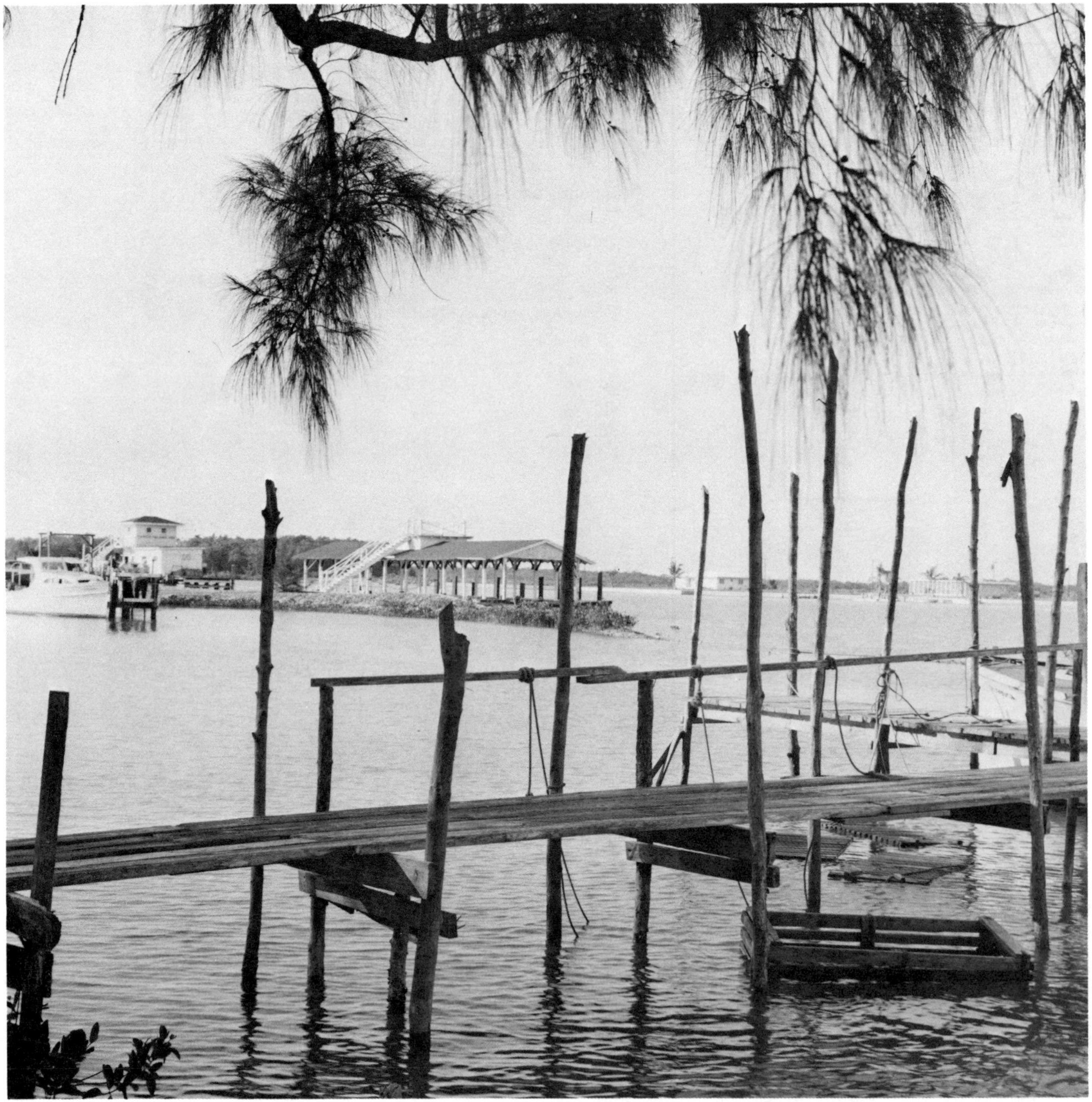

Indian River County

Indian River County is located along the mid-Florida east coast. The county seat, Vero Beach, is built on both the mainland and the barrier island. Avenues of coconut palms enhance the residential area of this quiet resort town. Across the Indian River Causeway lies the city's resort area that fronts a reef-protected beach.

History records the sinking of eleven Spanish ships carrying $14 million in gold and silver from Havana during a violent hurricane in July 1715. The loss of the ships cost the lives of 1,000 persons although 1,500 managed to reach the shore and set up camps. Offshore salvagers recovered $5 million of treasure, only to have it stolen by a swashbuckling English pirate.

Although early Indian tribes settled in the county, a trader named Barker was one of the first to start a business on top of an Indian shell mound. In 1865, August Park settled in the present-day Sebastian area, opening a store to sell food and stores to boating crews.

In April 1894 the S.S. *Breckenshire* ran onto the reef about ¾ mile off present-day Humiston Park. Its rusting remains are now sanctuary for multitudes of fish, which makes the area interesting for scuba divers.

Grapefruit and oranges ripen during the fall and winter and can be picked until summer and early fall. Vero Beach is the shipping center for the famous Indian River citrus produced throughout this area. The groves and packinghouses often welcome interested visitors.

Located at the northern county line is Sebastian Inlet State Recreation Area. It covers over 643 acres and is located on S.R. A1A about halfway between Vero Beach and Melbourne. This developed park area includes the McLarty State Museum on the site of an old Spanish salvage camp.

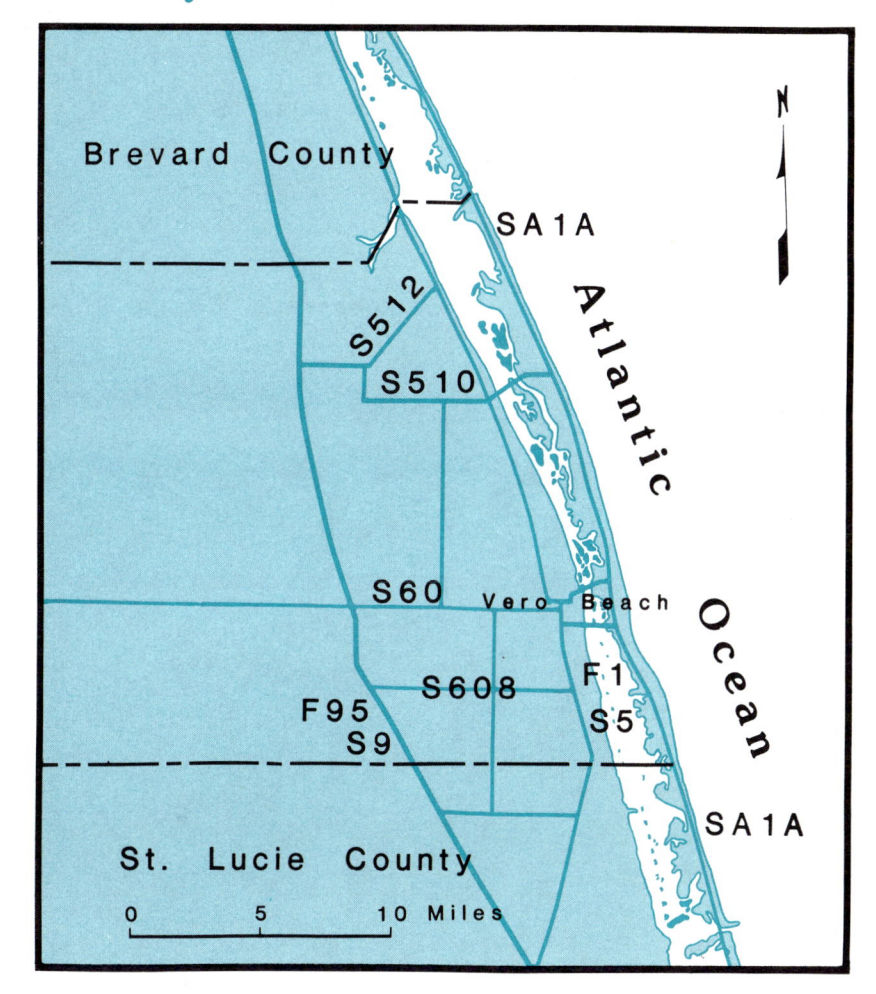

Sebastian Inlet State Recreation Area is an undeveloped beachfront state park about 9,500 feet in length. It is 5.5 miles north of Wabasso Beach on S.R. A1A with access and parking on both sides of the highway before Sebastian Inlet Bridge. Alcohol is not permitted. Overnight camping and a fishing pier.

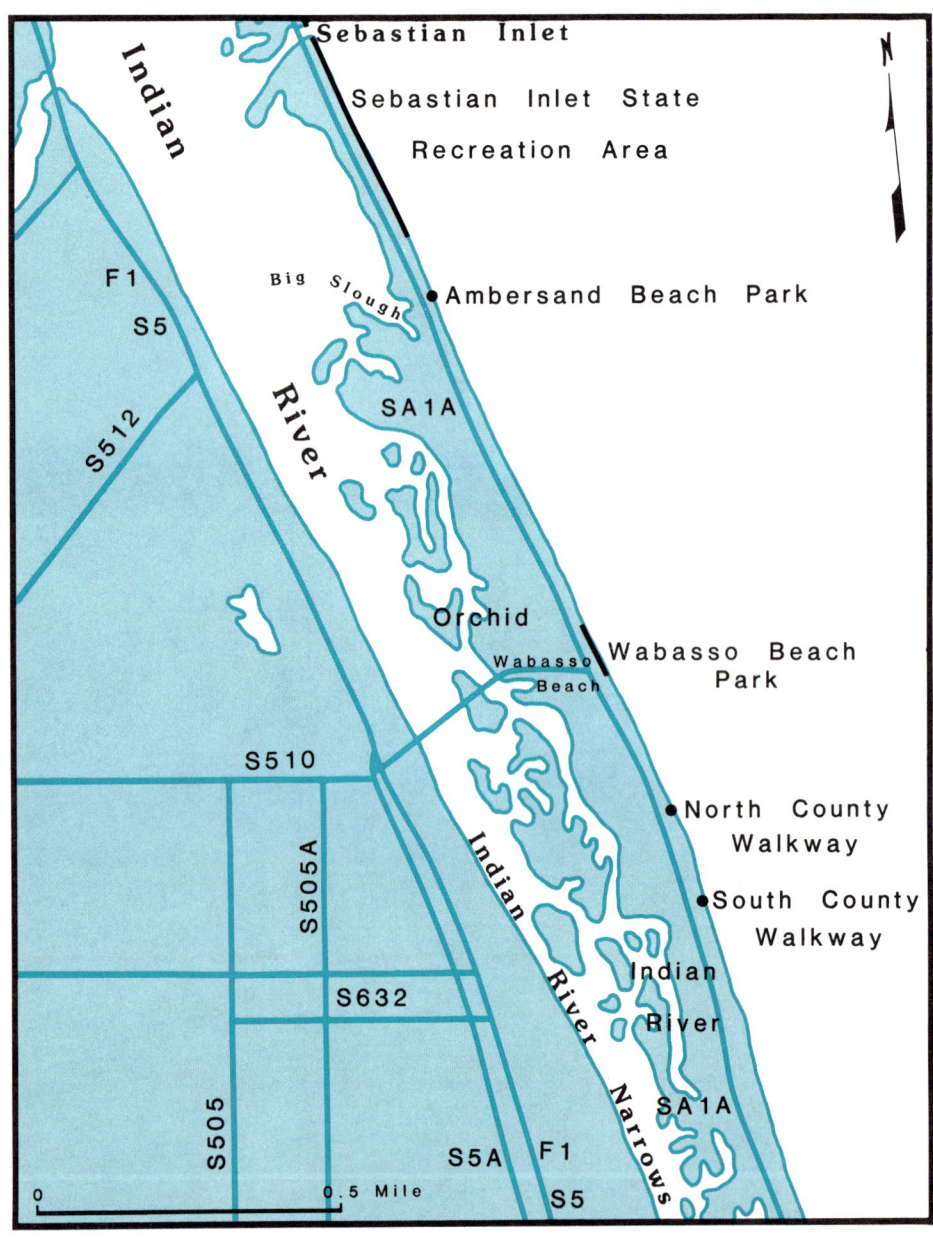

Ambersand Beach Park will open in summer 1985. It lies north of Orchid City about 5 miles from the intersection of S.R. 510 and S.R. A1A. There is a dune walkover. Development plans include a "sail-cat" ramp.

Wabasso Beach Park is located less than 100 feet north of the intersection of S.R. 510 and S.R. A1A in Wabasso Beach. There is access to 400 feet of beachfront at Caymen Rd., Sandpiper Rd., Sand Dollar Ln., and Pebble Path. It is a developed beach with lifeguards and is generally crowded on the weekends. Alcohol is not permitted.

North County Walkway lies 2 miles south of Wabasso Beach on S.R. A1A. There is a dune walkover to almost 400 feet of undeveloped beach.

South County Walkway provides access to 525 feet of undeveloped beach located 3 miles south of Wabasso Beach on S.R. A1A.

Northern district

Indian River County

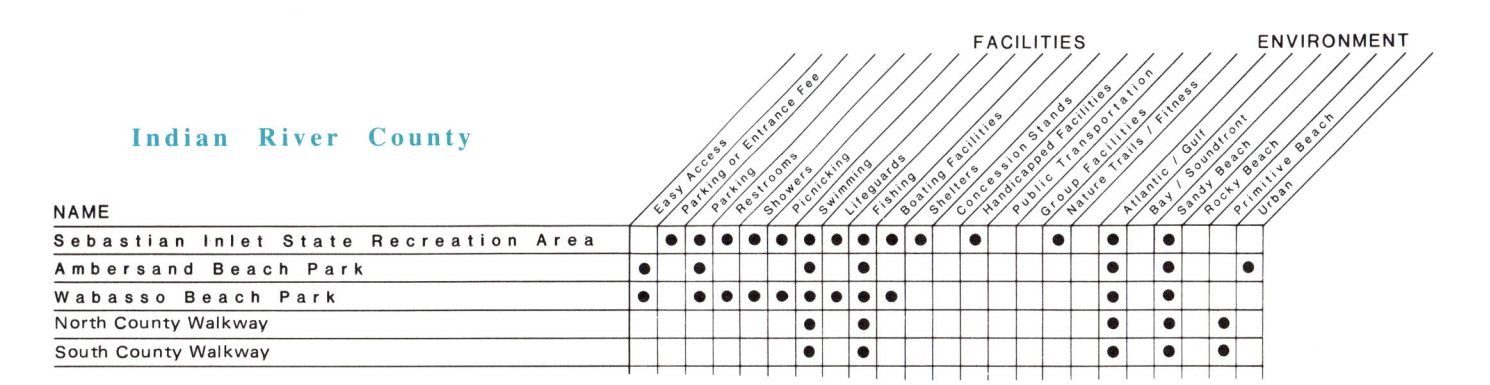

NAME	Easy Access	Parking or Entrance Fee	Parking	Restrooms	Showers	Picnicking	Swimming	Lifeguards	Fishing	Boating Facilities	Shelters	Concession Stands	Handicapped Facilities	Public Transportation	Group Facilities	Nature Trails / Fitness	Atlantic / Gulf	Bay / Soundfront	Sandy Beach	Rocky Beach	Primitive Beach	Urban
Sebastian Inlet State Recreation Area	●	●	●	●	●	●	●	●	●	●		●			●		●		●			
Ambersand Beach Park	●		●			●		●							●		●		●			
Wabasso Beach Park	●		●	●	●	●		●	●	●					●		●		●			
North County Walkway						●		●							●		●			●		
South County Walkway						●		●							●		●		●			

FACILITIES ENVIRONMENT

Tarpon fishing

Vero Beach

55

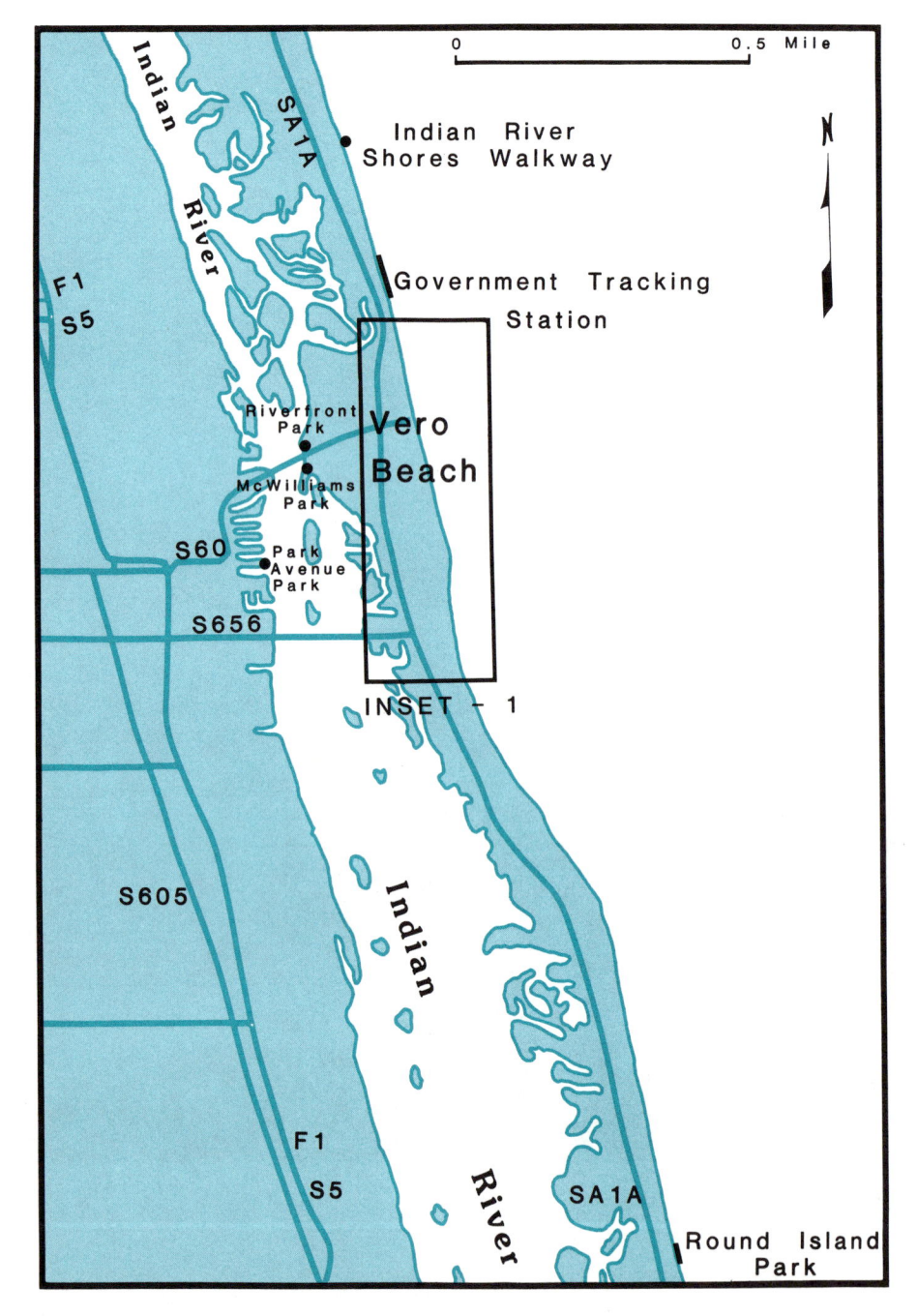

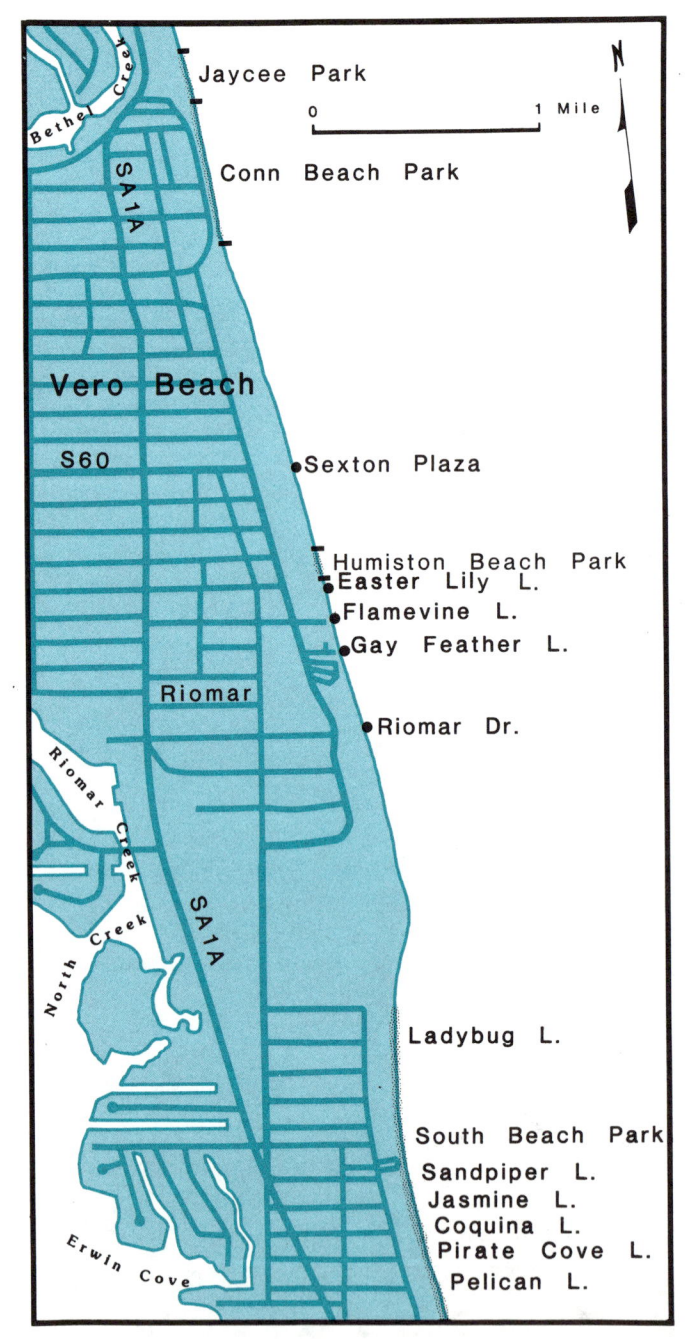

Southern district

Vero Beach (inset 1)

Indian River County

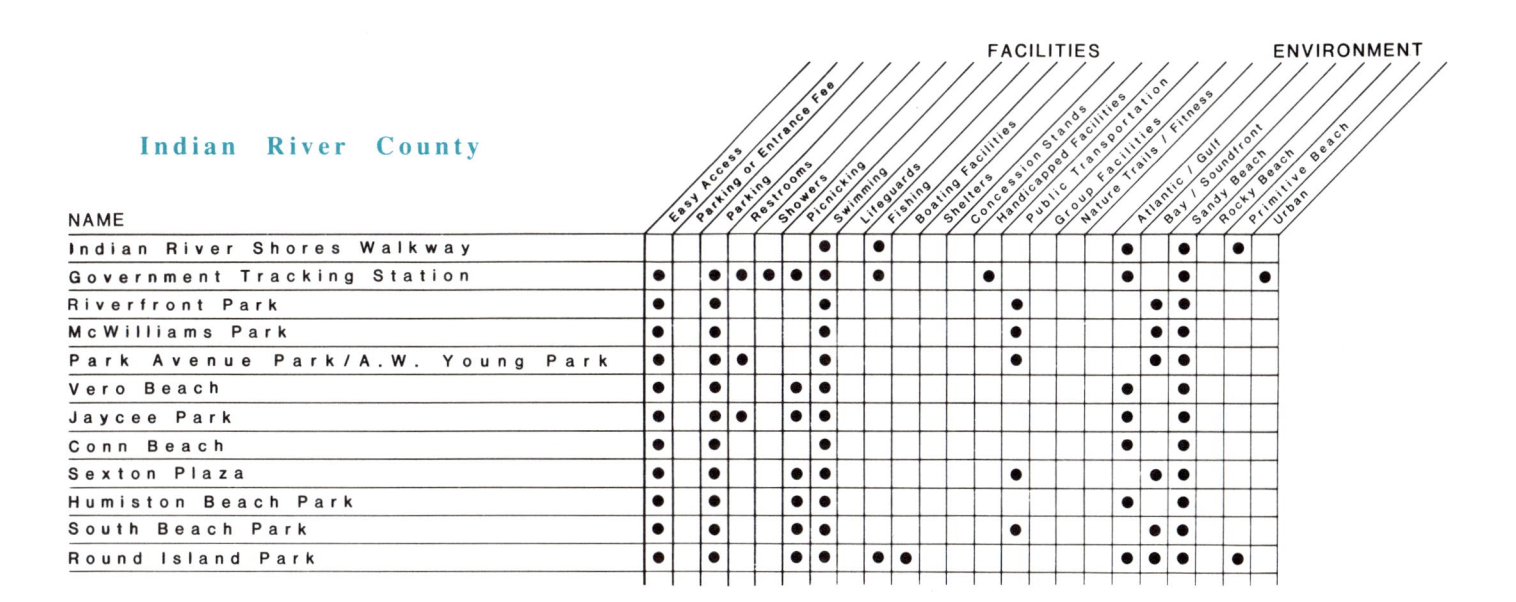

NAME	Easy Access	Parking or Entrance Fee	Parking	Restrooms	Showers	Picnicking	Swimming	Lifeguards	Fishing	Boating Facilities	Shelters	Concession Stands	Handicapped Facilities	Public Transportation	Group Facilities	Nature Trails/Fitness	Atlantic/Gulf	Bay/Soundfront	Sandy Beach	Rocky Beach	Primitive Beach	Urban
Indian River Shores Walkway						●		●									●		●	●		
Government Tracking Station	●		●	●	●	●	●		●			●					●		●			●
Riverfront Park	●		●			●						●					●	●				
McWilliams Park	●		●			●						●					●	●				
Park Avenue Park/A.W. Young Park	●		●	●		●						●					●	●				
Vero Beach	●		●				●	●									●		●			
Jaycee Park	●		●	●			●	●									●		●			
Conn Beach	●		●														●		●			
Sexton Plaza	●		●			●	●						●				●	●				
Humiston Beach Park	●		●			●	●										●		●			
South Beach Park	●		●			●	●						●				●	●				
Round Island Park	●		●			●	●		●	●							●	●	●			●

Indian River Shores Walkway is on Beachcomber's Lane about 3 miles north of Vero Beach from S.R. 510 on S.R. A1A or about 500 feet north of the Government Tracking Station.

Government Tracking Station, 9.5 acres of developed beach accessible at Reef Lane, Surf Lane, and Pebble Lane off S.R. A1A, is situated 2 miles north of the intersection of U.S. 60 and S.R. A1A in Vero Beach.

Riverfront Park is located riverside of Vero Beach. Access is marked on the first left off the Merrill Barber Bridge East.

McWilliams Park is a riverside recreation area. Access is 1,500 feet on the right after crossing the Merrill Barber Bridge East.

Park Avenue Park (A. W. Young Park) is an inland park located between Canal 5 and Canal 6 on Indian River. There is marked access on Park Ave. E.

Jaycee Park has 2,000 feet of developed beach with marked access on S.R. A1A about 1.5 miles north of the intersection of S.R. A1A and U.S. 60 in Vero Beach.

Conn Beach is on Phoenix Palm (S.R. A1A) about one mile north of U.S. 60 in Vero Beach. This undeveloped beach is accessible via Avenida Palm, Lilac Rd., and Conn Way.

Vero Beach is an undeveloped city beach. Access roads, in order of appearance on S.R. A1A south, are Easter Lily, Flamevine, Gay Feather, Riomar, Ladybug, Sand Piper, Jasmine, Coquina, Pirate Cove, Turtle Cove, and Sea Gull.

Sexton Plaza has 150 feet of developed beach in the Vero Beach central business district. Access to the beach is at the end of Beachland Blvd. E.

Humiston Beach Park provides 500 feet of developed beach. It is reached by taking U.S. 60 east to Ocean Drive South for about 0.5 mile to Dahlia Lane; access is marked.

South Beach Park lies south of Riomar Country Club. Take S.R. A1A east, turn left on Causeway Blvd. to Ocean Dr. There is marked access to about 300 feet of developed beachfront.

Round Island Park has about 400 feet of undeveloped beach; access is off the Charles Mitchell Highway (S.R. A1A) 12.5 miles south of U.S. 60 in Vero Beach.

Coastal Management Program

Florida's coast is a unique ecological resource with many competing uses. The coast is important for commerce and economic development, for tourism and recreation, and for residential purposes. To manage the competing demands on this resource, the 1978 Florida legislature passed the Coastal Management Act, which established a coastal management program based on existing laws and regulations to protect, maintain, and develop coastal resources. It stated that "the coastal zone is rich in a variety of natural, commercial, recreational, ecological, industrial, and aesthetic resources of immediate and potential value to the present and future residents of this state which will be irretrievably lost or damaged if not properly managed."

The key to an effective program in Florida is coordination among the state agencies charged with administering the laws and programs that affect the coastal zone. The act directed the Department of Environmental Regulation (DER) to develop a management framework for coastal programs. The Interagency Management Committee (IMC) was created to unify state activities by making recommendations to the governor and cabinet on new policy, legislation, and interagency agreements.

The daily implementation of key coastal programs is carried out primarily by three agencies—the DER, the Department of Natural Resources (DNR), and the Department of Community Affairs (DCA). Other agencies involved include the Departments of Agriculture, Commerce, Health and Rehabilitative Services, and Transportation, the Game and Fish Commission, and the Office of the Governor. The comprehensive management program also includes local regional agencies such as the water management districts and regional planning councils.

Florida's coastal management program involves and affects a number of people. Citizen participation is guaranteed by Florida Statutes in three ways: access to public information, participation in rule making, and participation in licensing and enforcement judgments. Florida's Coastal Resources Citizens' Advisory Committee (CAC) provides one avenue for public participation by advising the IMC on coastal management issues. Appointed by the governor, the committee is made up of persons representing government, industry, and environmental interests.

Because of the geographic and legal bases for the program, the entire state of Florida is considered to be within the coastal zone. The coastal management program is based on existing statutes and separate statewide programs. The IMC coordinates these authorities and assures that they are consistently and efficiently applied throughout the coastal area.

The essential purpose of the program is to balance development demands with environmental considerations. Reconciliation of competing resource demands is often a difficult task. The statutes and guidelines contained in the program are designed to ensure that public resources are used wisely. But these are not enough. Ultimately, it will be the public's awareness, understanding, and appreciation of limited resources that will ensure the continuation of these resources for future generations.

St. Andrews State Recreation Area

Fort Pierce and area

St. Lucie County

St. Lucie County's name honors Saint Lucie of Syracuse, a name the Spanish also bestowed on Fort Santa Lucia in 1565. The county seat is Fort Pierce, known as the "Sunrise City." Commercial and pleasure fishing in the ocean, inlet, and river channel are important attractions around Fort Pierce. Treasure hunting is also a big business: more than $2 million in gold, silver, and historic artifacts have been recovered by salvage companies working off the Fort Pierce–Vero Beach coast. Recently, professional treasure hunters located the sand-covered wrecks of two of the eleven ships of the Silver Fleet that sank in a hurricane in 1715.

The earliest inhabitants of the Fort Pierce–St. Lucie County area were the Ais Indians. In 1565, this tribe resisted the attempt by Pedro Menéndez de Avilés to settle the area. Later the Ais were driven out of the area by the Seminoles. The first European settlement was established by James Hutchinson on the beautiful barrier island now named for him.

Within the city limits of Fort Pierce is "The Savannas," a unique wilderness area purchased in part by the state under the Land Conservation Act of 1972. Over 550 acres are managed by the county, the rest by the State Game and Freshwater Fish Commission. "The Savannas" has widespread fame as a well-managed fishing area in addition to its other outdoor recreation opportunities.

Jack Island State Preserve, formerly known as the Indian River Inlet Area, is a 950-acre island that is a sanctuary to more than 100 varieties of birds. Fishing around the island is excellent. An observation tower allows visitors to view a large expanse of the island and the surrounding Indian River. A four-mile trail is maintained along perimeter dikes.

St. Lucie County has almost 22 miles of oceanfront and 50 miles of Indian River shoreline. Located in the heart of the world-famous Indian River citrus area, the county offers visitors ocean breezes and mild climate in addition to luscious fruit.

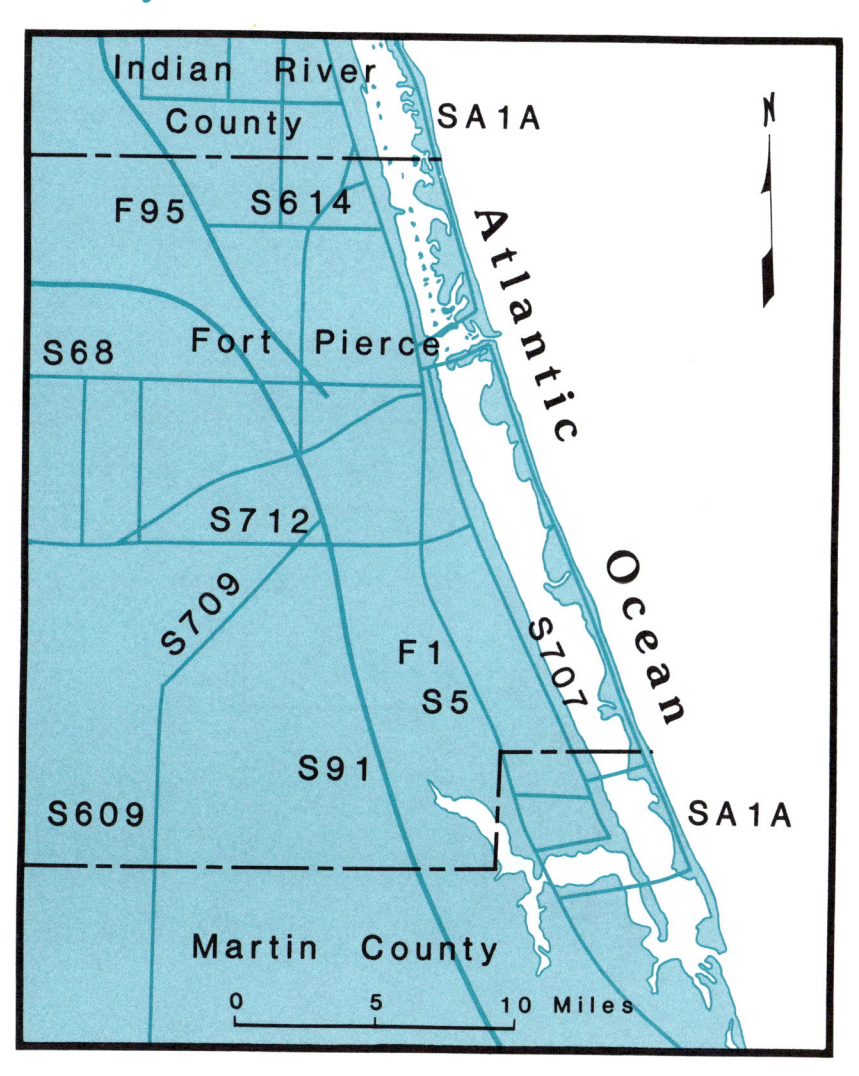

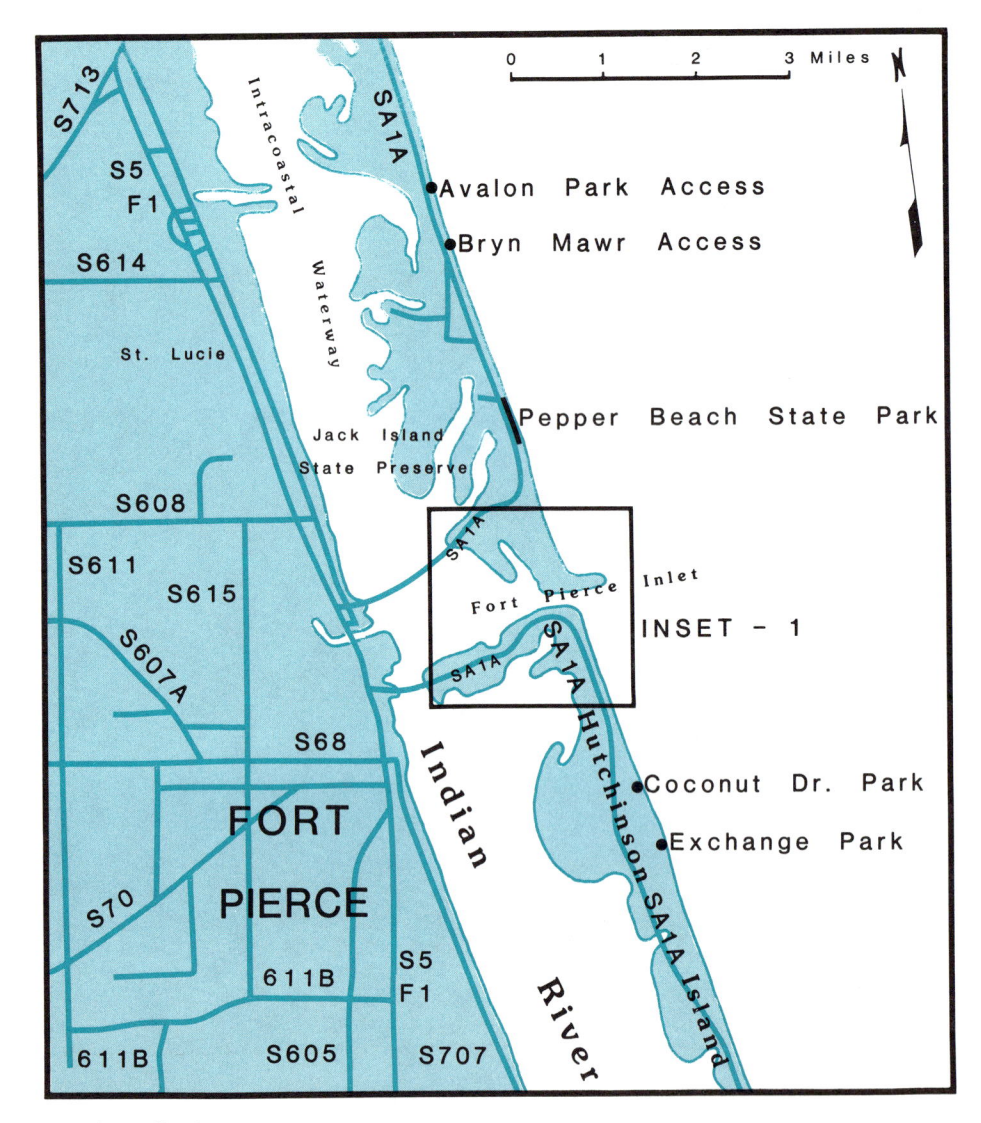

Northern district

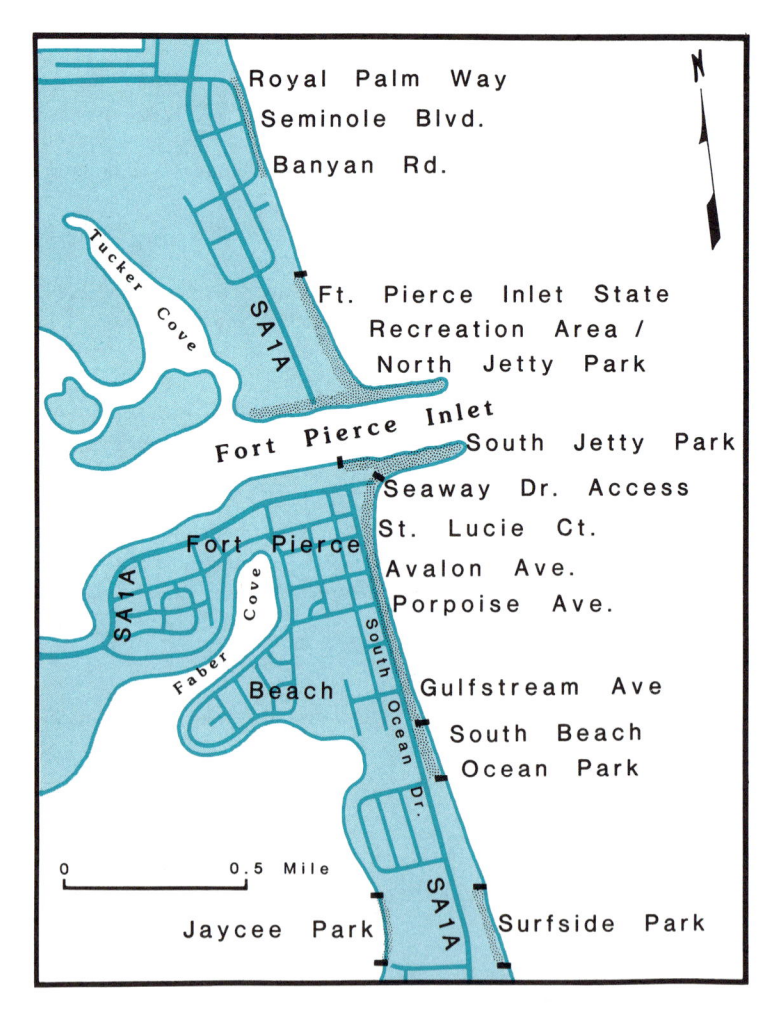

Fort Pierce (inset 1)

St. Lucie County

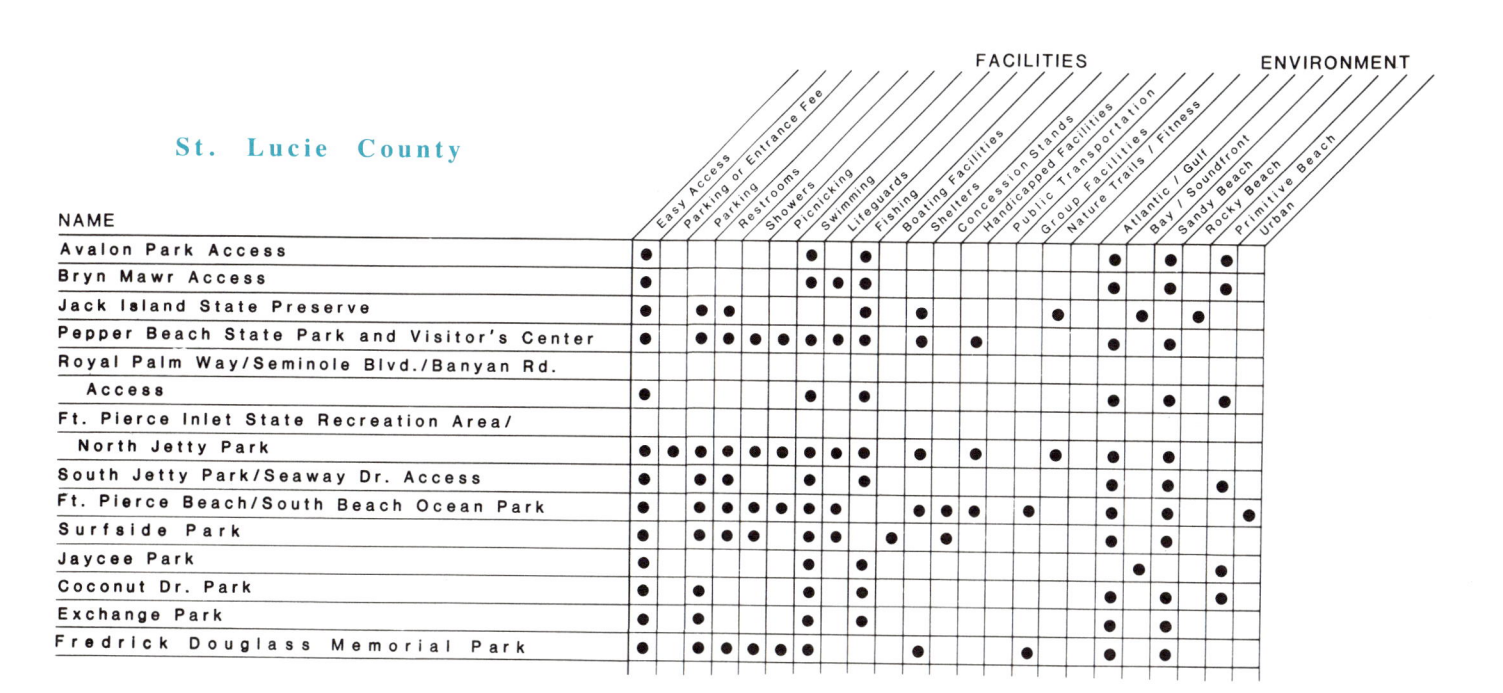

FACILITIES / ENVIRONMENT

NAME	Easy Access	Parking or Entrance Fee	Parking	Restrooms	Showers	Picnicking	Swimming	Lifeguards	Fishing	Boating Facilities	Shelters	Concession Stands	Handicapped Facilities	Public Transportation	Group Facilities	Nature Trails/Fitness	Atlantic/Gulf	Bay/Soundfront	Sandy Beach	Rocky Beach	Primitive Beach	Urban
Avalon Park Access	●					●	●										●		●		●	
Bryn Mawr Access	●					●	●	●									●		●		●	
Jack Island State Preserve	●	●	●				●		●					●			●		●			
Pepper Beach State Park and Visitor's Center	●	●	●	●	●	●	●	●		●	●						●		●			
Royal Palm Way/Seminole Blvd./Banyan Rd. Access	●					●	●										●		●		●	
Ft. Pierce Inlet State Recreation Area/North Jetty Park	●	●	●	●	●	●	●	●	●		●		●		●		●		●			
South Jetty Park/Seaway Dr. Access	●		●	●		●		●									●		●			
Ft. Pierce Beach/South Beach Ocean Park	●		●	●	●	●	●			●	●	●	●				●		●			●
Surfside Park	●		●	●	●		●	●			●	●					●		●			
Jaycee Park	●					●	●											●		●		
Coconut Dr. Park	●		●			●	●										●		●			
Exchange Park	●					●	●										●		●			
Fredrick Douglass Memorial Park	●		●	●	●	●			●			●					●		●			

Avalon Park Access: 60 feet of undeveloped beach, 1.5 miles south of the north county line on S.R. A1A.

Bryn Mawr Access: 300 feet of undeveloped beach, 2 miles south of the north county line on S.R. A1A.

Jack Island State Preserve is 631 acres of mangrove island on the Indian River about 2.5 miles north of Fort Pierce Inlet. Bird sanctuary with nature trails.

Pepper Beach State Park and Visitor's Center: 2,000 feet of developed beach and a local history museum north of Fort Pierce Inlet.

Royal Palm Way/Seminole Blvd./Banyan Rd. access: 5 miles south of the north county line; access to 180 feet of undeveloped beach.

Fort Pierce Inlet State Recreation Area/North Jetty Park covers 340 acres just north of the Fort Pierce Inlet on S.R. A1A. The area has an abundant variety of birdlife.

South Jetty Park/Seaway Dr. Access is located at the north end of Hutchinson Island. It has 1.5 acres of developed ocean park with a fishing pier. No alcohol.

Fort Pierce Beach/South Beach Ocean Park: 1,240 feet. A developed beach with a boardwalk.

Surfside Park 6,000 feet south of Fort Pierce Inlet; 2.8 acres of developed, urban beach park. Parking for 60 cars. No alcohol permitted.

Jaycee Park is across South Ocean Dr. (S.R. A1A) from Surfside Park; access west toward the Indian River.

Coconut Dr. Park has 81 feet of undeveloped beach at the south end of Surfside Dr. on Blue Heron Ave. No alcohol.

Exchange Park: 1.5 undeveloped acres about 2.25 miles south of Fort Pierce Inlet.

Southern district

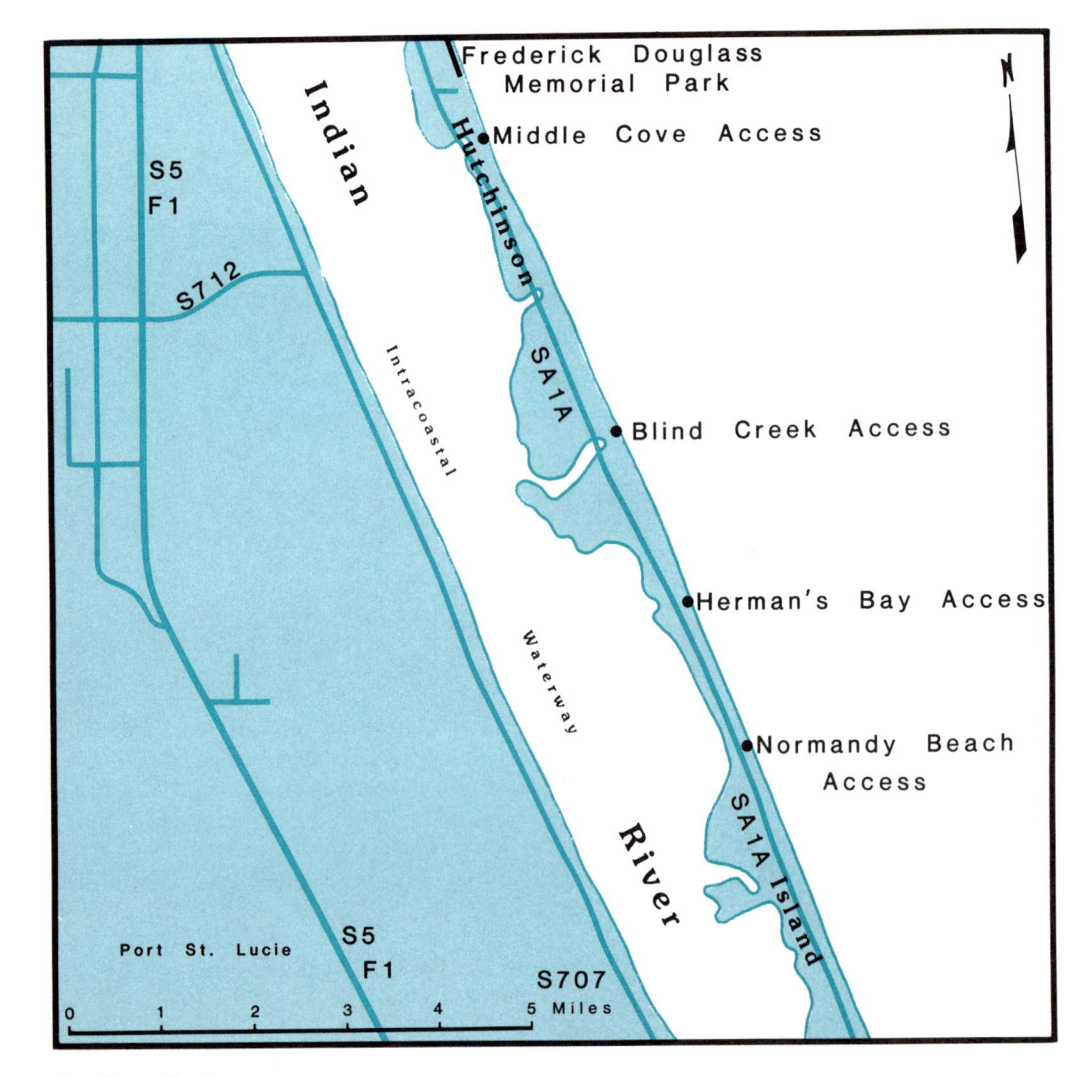

Hutchinson Island

64

St. Lucie County

FACILITIES / ENVIRONMENT

NAME	Easy Access	Parking or Entrance Fee	Parking	Restrooms	Showers	Picnicking	Swimming	Lifeguards	Fishing	Boating Facilities	Shelters	Concession Stands	Handicapped Facilities	Public Transportation	Group Facilities	Nature Trails / Fitness	Atlantic / Gulf	Bay / Soundfront	Sandy Beach	Rocky Beach	Primitive Beach	Urban
Middle Cove Access	●	●		●		●											●		●		●	
Blind Creek Access	●	●		●		●											●		●		●	
Herman's Bay Access	●	●		●		●											●		●		●	
Normandy Beach Access	●	●		●		●											●		●		●	

Frederick Douglass Memorial Park: 1,040 feet, this developed park is located 4 miles south of Fort Pierce Inlet on S.R. A1A.

Middle Cove Access: 110 feet of undeveloped beach, 5 miles south of Fort Pierce Inlet.

Blind Creek Access: 335 feet of undeveloped beach, 7 miles south of Fort Pierce Inlet.

Herman's Bay Access: 110 feet of undeveloped beach, 10 miles south of Fort Pierce Inlet.

Normandy Beach Access: 110 feet of secluded, undeveloped beach, 11 miles south of Fort Pierce Inlet.

Fort Pierce area showing dune walkover

Florida Seafood

The waters of the Atlantic Ocean and the Gulf of Mexico yield a variety of finfish and shellfish—tilefish, shrimp, lobster, mackerel, snapper, oysters, flounder, scallops, bluefish, and crab, to name a few of the most popular delicacies. Most of Florida's coastal cities have at least one trait in common: fresh Florida seafood year round.

Fish are classified as lean or fat, depending on their fat content. Lean fish have a fat content of less than 5 percent, with the oil concentrated in their liver. Fat fish, with over 5 percent fat content, are darker in color because the fat is distributed throughout the flesh. As a rule, the species that contain higher percentages of oil have more flavor. Fat fish, however, do not freeze as well as lean fish. It is recommended that after freezing they be used within three months. In preparing a recipe, lean fish may be substituted for fat fish, but the flavor of the fish may be masked and more basting is required to prevent drying. For recipes that require frequent handling of the fish, such as chowders, soups, or pickling, firm-textured fish such as grouper, red drum, or tilefish are preferred. These fish retain their shape and have a more pleasing finished appearance.

When buying a whole fresh fish, look for these signs of freshness: eyes, bright, clear, and bulging; gills, bright red in color and free of slime; flesh, firm and elastic with no trace of browning or drying out; odor, fresh and mild, not fishy; skin, iridescent and unfaded. The color should be characteristic of the species. Fresh steaks and fillets should be judged by their flesh, skin, and odor. Frozen fish of good quality show no discoloration of the flesh, no freezer burn (white, dry appearance on the edge), no ice crystals, and little or no odor.

Shrimp. Due to its distinctive flavor, shrimp is a popular shellfish. It is sold according to size or grade, based on the number of headless shrimp per pound. For maximum quality, cook fresh shrimp within one or two days of purchase. Rock shrimp is a unique variety easily mistaken for a miniature lobster tail. Not only the hard shell, from which it derives its name, but also the texture of the meat is like that of a lobster. The flavor is a blend of both. This delicious creature is more perishable than other shrimp; therefore, most are marketed in a raw or frozen state.

For either type of shrimp, two pounds of green tails (raw shrimp) will yield one pound of cooked, peeled, and deveined shrimp, enough to serve six people. Shrimp can be simmered, broiled, smoked (hickory chips are a favorite), and pan- or deep-fried, breaded or battered.

Shark. While most Americans are not familiar with shark as a food, other countries consider it a delicacy, such as Oriental shark-fin soup. Because of its abundance, shark is considered an economical food for family consumption. Consumer demand is increasing its availability in supermarkets, retail seafood outlets, and restaurants. The qualities that make shark appealing are a firm texture and versatility in methods of preparation. The meat can be broiled, baked, fried, poached, barbecued, sauteed, or used in soups or casseroles. Cooked shark meat is also delicious when flaked and added to salads and dips. For the angler, freshly caught shark should be soaked in icewater with ½ tablespoon of lemon juice or one tablespoon of cider vinegar per pound of fish. Refrigerate and allow to soak for four hours to neutralize any ammonia in the flesh of the shark.

Mullet. Mullet is prized for its delectable nutlike flavor. The most prevalent and abundant species caught in the United States is the gray or striped mullet. Considered a fat fish, mullet can be prepared in a variety of ways—barbecued, baked, broiled, smoked, and fried—without becoming dry.

Florida has an abundance of fresh seafood waiting to be sampled in local markets and restaurants. For information on preparing various species or for recipes on 20-minute seafood dishes, outdoor grilling, seafood hors d'oeuvres, seafood slimmers, or smoked seafood, contact Florida Department of Natural Resources, Bureau of Marketing and Extension Services, 3900 Commonwealth Blvd., Tallahassee, FL 32303.

Fresh Florida seafood

Forest Zone

An adequate supply of fresh water, soil with sufficient organic content, and protection from harmful salt spray landward of the highly stabilized dunefields are the conditions necessary for the development of a "forest zone."

The slash pine (*Pinus ellioti*) is the primary species in the coastal woodland areas of North Florida. Distinguished by long needles clumped in twos and threes, it is a rapidly growing tree often used in reforestation development. In South Florida its relative, the south Florida slash pine, occurs on the coast as far north as Daytona Beach and Tampa and is the only pine on Big Pine Key. Physically similar to the slash pine, this tree prefers a drier environment and has thicker stemmed seedlings. The sand pine (*Pinus clausa*) occurs on coastal sand dunes as far south as Dade and Lee counties. Its most remarkable characteristic is its ability to reseed in burned-over areas.

Also found in the coastal woodlands, although mainly in South Florida, are hardwood hammock areas or "tree islands." These hammocks occur on higher rocky land usually surrounded by a lower marsh. Composed primarily of tropical hardwoods, the hammock has a dense leafy canopy that allows little light to filter through, thus creating an open, sheltered area. Characteristic trees of the hammock include live oak (*Quercus virginiana*), gumbo-limbo (*Bursera simaruba*), pond apple (*Annona glabra*), wild lime, magnolia (*Magnolia spp.*), and cabbage palm (*Sabal palmetto*). Orchids, Spanish moss, and other bromeliads are abundant. Vines present are devils claw, poison ivy, wild bamboo, possum grape, and Virginia creeper. A dense bank of saw palmettos (*Serenoa repens*) forms the perimeter of many hammocks. Between 20 and 100 feet wide, the palmetto stand serves as an excellent firebreak, a shelter for rattlesnakes, and an obstacle for hikers.

Forest edge

Martin County

Bordered by the Atlantic Ocean on the east and Lake Okeechobee on the west, Martin County, named for Governor John W. Martin, is famed for sailfishing and all other types of sport fishing found in Florida. Ocean fishing, river and bay fishing, freshwater fishing in the North and South Forks of the peaceful St. Lucie River, bridge fishing, and surf casting attract anglers from all over the United States.

The county seat, Stuart, is the eastern terminus of the 140-mile South Florida Cross-State Waterway that connects the Gulf of Mexico near Fort Myers with the Atlantic Ocean. The waterway was intended as a drainage artery for Lake Okeechobee; however, boaters soon discovered it to be a sheltered cruise through the center of the state.

A favorite fishing and recreation area is Jonathan Dickinson State Park, covering over 9,500 acres. The park is named for a Quaker who survived a shipwreck in this area in 1696. Captured by the Hoe-Bay Indians (from which Hobe Sound gets its name), Dickinson was later released to endure a difficult journey to St. Augustine. Swimming in the Loxahatchee River, freshwater fishing in Kitching Creek, both salt- and freshwater fishing in the Loxahatchee, a boat dock and launching ramp, and a jungle cruise on the "Loxahatchee Queen" make this waterfront park popular with residents and visitors.

Hobe Sound National Wildlife Refuge has about 400 acres dedicated to the preservation of natural conditions, providing protection to the diminishing population of manatees and green turtles that often frequent Hobe Sound. Wildlife populations in the refuge include armadillos, raccoons, whitetail deer, and bobcats. Bobwhite quail and mourning doves, as well as bald eagles, brown pelicans, and many types of wading and shore birds, have been observed in the refuge. Although small, this habitat is typical of a rapidly diminishing type in Florida. High dunes reaching up to 30 feet above sea level drop sharply to the shoreline of Hobe Sound. Along the shore, Australian pines and mangroves create a tranquil atmosphere. Inland the high dunes are covered with palmettos and scrub oaks, vines and wildflowers. Hobe Sound Refuge is the last of its kind in South Florida.

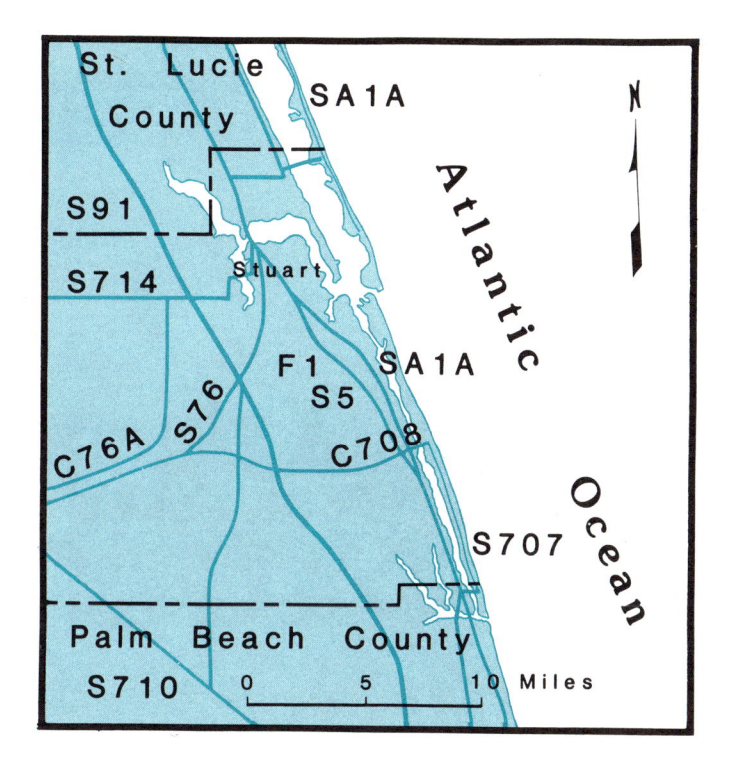

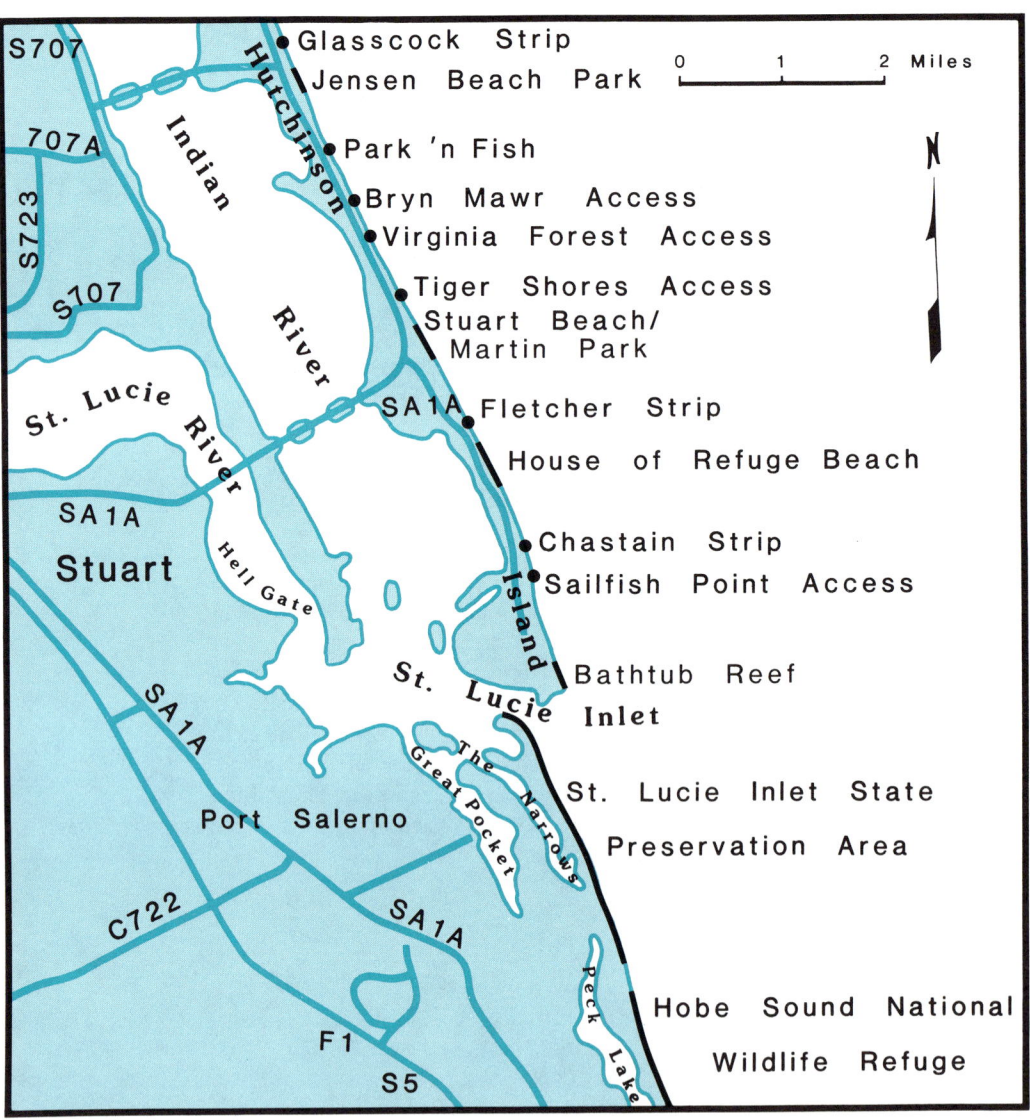

Northern district

Martin County

NAME	Easy Access	Parking or Entrance Fee	Parking	Restrooms	Showers	Picnicking	Swimming	Lifeguards	Fishing	Boating Facilities	Shelters	Concession Stands	Handicapped Facilities	Public Transportation	Group Facilities	Nature Trails / Fitness	Atlantic / Gulf	Bay / Soundfront	Sandy Beach	Rocky Beach	Primitive Beach	Urban
NORTHERN DISTRICT																						
Jensen Beach Park	●		●	●	●	●	●	●	●		●	●	●				●		●			
Bob Graham Beach	●		●				●		●								●		●		●	
Stuart Beach/Martin Park	●		●	●	●	●	●	●	●		●	●	●				●		●			
House of Refuge Beach	●		●				●										●		●		●	
Bathtub Reef							●		●								●		●		●	
St. Lucie Inlet State Preservation Area							●		●								●		●		●	
SOUTHERN DISTRICT																						
Hobe Sound National Wildlife Refuge							●		●								●		●		●	
Hobe Sound Beach	●		●	●	●	●	●	●	●		●	●					●		●			
Jupiter Island Park	●		●				●		●								●		●		●	

Jensen Beach Park has 3,000 feet of developed beach. Access is at the intersection of S.R. 707A and S.R. A1A. Dune walkovers, bicycle facilities, a boardwalk, and a concession stand are available.

Bob Graham Beach is located 3,500 feet south of Jensen Beach on S.R. A1A. It has 2,000 feet of undeveloped beach with dune crossovers.

Stuart Beach/Martin Co. Park is 5 acres of developed oceanfront park on MacArthur Blvd. off S.R. A1A. It has bicycle facilities, 150 parking spaces, a 250-foot boardwalk, and dune crossovers. On the property is the Elliot Museum, housing a collection of Americana dating as far back as 1750.

Access roads to undeveloped ocean beach off S.R. A1A: Glasscock Strip, Bryn Mawr, Virginia Forest, Tiger Shores.

Access roads to undeveloped ocean beach off MacArthur Blvd.: Fletcher Strip, Chastain Strip, Sailfish Point.

House of Refuge Beach lies one mile south of Stuart Beach on Gilbert's Bar, Hutchinson Island. The oldest standing structure in Martin County, built in 1875, served as a rescue and refuge center for shipwrecked sailors. Visitors' museum open 1–5 P.M.; observation tower. Current site of sea turtle research.

Bathtub Reef is 1,100 feet of undeveloped beach at the south end of Hutchinson Island, just north of St. Lucie Inlet. An offshore reef forms a shallow "bathtub" where snorkeling is excellent.

St. Lucie Inlet State Recreation Area has over 2.5 miles of remote beach on the north end of Jupiter Island just south of St. Lucie Inlet. Access to this area (under development) is by boat.

Hobe Sound National Wildlife Refuge has 4 miles of ocean beach on the northern end of MacArthur Blvd. on Jupiter Island. Access to this remote and undeveloped area is by boat or by foot from a small parking area.

Hobe Sound Beach lies approximately 2 miles north of Bridge Road on Jupiter Island.

Jupiter Island Park has 300 feet of developed beach at the eastern terminus of Bridge Road (S.R. 70) in Hobe Sound.

71

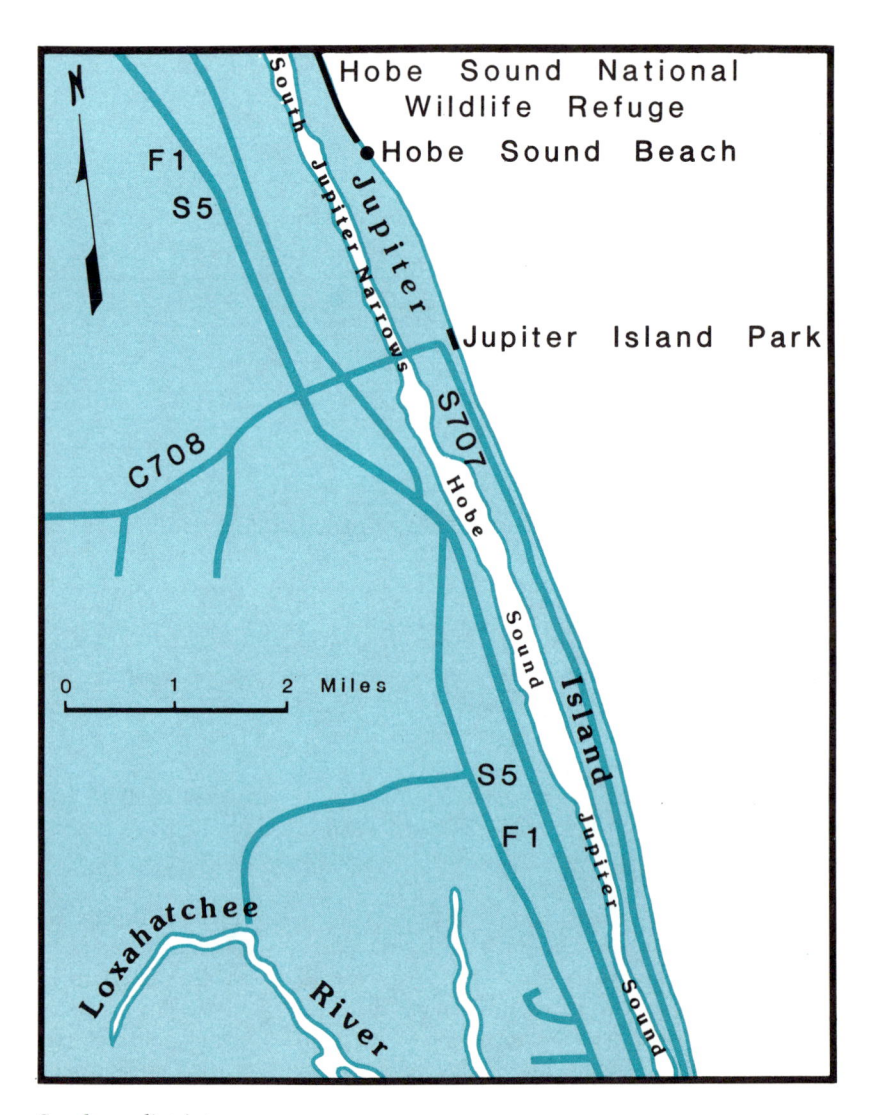

Southern district

Underwater Archaeology

Below the surface of the warm waters off the Florida coast lies a different world, one that contains many artifacts associated with the past. Some of these relics are ancient, such as cannons from old "men-of-war"; others are more recent, such as a newly purchased anchor.

Much of the history of early Florida lies submerged off both its coasts. Glaciers formed during the last Ice Age (Pleistocene Epoch) have been gradually melting. As a result the sea level surrounding the Florida peninsula has risen an estimated 115 feet and in the process has "buried" prehistoric and early historic coastal sites of habitation.

One method of retrieving this lost history is through underwater archaeology, although such research is limited because equipment is expensive to maintain. The proper cleaning and preservation of recovered objects that have been submerged are critical, since sea salts cause severe deterioration that continues after the artifact has been removed from the underwater site.

Florida has the oldest program of conservation and preservation of underwater artifacts in the United States. The first known underwater work was done in 1952, when the National Park Service was trying to find evidence of French Huguenot occupation at Fort Caroline on the St. Johns River in Duval County.

In addition to investigating former land sites now submerged, the state has also supervised private salvage operations on Spanish treasure ships known to have been lost to summer and fall hurricanes. Florida's share of recovered materials is preserved at the Division of Archives, History and Records Management in Tallahassee. Of course, artifacts from the most recent past—such as that anchor—may be kept. Valuable items from historical times should be turned over to the state for evaluation.

Palm Beach

Palm Beach County

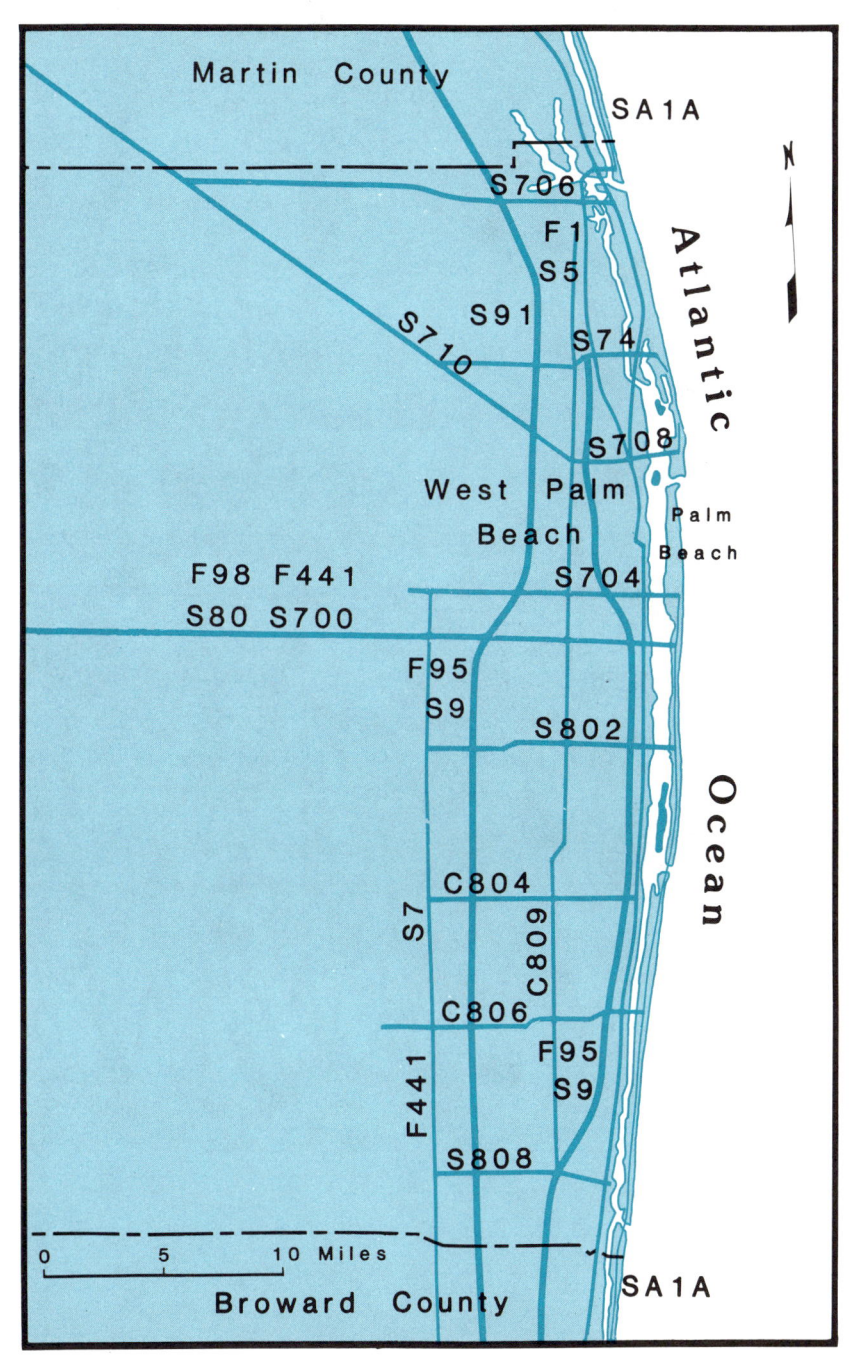

The graceful palms that inspired a county's name create a tropical setting for a world-renowned resort area. Palm Beach was South Florida's first luxury wintering spot for northern tourists, and it is still frequented by the famous and the wealthy.

Palm Beach County has 46 miles of frontage on the Atlantic Ocean and is one of the largest counties in Florida in both area and population. The Gulf Stream, originating near the equator off the coast of Africa, comes closer to the shores of Lake Worth than to any other point in the United States. Its constant temperature of 73 degrees in winter and 82 degrees in summer give the area one of the most delightful climates in the continental United States.

Its history of settlement begins in the recent past, as Spanish and British occupations of Florida left no imprint on the lower east coast. The settlement of Palm Beach County and its eventual development as a tropical resort are due to the wreck of a Spanish ship, the *Providencia*, in 1878 on the beach at Lake Worth. The early pioneer settlers were offered the cargo of 20,000 coconuts, of which 14,000 were planted on the wilderness beach. With time the barrier sand key was transformed to Palm Beach.

The last remaining significant stretch of undeveloped barrier island in South Florida has recently been acquired by the State of Florida and Palm Beach County for conservation and recreational development. Located on Singer Island in North Palm Beach, the proposed John D. MacArthur State Recreation Area has over 225 acres of a wide range of natural coastal ecosystems including a wide beach, dunes, tropical hammock, mangroves, and upland hammock. Also interesting is Munyon Island, a 48-acre spoil island in Lake Worth almost entirely covered with Australian pines. This site with its various environments in close proximity to each other represents one of the finest coastal areas in the state. Palm Beach County is unique in that it offers both the wilderness beauty of the Everglades and the magnificence of a luxurious tropical resort area having some of the finest coastline in the state.

Northern district

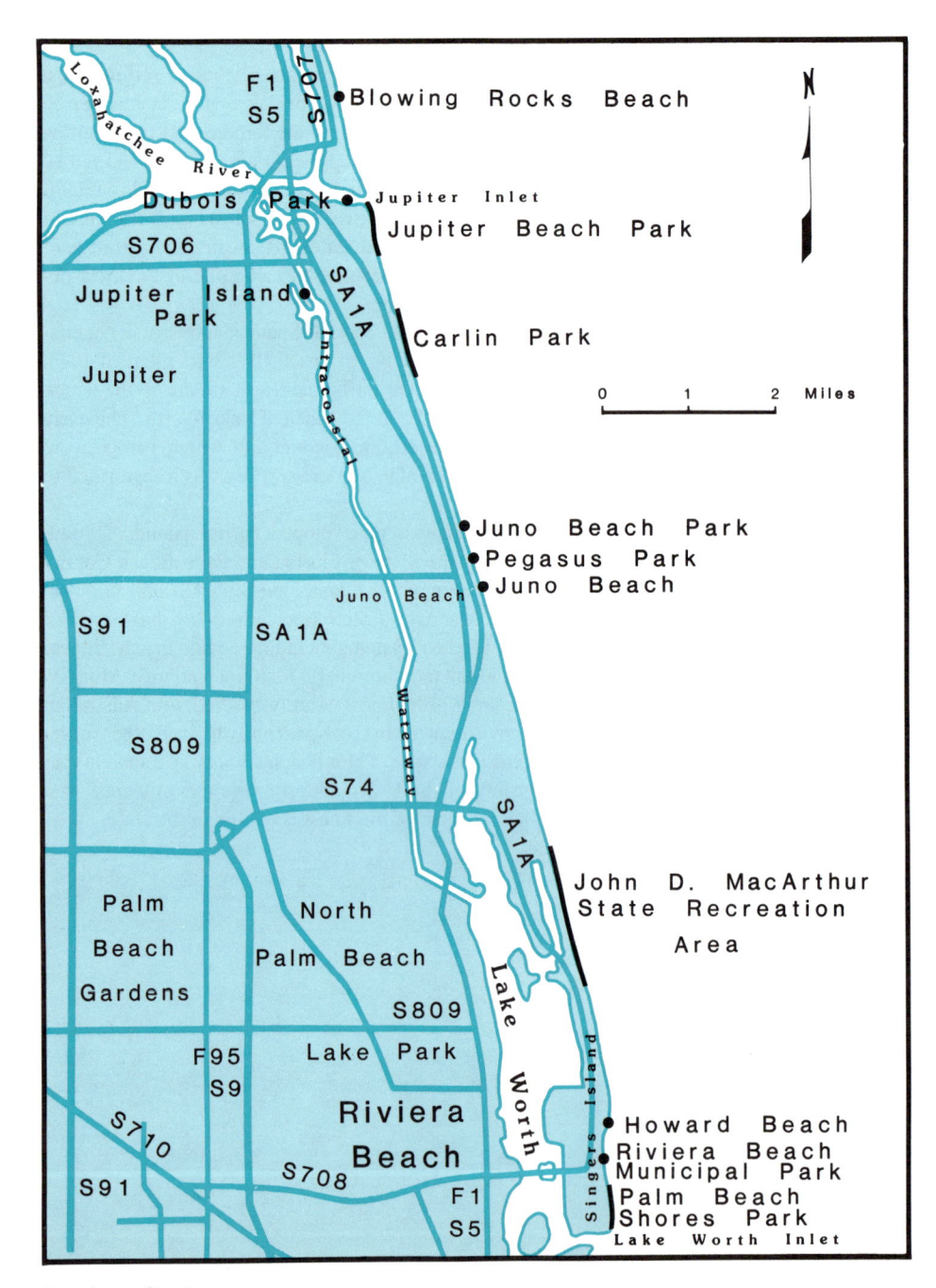

Palm Beach

Palm Beach County

NAME	Easy Access	Parking or Entrance Fee	Parking	Restrooms	Showers	Picnicking	Swimming	Lifeguards	Fishing	Boating Facilities	Shelters	Concession Stands	Handicapped Facilities	Public Transportation	Group Facilities	Nature Trails / Fitness	Atlantic / Gulf	Bay / Soundfront	Sandy Beach	Rocky Beach	Primitive Beach	Urban
Blowing Rocks Beach (Coral Cove Park)	●		●	●	●	●	●	●			●			●			●		●			
Jupiter Beach Park	●		●	●	●	●	●	●	●								●		●			
DuBois Park	●		●	●	●	●	●	●			●							●	●			
Jupiter Island Park	●		●	●	●	●	●	●	●								●		●			
Carlin Park	●		●	●	●	●	●	●	●		●	●	●	●			●		●	●		
Juno Beach Park	●		●	●	●	●	●	●				●	●				●		●			
Pegasus Park (Loggerhead Park)						●		●									●		●		●	
Juno Beach			●														●		●		●	
McArthur State Recreation Area			●														●		●			
Howard Beach	●		●	●	●	●	●	●	●			●	●				●		●			
Riviera Beach Municipal Park	●		●	●	●	●	●	●			●	●	●	●			●		●			●
Palm Beach Shores Park	●		●	●	●			●	●			●					●		●			●

Coral Cove Park (Blowing Rocks Beach) is a developed beach with rock formations one mile south of Martin Co. on Jupiter Island. It has 600 feet of beachfront accessible via S.R. A1A. Lifeguards on weekends and holidays.

Jupiter Beach Park is a developed beach (1,700 feet) with lifeguards. Access is off S.R. A1A on Beach Dr. just south of Jupiter Inlet on the ocean.

DuBois Park has 20 acres of developed beach with lifeguards; accessible via S.R. A1A on DuBois Rd. south of Jupiter Inlet.

Jupiter Island Park in North Palm Beach Co. is surrounded by the Intracoastal Waterway. This rocky beach is reached via U.S. 1, north of Indiantown Rd. (S.R. 706), south of Jupiter Inlet and S.R. A1A.

Carlin Park is a developed recreation area with 3,000 feet of rocky beach on the Atlantic Ocean with lifeguards. Access is on S.R. A1A, south of Indiantown Rd. (S.R. 706) in Jupiter.

Juno Beach Park lies one mile north of Juno Beach city limits on S.R. A1A. It has 300 feet of undeveloped beach with lifeguards. Fishing pier.

Loggerhead Park (Pegasus Park) has 900 feet of beachfront to open in summer 1985. It is located on S.R. A1A and U.S. 1 about ⅛ mile north of Donald Ross Rd. north of Juno Beach.

Juno Beach has a small (2,600 feet), undeveloped oceanfront beach that is difficult to find as it is not marked (1983). It lies about ¾ mile south of Donald Ross Rd. on S.R. A1A. No alcohol allowed. Fishing pier available.

John D. McArthur State Recreation Area (formerly Air Force Beach) is located on Singer Island between Lake Worth and the Atlantic Ocean. This is a new state park with plans for development of 8,000 feet of oceanfront beach in 1985. Access is off S.R. A1A, 2.5 miles north of Riviera Beach. No alcohol is allowed.

Howard Beach (Diamond Head) is 700 feet of undeveloped beach on S.R. A1A, directly north of the Hilton Inn on Singer Island.

Riviera Beach Municipal Park has three marked access points off Blue Heron Blvd. in Riviera Beach east of I-95. This is a developed beach with lifeguards.

Palm Beach Shores Park is a developed beach north of Lake Worth Inlet. It has 3,000 feet of beachfront with lifeguards.

77

Palm Beach Municipal Beach has unlimited access to 6 miles of undeveloped beach with lifeguards. A large portion of the beach is bulkheaded. Parking is very limited. The beach runs about 3 miles both north and south of Royal Palm Way off S.R. A1A. No alcohol is allowed.

Phipps Ocean Park has 1,300 feet of developed beach. Marked access is on Ocean Blvd., 2.5 miles south of Southern Blvd. Bridge. No alcohol allowed.

Richard G. Kreusler Park has 450 feet of developed, rocky beach with lifeguards. Access is on S.R. A1A in Palm Beach, directly north of Lake Worth Public Beach.

Lake Worth Beach, in Palm Beach, is accessible at Ocean Blvd. (S.R. A1A) and Lake Ave. (S.R. 802). This is a developed beach (1,200 feet) with a pool and lifeguards. No alcohol allowed.

Lantana Park lies between Lake Worth and the Atlantic Ocean one mile east of Lantana at the intersection of S.R. 12 and Ocean Blvd. (S.R. A1A). This is an undeveloped beach with lifeguards. No alcohol allowed.

Manalapan Beach is an undeveloped beach about 1,350 feet south of Lantana Ave.

Boynton Inlet Park has access to 600 feet of beach, in the town of Ocean Ridge, on S.R. A1A, directly south of the South Boynton Inlet and north of Island Dr.

Ocean Ridge Hammock Park lies north of Boynton Public Beach. It has 1,100 feet of undeveloped beach accessible from S.R. A1A, in Ocean Ridge.

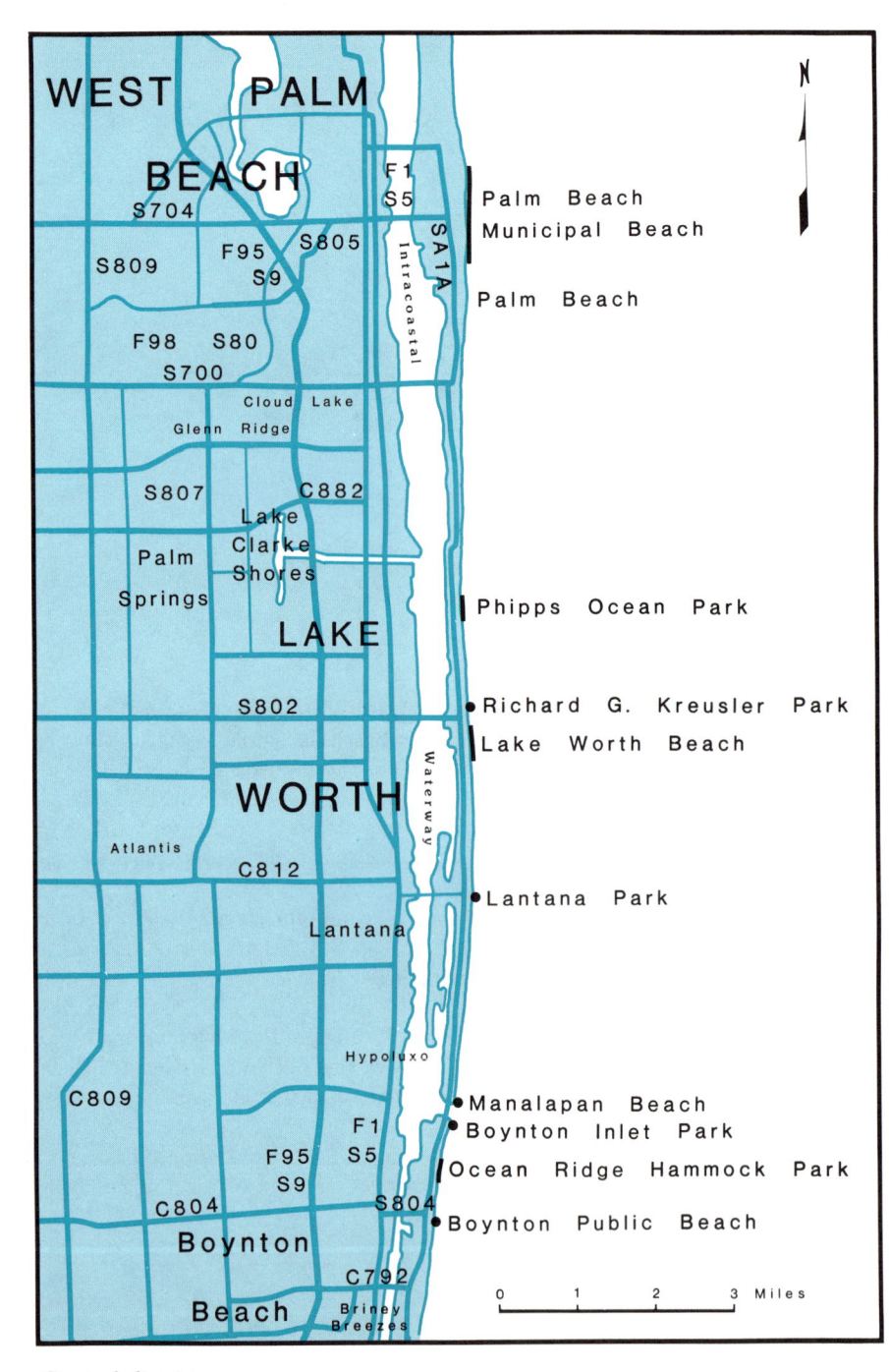

Central district

Palm Beach County

NAME	Easy Access	Parking or Entrance Fee	Parking	Restrooms	Showers	Picnicking	Swimming	Lifeguards	Fishing	Boating Facilities	Shelters	Concession Stands	Handicapped Facilities	Public Transportation	Group Facilities	Nature Trails / Fitness	Atlantic / Gulf	Bay / Soundfront	Sandy Beach	Rocky Beach	Primitive Beach	Urban
CENTRAL DISTRICT																						
Palm Beach Municipal Beach	●		●		●		● ●					●			●		●		●	●		
Phipps Ocean Park	●	●	●	●	●	●	● ●					●	●	●			●		●			●
Richard G. Kreusler Park	●	●	●	●	●		● ●					●	●				●		●			●
Lake Worth Beach	●		●	●	●	●	● ●		●	●	●	●	●				●					●
Lantana Park	●	●	●	●	●	●	● ●		●	●	●	●	●				●		●			●
Manapalan Beach			●				●	●									●		●	●		
Boynton Inlet Park			●				●	●									●		●			
Ocean Ridge Hammock Park			●				●	●									●		●			
Boynton Public Beach	●	●	●	●	●	●	●				●	●	●	●	●	●	●		●			●
SOUTHERN DISTRICT																						
Gulf Stream County Park	●	●	●	●	●	●	● ●			●		●	●				●		●			●
Delray Public Beach		●	●			●	● ●	●	● ●		●			●			●		●			
Atlantic Dunes Park	●	●	●	●	●	●	●			●		●	●		●		●		●			●
Spanish River Park		●	●	●	●		● ●		●	●	●		● ●				●		●			●
Red Reef Park	●	●	●	●	●	●	●		●			●					●		●			
South Beach Park			●	●	●		● ●					●					●		●			●
South Inlet Park		●	●	●	●	●	● ●	●		●			● ●		●		●		●			

Boynton Public Beach is 2 blocks north of the I-95 exit on S.R. A1A in Ocean Ridge.

Gulf Stream Co. Park has 600 feet of developed beach on S.R. A1A, south of Briny Breezes mobile home community, north and east of the St. Andrews Club.

Delray Public Beach is accessible along the entire 7,000-foot length bordering the east side of Ocean Blvd. in Delray Beach. This is a developed beach with lifeguards. No alcohol is allowed.

Atlantic Dunes Park, on S.R. A1A approximately one mile south of Delray Public Beach, has 7 acres of developed beach with lifeguards. No alcohol is allowed.

Spanish River Park lies one mile east of Boca Raton. Access points are on Ocean Blvd. (S.R. A1A) just south of NE Spanish River Blvd. in Boca Raton. This developed beach covers 49 acres of the island from the Atlantic Ocean to the Intracoastal Waterway. It has over a half mile of beachfront with lifeguards. No alcohol is allowed.

Red Reef Park, about one mile south of Spanish River Park on S.R. A1A, has access to 67 acres of developed beach with lifeguards. No alcohol is allowed.

South Beach Park is on S.R. A1A (Ocean Blvd.) at NE 4th St. The beach and ocean can be reached from three locations in the park. This is an undeveloped beach with lifeguards. No alcohol is allowed.

South Inlet Park has 850 feet of developed beach with lifeguards on S.R. A1A, directly south of Boca Inlet.

Southern district

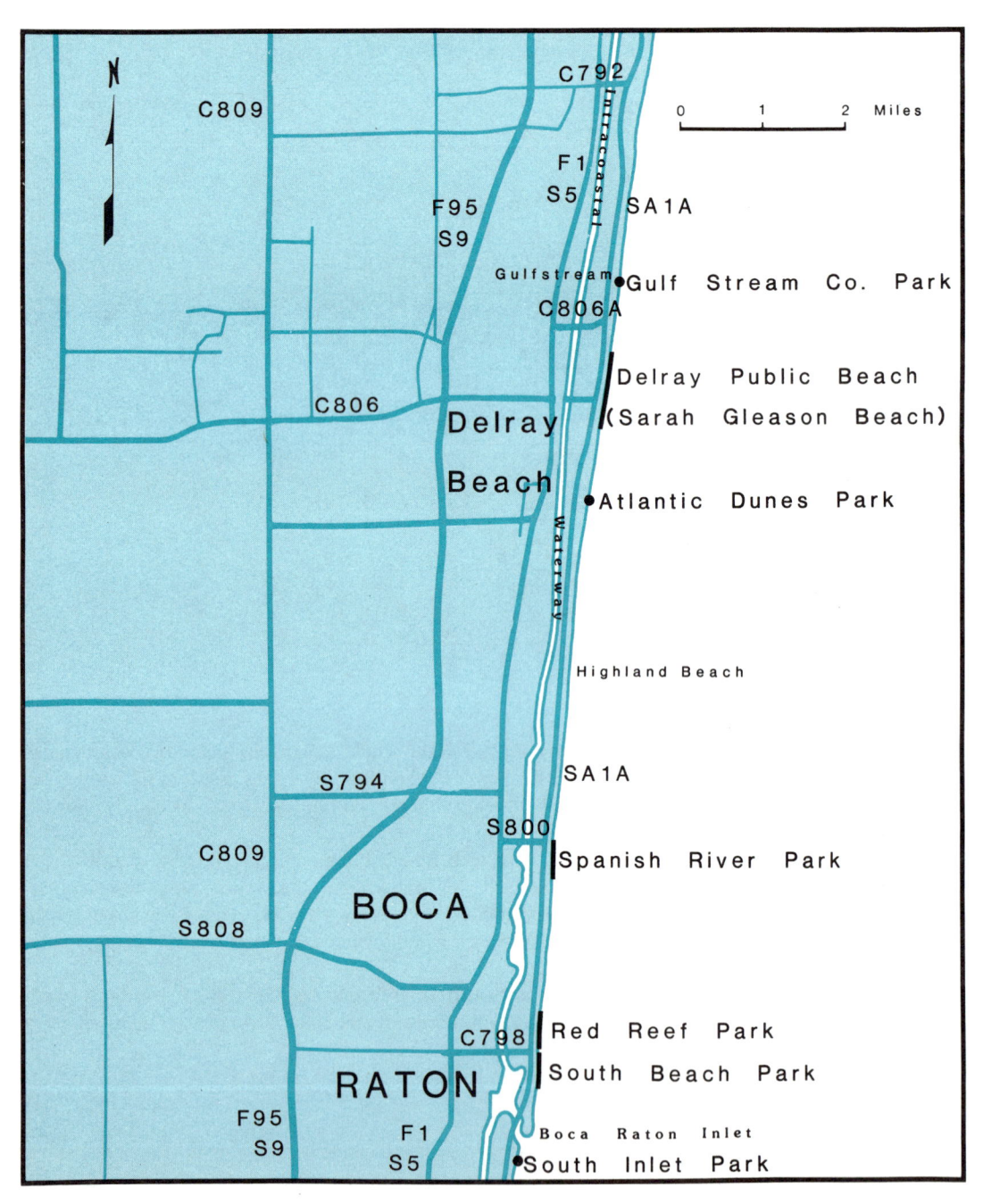

- C792
- C809
- F1
- S5
- SA1A
- F95
- S9
- Gulfstream — Gulf Stream Co. Park
- C806A
- C806
- Delray Public Beach (Sarah Gleason Beach)
- Delray Beach
- Atlantic Dunes Park
- Intracoastal Waterway
- Highland Beach
- S794
- SA1A
- S800
- C809
- Spanish River Park
- BOCA
- S808
- C798 — Red Reef Park
- RATON — South Beach Park
- F95
- S9
- F1
- S5
- Boca Raton Inlet
- South Inlet Park

0 1 2 Miles

White pelican

Southern district

Endangered Species Act of 1973

To conserve species of fish, wildlife, and plants that are in danger of extinction, the Endangered Species Act of 1973 makes it unlawful (with certain limited exceptions) to take, to import into, or to export from the United States any species designated as "endangered" on the official list published by the secretary of the interior, or any part of or product made from such species. A second category of species, "threatened," is provided for in the act and may also be subject to import and export limitations. An up-to-date list of the species affected will be provided to the traveler upon request. Such listed species, or any parts of or products made from such species, that are being imported or exported will be subject to seizure by agents of the U.S. Government.

The secretary of commerce has responsibility for certain endangered and threatened species, particularly certain species that reside in the marine environment. The Department of the Interior has responsibility for other endangered and threatened species. Among the animals that are endangered are some species of whales and sea turtles. Some of the products made from these sea turtles are combs, polished shells, jewelry, leather goods, meat, oil, cosmetics, and preserved animals.

To facilitate enforcement of this law, all fish and wildlife, other than some exempted personal and household effects, that are imported into, or exported from, the United States must pass through certain designated ports in accordance with Title 50 of the Code of Federal Regulations, Part 14. The following ports are currently designated as ports of entry for all fish and wildlife: New York, N.Y.; Miami, Florida; Chicago, Illinois; San Francisco and Los Angeles, California; New Orleans, Louisiana; Seattle, Washington; Honolulu, Hawaii.

Complete information may be obtained from Director, U.S. Fish and Wildlife Service, Department of the Interior, Washington, DC 20240; Director, National Marine Fisheries Service, Washington, DC 20235; Regional Director, Southeast Region, 9450 Gandy Boulevard, Duval Building, St. Petersburg, Florida 33702. Marine Patrol: (904) 488-5757.

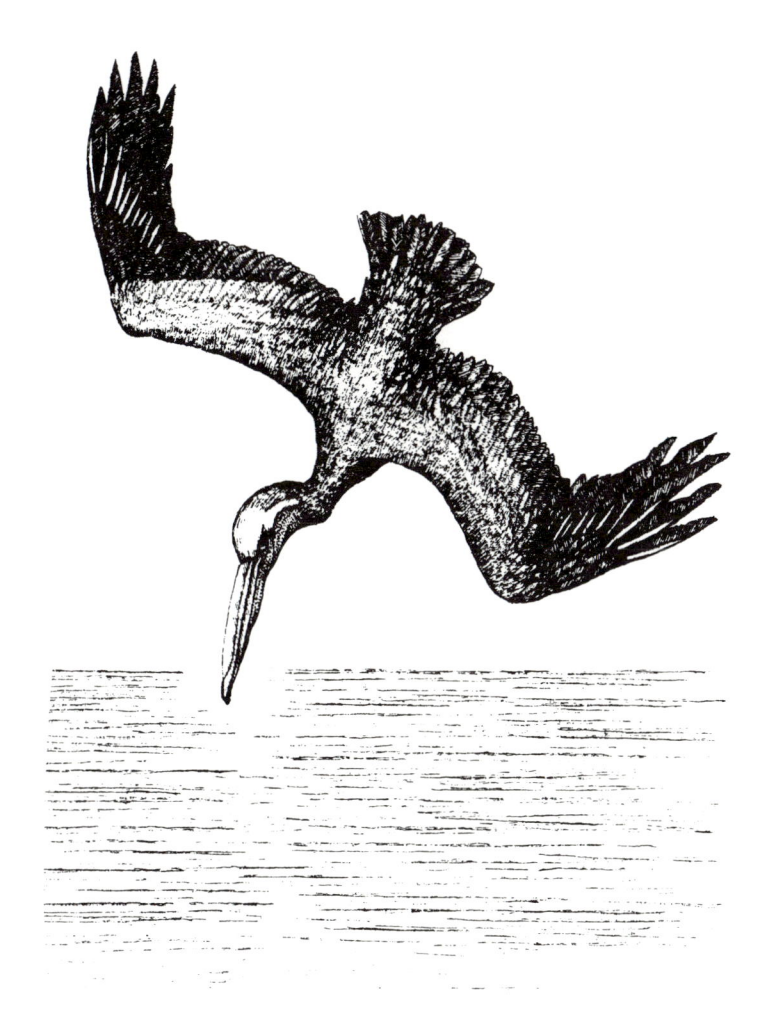

Brown pelican

Pompano Beach

Broward County

Broward County, named for Governor Napoleon B. Broward, is located in southeast Florida between Dade and Palm Beach counties in the heart of the Gold Coast. Lavish tropical beauty and miles of palm-lined lagoons, canals, and rivers honeycomb the exotic resort city of Fort Lauderdale. Plush waterfront homes face canals, and much of the public beach is lined with towering luxury hotels and high-rise condominium apartments. Each spring college students ritually flock to the Fort Lauderdale area to escape the chill of winter and the stress of studying. Fort Lauderdale was named for an officer of an army fort built in 1838, during the Seminole War.

Fort Lauderdale Beach begins at the north end of what is known as the Galt Ocean Mile. This area represents a one-mile stretch of hotels, high-rises, and condominiums. There is little or no public access to the beach from the west. However, north-south foot traffic or access is available.

From the Galt Ocean Mile, the beach continues south, bordered by smaller motels and private residences, with limited public access for approximately one and a quarter miles. The portion of coastline that is referred to as the Public Beach begins at the 1800 block of North Atlantic Blvd. and extends 3.5 miles south. This portion has unlimited public access: metered parking, immediately adjacent to the beach, is available. The city provides surfing and windsurfing areas and a launching area for nonpowered small craft under 18 feet. Lifeguards are on duty from 9:00 A.M. to 5:00 P.M. All beach rules are enforced via city ordinances. Food and beverages are available across the street from the beach in local restaurants and hotels.

Hugh Taylor Birch State Recreation Area covers 188 acres in downtown Fort Lauderdale. The tropical atmosphere is a result of the 70-degree Gulf Stream that extends easterly along the ocean and borders the Intracoastal Waterway on the west. Two unusual features of the park are a youth camp, which can accommodate 70 people by reservation, and a scenic railroad, which winds for 3 miles through the park, along the ocean and waterway and over two freshwater lagoons.

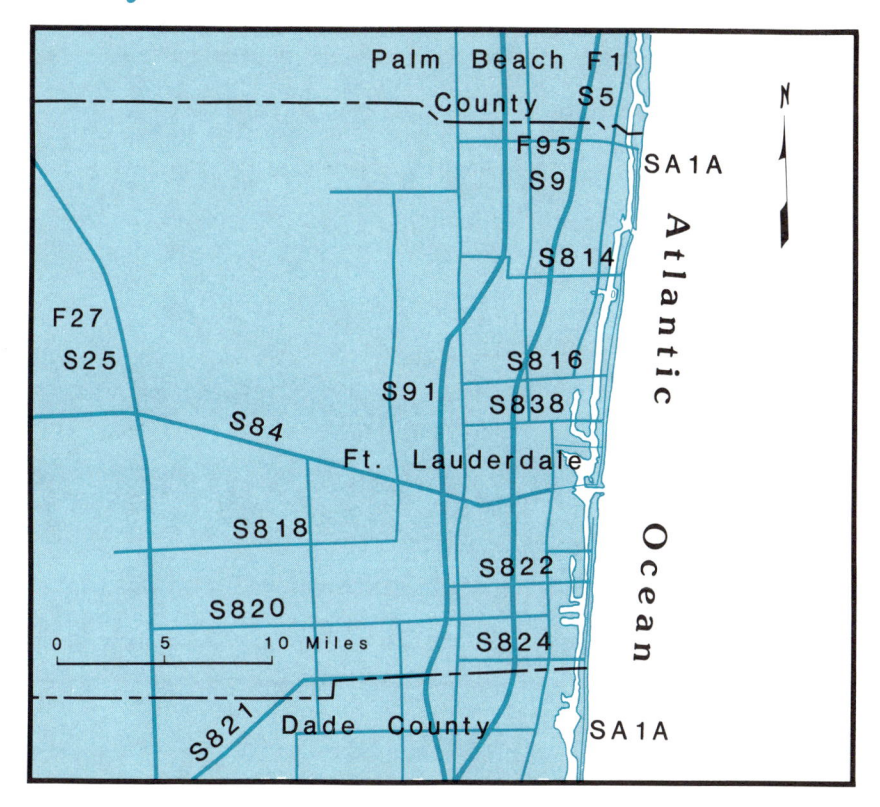

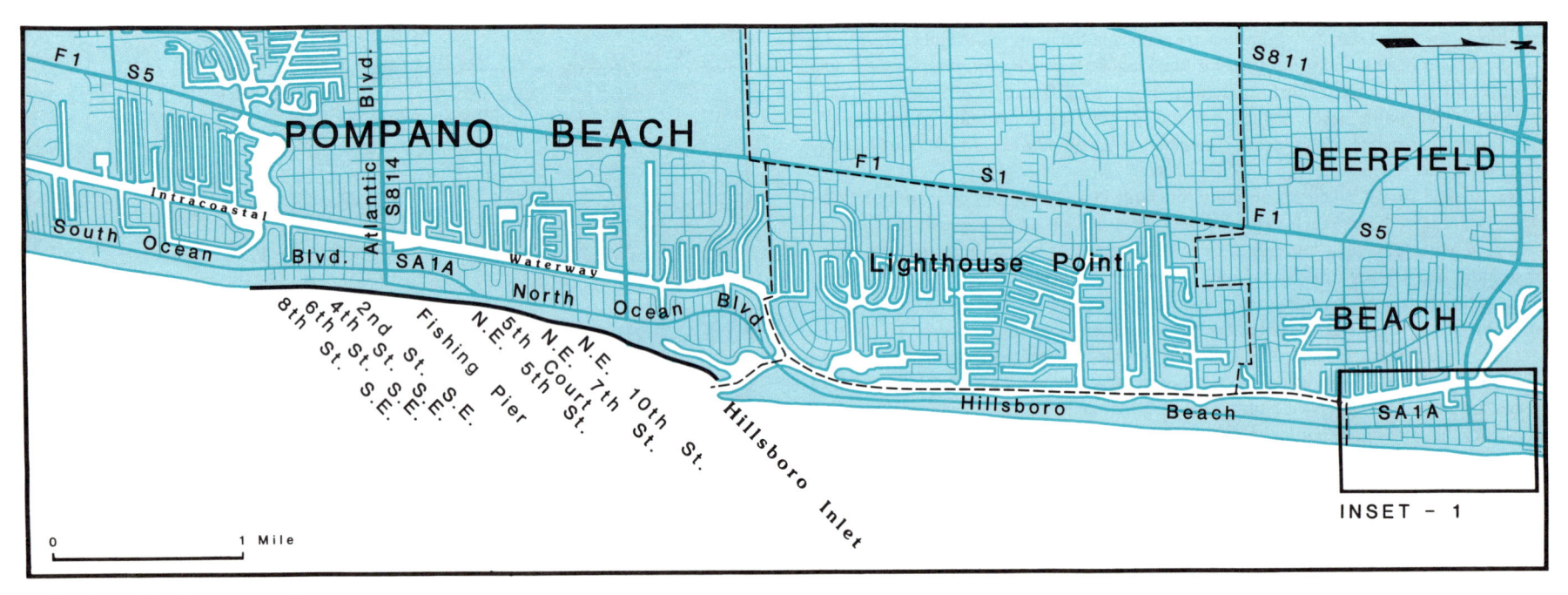

Northern district

Deerfield Beach is an undeveloped beach with 3,030 feet of beach frontage. Access is off S.R. A1A. North Beach: at N.E. 7th St., 4th Ct., and 2d St. Central Beach: at Access Way, S.E. 1st St.–4th St. South Beach: at S.E. 6th, 7th, 9th, and 10th streets.

Deerfield Island Park is a destination island park for boaters. These 56 acres are accessible only by water.

Pompano Beach has 3,000 feet of developed beach with lifeguards. Access is off North Ocean Blvd. (S.R. A1A) at N.E. 10th St., 7th St., 5th Ct., 5th St.; at 2d St. S.E., 4th St., 6th St., and 8th St. A public parking lot is located between N.E. 5th St. and Atlantic Blvd. (S.R. 814). Fishing pier.

Broward County

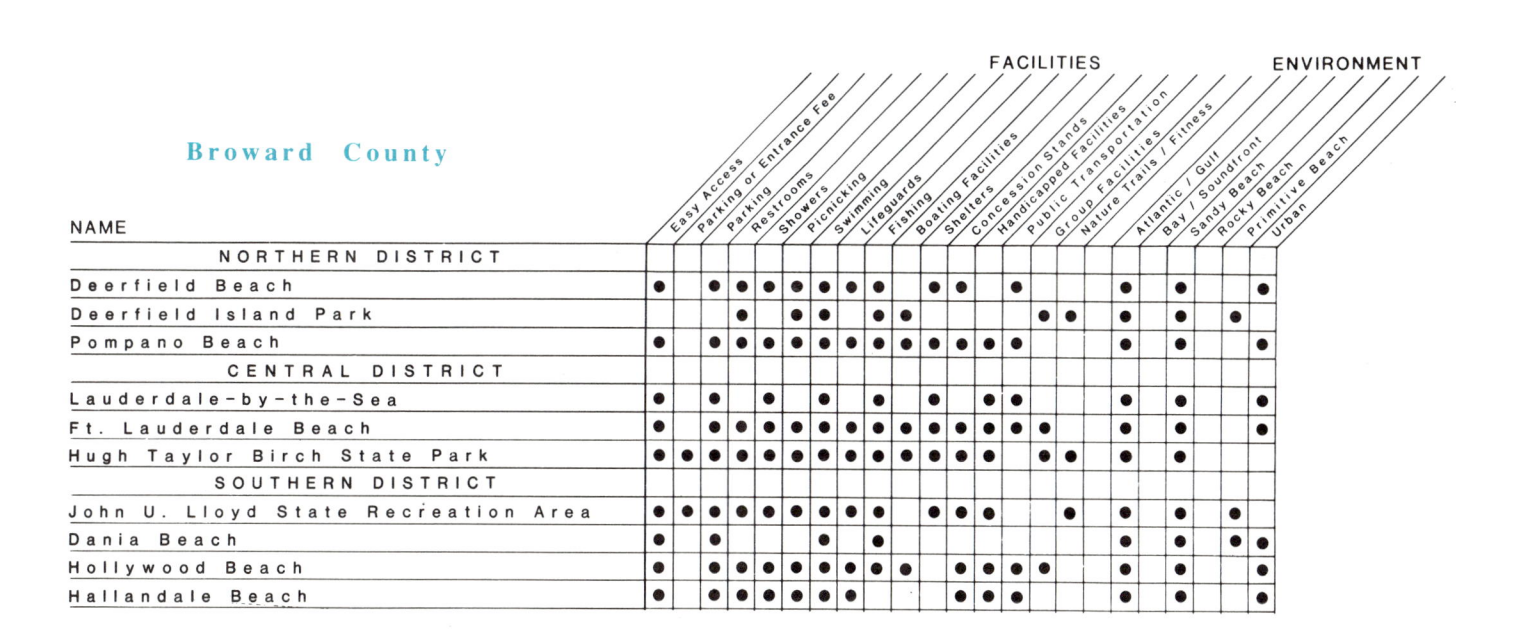

NAME	Easy Access	Parking or Entrance Fee	Parking	Restrooms	Showers	Picnicking	Swimming	Lifeguards	Fishing	Boating Facilities	Shelters	Concession Stands	Handicapped Facilities	Public Transportation	Group Facilities	Nature Trails / Fitness	Atlantic / Gulf	Bay / Soundfront	Sandy Beach	Rocky Beach	Primitive Beach	Urban
NORTHERN DISTRICT																						
Deerfield Beach	•		•	•	•	•	•	•			•	•	•				•		•			•
Deerfield Island Park			•	•		•		•	•					•	•		•	•	•			
Pompano Beach	•		•	•	•	•	•	•	•		•	•	•				•		•			•
CENTRAL DISTRICT																						
Lauderdale-by-the-Sea	•		•		•		•	•			•	•					•		•			•
Ft. Lauderdale Beach	•		•	•	•	•	•	•	•		•	•	•				•		•			•
Hugh Taylor Birch State Park	•	•	•	•		•	•		•		•	•	•		•	•	•		•			
SOUTHERN DISTRICT																						
John U. Lloyd State Recreation Area	•	•	•	•	•	•	•	•	•		•	•			•		•		•	•		
Dania Beach	•		•			•	•										•		•	•		•
Hollywood Beach	•		•	•	•	•	•	•	•		•	•	•				•		•			•
Hallandale Beach	•		•	•	•	•	•	•			•	•	•				•		•			•

Pompano Beach

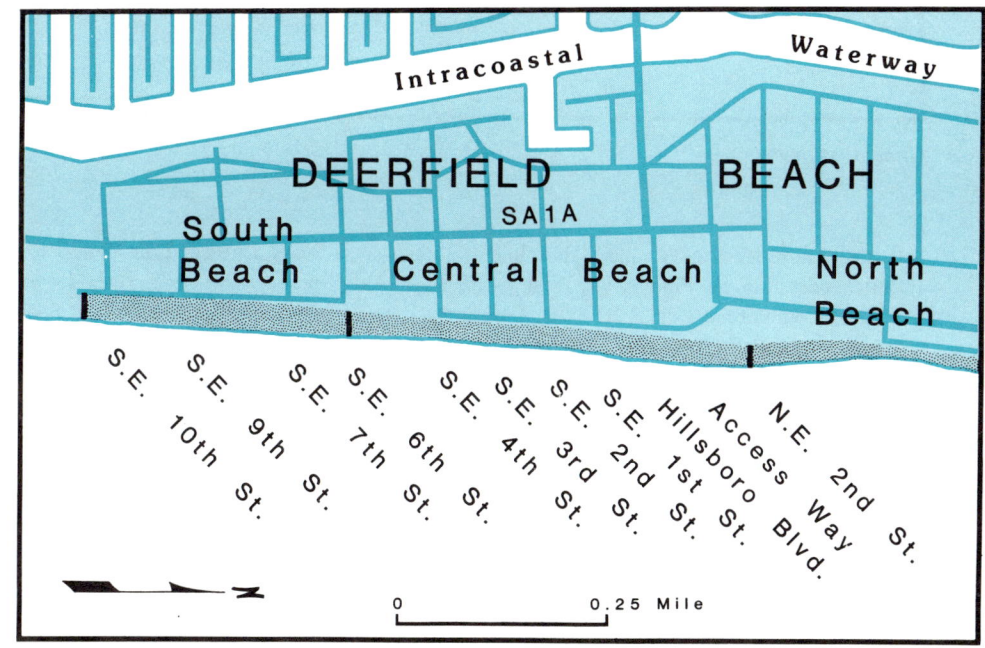

Deerfield Beach (inset 1)

85

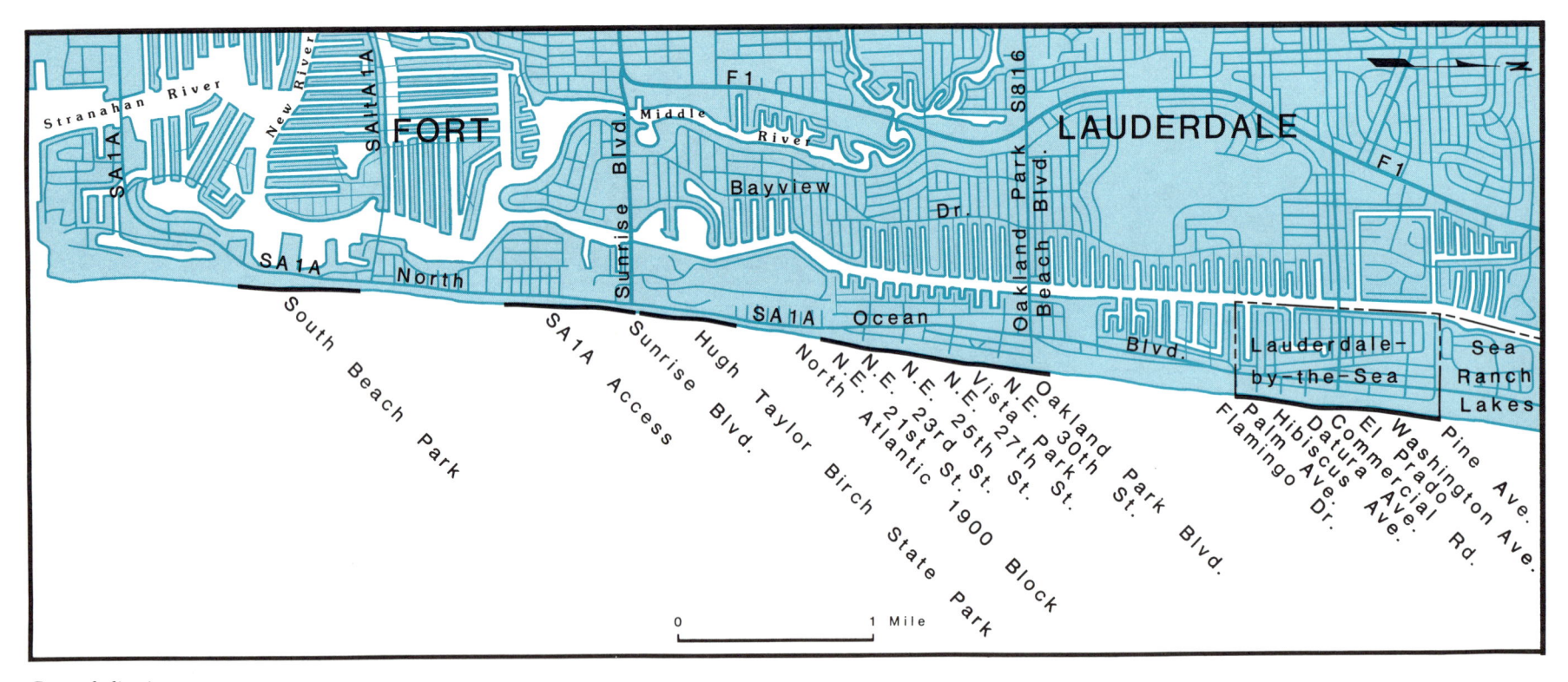

Central district

Lauderdale-by-the-Sea has a shoreline of about one mile accessible on Pine Ave., Washington Ave., El Prado, Commercial Rd. (S.R. 870), Datura Ave., Hibiscus Ave., Palm Ave., and Flamingo Dr. No alcohol allowed. Fishing pier.

Fort Lauderdale Beach has access to over 21,000 feet of developed crushed shell/coarse sand beach. Access street ends: Oakland Park Blvd. (S.R. 816), N.E. 30th St., Vista Park, N.E. 27th St., 25th St., 23d St., 22d St., 21st St., North Atlantic 1900 block, and Sunrise Blvd. (U.S. 1). Additional open access may be had along S.R. A1A where it parallels the beach. No alcohol allowed.

Hugh Taylor Birch State Park has access to 400 feet of beach. It is a developed state recreation area that extends from the Atlantic Ocean to the Intracoastal Waterway. Access from Fort Lauderdale is east on Sunrise Blvd. (U.S. 1) to North Atlantic Blvd. (S.R. A1A). No alcohol allowed.

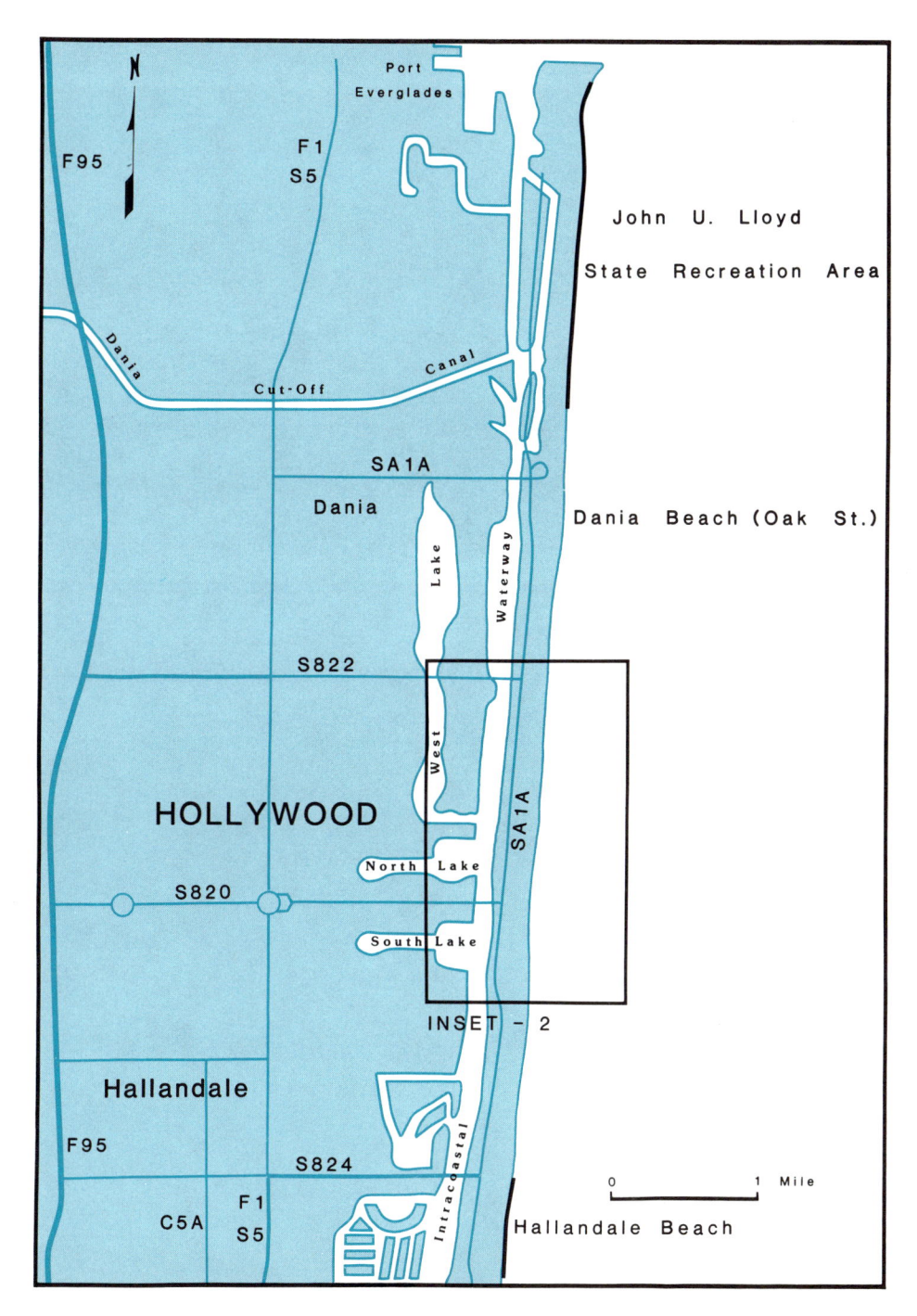

John U. Lloyd State Recreation Area covers 244 acres of barrier island north of Dania. Access is on Dania Beach Blvd. (S.R. A1A). This developed state recreation area has over 11,000 feet of beach frontage with lifeguarded areas. No alcohol permitted.

Dania Beach has 1,000 feet of undeveloped beach. Access is east off S.R. A1A on Oak St.

Hollywood Beach has over 5 miles of developed beach with lifeguards. Street-end access off North Ocean Dr. See map on next page for access streets, from Sherman in the north through Greenbriar at the south. No alcohol is allowed. There is a catamaran launch and community center. Hollywood Boardwalk is a wide sidewalk area bordered to the east by beach and the west by rows of shops, restaurants, and hotels.

Hallandale Beach is a developed beach with five dune walkways off Ocean Blvd., south of S.R. 824. No alcohol allowed.

Southern district

HOLLYWOOD

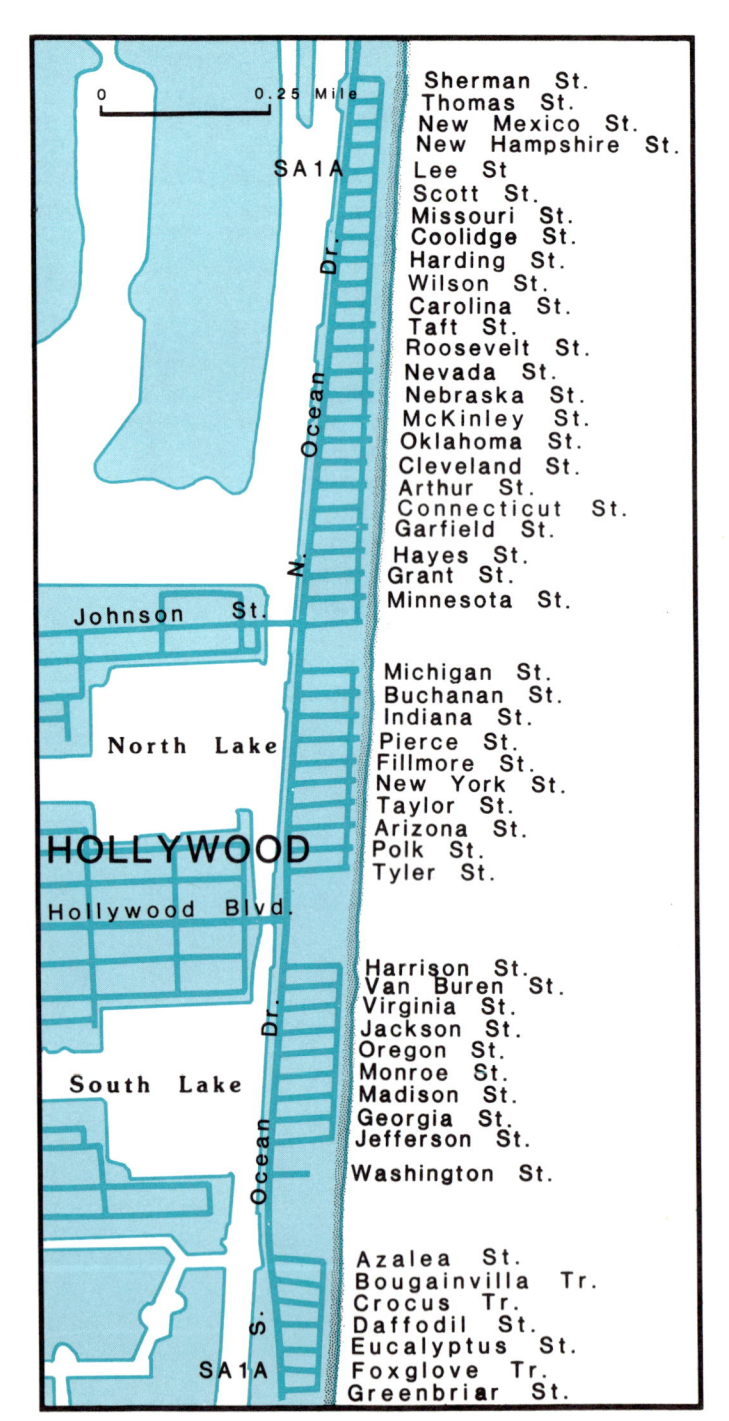

0.25 Mile

SA1A
N. Ocean Dr.
Johnson St.
North Lake
Hollywood Blvd.
South Lake
S. Ocean Dr.
SA1A

Sherman St.
Thomas St.
New Mexico St.
New Hampshire St.
Lee St.
Scott St.
Missouri St.
Coolidge St.
Harding St.
Wilson St.
Carolina St.
Taft St.
Roosevelt St.
Nevada St.
Nebraska St.
McKinley St.
Oklahoma St.
Cleveland St.
Arthur St.
Connecticut St.
Garfield St.
Hayes St.
Grant St.
Minnesota St.

Michigan St.
Buchanan St.
Indiana St.
Pierce St.
Fillmore St.
New York St.
Taylor St.
Arizona St.
Polk St.
Tyler St.

Harrison St.
Van Buren St.
Virginia St.
Jackson St.
Oregon St.
Monroe St.
Madison St.
Georgia St.
Jefferson St.
Washington St.

Azalea St.
Bougainvilla Tr.
Crocus Tr.
Daffodil St.
Eucalyptus St.
Foxglove Tr.
Greenbriar St.

Fort Lauderdale Beach

Hollywood Beach (inset 2)

Beach Erosion and Restoration

Sandy beaches, those areas valued by tourist and resident alike, are subject to continuing erosion, for a variety of reasons: wave action, offshore slope, sea-level rise, climatic conditions, type and source of sand. In addition, manmade structures (seawalls, groins, jetties) can accelerate erosion by intensifying and changing wave patterns along the shore. Given these twin forces, the rate of deposition of new sand on a beach is generally less than the rate at which these forces remove existing sand. Erosion results.

Because beaches are valued so highly for recreation, people have built structures on or near them. Erosion gradually exposes these structures to wave action and storms, and it reduces the visible beach area. The use of seawalls and groins does not generally protect beaches and associated property for long because the structures may add to erosion. The increase in beach construction and shoreline devices has increased erosion pressure on public and private beaches. Also, the loss of sand dunes through thoughtless construction practices has reduced a source of sand for replenishment of beaches as well as storm protection to structures.

Beach restoration is one way to replenish a beach with sand to add to its recreational and protective value. This process consists of finding, transporting, and placing acceptable sand onto the beach. While restoration cannot stop erosion, it can restore the beach to its original contours. This process must be done periodically to keep the same beach width.

The largest beach restoration project in the United States was begun in 1978–79 by the Army Corps of Engineers at Miami Beach in Dade County. This project consisted of placing 13.5 million cubic yards of sand on 9.3 miles of shore to create a new beach 300 feet wide where erosion had all but taken the sand off Miami Beach. In addition, 211,000 cubic yards are required annually to maintain the beach at its desired width. Naturally, the cost is high (over $62 million). The photographs on page 97 show the dramatic change brought about at Miami Beach through this project.

One alternative to periodic beach restoration is protection of existing sand dunes by preventing the construction of nearby buildings. Also, walkovers can be constructed over the dunes, grasses planted to "grow" new dunes, and fences placed to "capture" sand for new dunes. Dune protection, stabilization, and formation are important programs of the Florida Division of Beaches and Shores, which oversees shoreline construction and promotes dune protection. Further information can be obtained from Division of Beaches and Shores, Department of Natural Resources, Commonwealth Building, Tallahassee, FL 32303.

Biscayne Bay

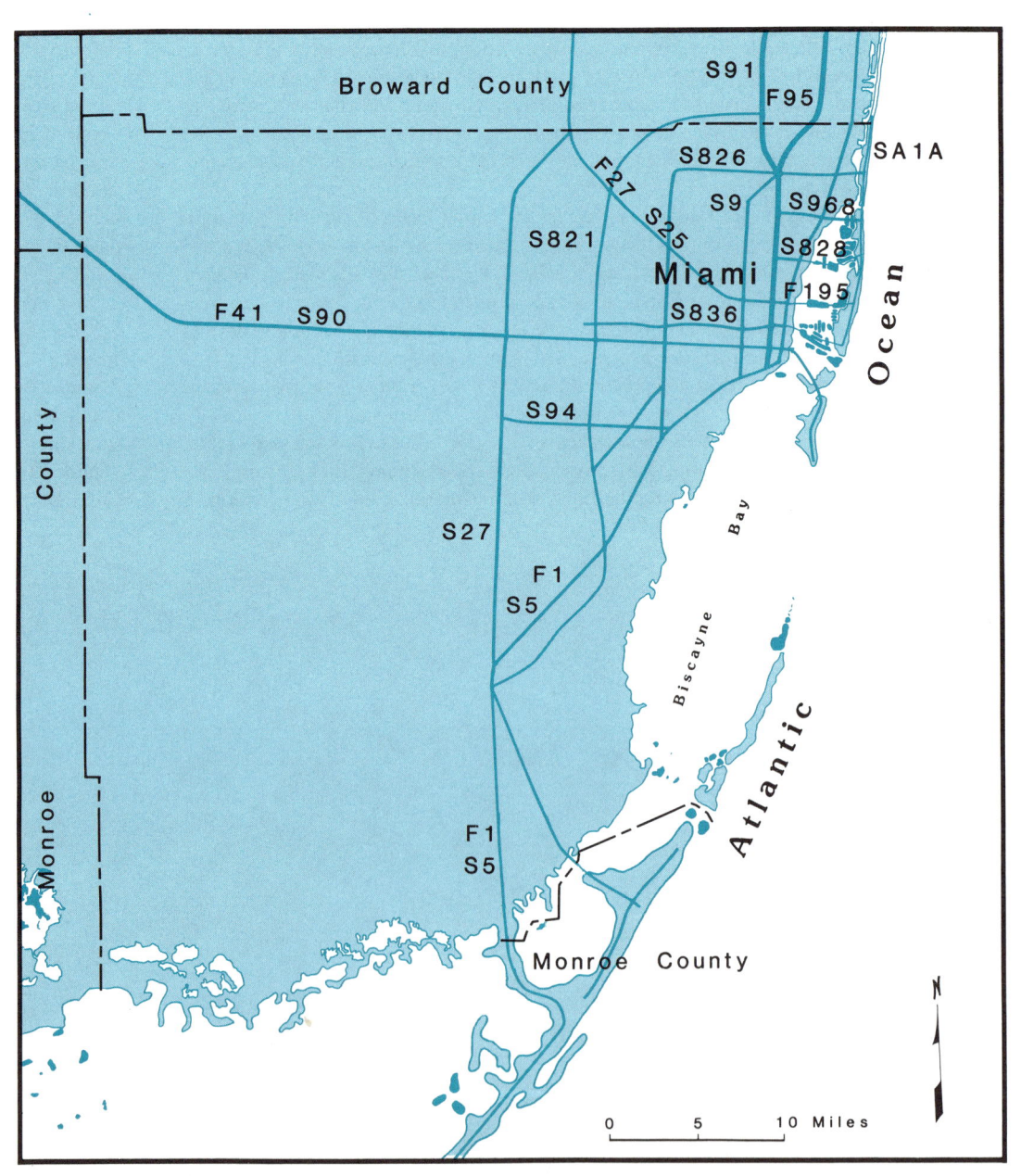

Dade County

Dade County lies on the southeastern tip of Florida. Established in 1836, this county was named after Major Francis L. Dade, the central figure in the Dade Massacre, near present-day Ocala, in 1835. Early settlers perched on the few high limestone ridges where dwellings could be constructed. It was not until 1910, when Carl Fisher migrated from Indiana, that landfill began in present-day Miami. Several years later he took title to what is now South Miami Beach. John Collins, a farmer, was attempting to grow avocados on the narrow barrier island, now developed with hotels and motels. Both Collins and Fisher began dredging and filling the landward side with silt that was replacing dense mangrove and tropical vegetation areas. Miami Beach was in its beginning stage, to be settled gradually by hundreds of northerners escaping frigid winters, carried south on Flagler's railroad.

To the west were the vast Everglades swamps and to the south a string of fragile islands named the Florida Keys. When Miami Beach vacationers stand on the 9.3 miles of restored beach and watch the construction of the Beachfront Park and the 1.8-mile Promenade scheduled for completion in 1986, many may not be aware of the serious beach erosion that had all but depleted most of the beach. The boardwalk and storm berm on which it is being constructed are expected to provide an excellent beach attraction.

The City of Miami Beach has many buildings constructed during the 1920s and 1930s. The architectural styles of this era, known as Art Deco and Mediterranean, can be seen as renovation of the buildings progresses.

Dade County has become a dynamic intercultural area, blending many nationalities and opening a vast market for international business. A major Cuban influence has added diversity to the rapidly growing area. Extensive port and airport facilities support the movement of people and goods to other areas.

The 406-acre park at Cape Florida State Recreation Area, located off U.S. 1 (via Rickenbacker Causeway to the lower tip of Key Biscayne near Miami), has a lush profusion of palm trees and tropical plants bordering the swimming beach. Cape Florida Lighthouse, one of South Florida's oldest structures, is the focal point of the park. Built in 1825 the lighthouse has endured marauding Indians and other dramatic incidents by pirates and "lawless persons." Crandon Park, on the northern end of Key Biscayne, has 698 acres of beach with facilities. A public golf course is located within the park boundary.

On Virginia Key, just a short distance after crossing the Rickenbacker Causeway, is the Miami Seaquarium and Planet Ocean. On the mainland, just south of Miami, is the Vizcaya Art Museum. To the south of Coconut Grove, off Old Cutler Road, is the Fairchild Tropical Gardens, where tropical plants can be viewed in a parklike setting.

Prior to 1900, the pineland area south of Miami was known as the "homestead country," federally owned and open to homesteaders. Because of an abbreviated designation on a Florida East Coast Railway freight car, the area station was subsequently named Homestead. Many winter vegetables are grown here and shipped to northern markets.

Biscayne National Park lies to the east of Homestead Air Force Base and encompasses Sands Key, Elliot Key, and Old Rhodes Key. The 175,000 acres consist of islands, coral reefs, turtle grass beds, and hardwoods, of which mahogany, Jamaica dogwood, and lignum vitae are of interest to naturalists. The abundant variety of birds and fish seen in the shallow waters may capture the interest of many.

Miami Beach

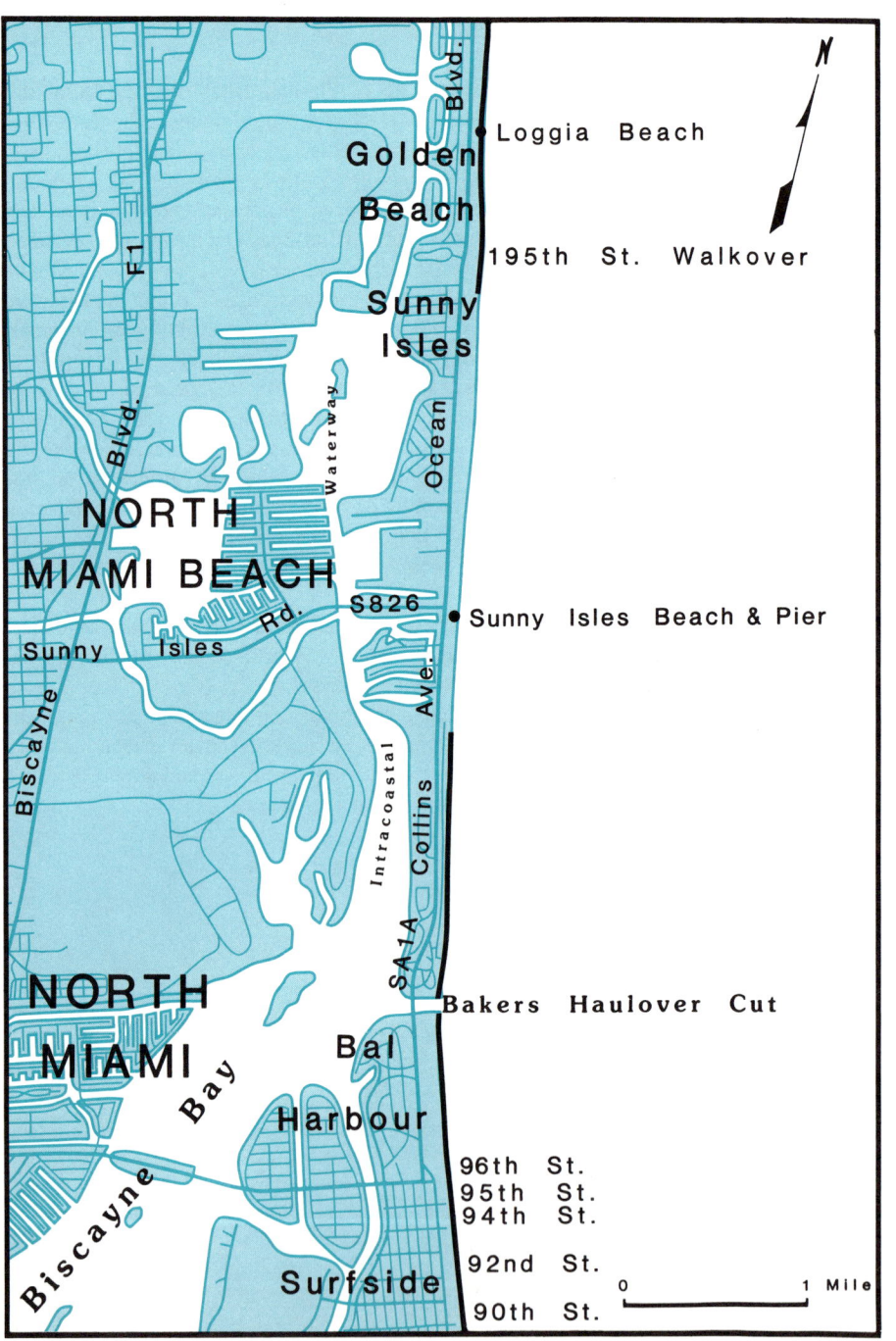

Northern district

Dade County

NAME	Easy Access	Parking or Entrance Fee	Parking	Restrooms	Showers	Picnicking	Swimming	Lifeguards	Fishing	Boating Facilities	Shelters	Concession Stands	Handicapped Facilities	Public Transportation	Group Facilities	Nature Trails / Fitness	Atlantic / Gulf	Bay / Soundfront	Sandy Beach	Rocky Beach	Primitive Beach	Urban
Golden Beach	•	•	•	•		•	•	•		•	•	•					•		•			•
Loggia Beach		•				•	•										•		•	•		
Sunny Isles Beach & Pier		•		•		•						•					•		•			•
Haulover Beach	•	•	•	•	•	•	•	•	•	•	•	•	•	•	•		•		•			•
Bal Harbor/Surfside	•		•	•	•		•	•	•			•	•				•		•			•

Golden Beach, 1.5 miles of undeveloped beach, is difficult to enter. There is a parking lot at 196th St. and a dune walkover at 195th St.; both are on Ocean Blvd. (S.R. A1A). No alcohol allowed.

Loggia Beach lies off S.R. A1A in Golden Beach. Access to this undeveloped beach is on Ocean Blvd. opposite the entrance to Central Island. No alcohol allowed.

Sunny Isles Beach and Pier lies on the barrier island off North Miami Beach. Access to this developed recreation area is at the intersection of Sunny Isles Causeway (S.R. 826) and Collins Ave. (S.R. A1A).

Haulover Beach and Pier has 172 acres of developed beach with lifeguards. Access is one-half mile south of S.R. 826 (Sunny Isles Rd.) on Collins Ave. (S.R. A1A). No alcohol allowed.

Bal Harbour/Surfside have 2 miles of undeveloped beach. Public access along Collins Ave. (S.R. A1A) at 96th, 95th, 94th, 92d, 90th, 89th, and 88th streets.

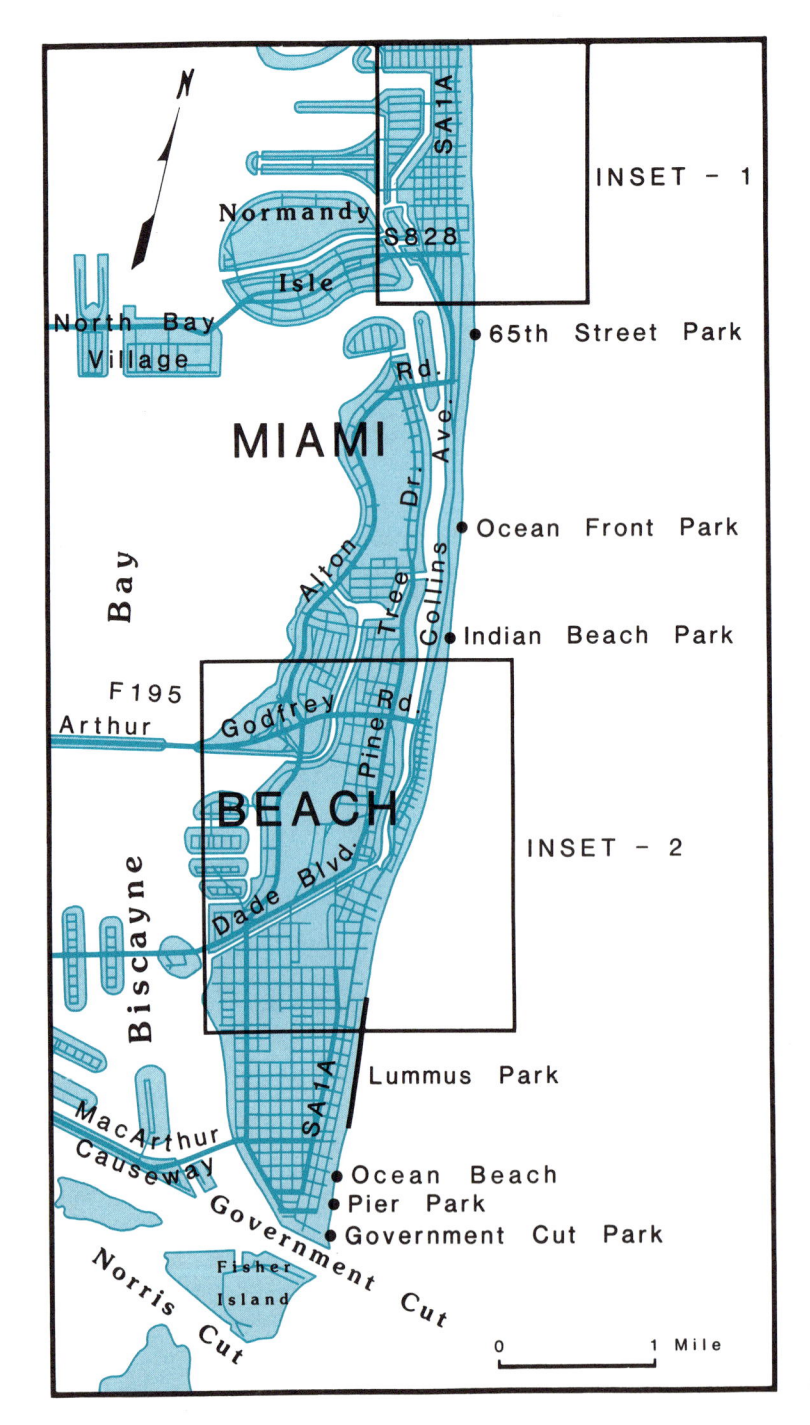

Crandon Park

Central district

Dade County

NAME	Easy Access	Parking or Entrance Fee	Parking	Restrooms	Showers	Picnicking	Swimming	Lifeguards	Fishing	Boating Facilities	Shelters	Concession Stands	Handicapped Facilities	Public Transportation	Group Facilities	Nature Trails / Fitness	Atlantic / Gulf	Bay / Soundfront	Sandy Beach	Rocky Beach	Primitive Beach	Urban
North Shore Ocean Front Park	●		●	●	●	●	●	●		●	●	●	●		●		●	●				●
North Shore Park	●	●	●	●	●		●	●			●	●	●				●	●				●
65th St. Park			●	●	●		●	●	●								●	●				●
Ocean Front Park			●	●	●		●	●	●					●			●	●				●
Indian Beach Park			●	●	●		●	●	●								●	●				
Miami Beach				●			●										●	●	●	●		
Collins Park	●		●	●	●		●	●				●	●				●	●				●
Lummus Park	●		●	●	●		●	●				●	●				●	●				●
Ocean Beach	●			●	●		●	●				●	●				●	●				●
Pier Park	●		●	●	●		●	●	●			●	●				●	●				●
Government Cut Park							●		●								●	●				●

North Shore Ocean Front Park has 24 acres of developed beach with lifeguards. Access and parking are available east on 79th St. Additional parking is between 84th/85th St. and 81st/80th St., west off Collins Ave. (S.R. A1A). No alcohol allowed.

North Shore Park lies on the barrier island between Biscayne Bay and the Atlantic Ocean. It has 36 acres of developed beach with lifeguard areas. Access points along Collins Ave. at 77th St., 74th St., 73d St., 71st St. (S.R. 828), and 69th St. Parking lots are off Collins Ave. at 76th/75th St., Ocean Terrace, and 73d/72d St.

65th St. Park, off Collins Ave. about 100 yards south of 65th St., has a small (1.5 acres), undeveloped, ocean-front beach with lifeguards.

Ocean Front Park lies on the barrier island between Indian Creek and the Atlantic Ocean. It is a developed beach with lifeguards. Access is from Collins Ave., about 2 miles south of 71st St. (S.R. 828).

Indian Beach Park is on Collins Ave. about one-half mile north of Arthur Godfrey Rd. The park is the main access to almost 2 miles of boardwalk and beach with lifeguard areas.

Miami Beach provides public access to undeveloped beach east of Collins Ave. at street ends from 43d St. through 29th St.

Collins Park is a developed beach with lifeguards and access to the boardwalk. The park is situated between 21st St. and 22d St.

Lummus Park has 48 acres of developed beachfront park with lifeguards. It is between 5th St. and 14th St. on Ocean Dr. in Miami Beach.

Ocean Beach on Miami Beach is an undeveloped recreation area with lifeguards. This area of 6 acres is accessible at 2d St. and 3d St. on Ocean Dr., south of Lummus Park.

Pier Park is at the intersection of Biscayne St. and Ocean Dr. at the south end of Miami Beach. This developed beach has more than 4 acres of beach with lifeguards.

Government Cut Park (South Point Park) at the southern tip of Miami Beach is under development and will be partly open in summer 1985.

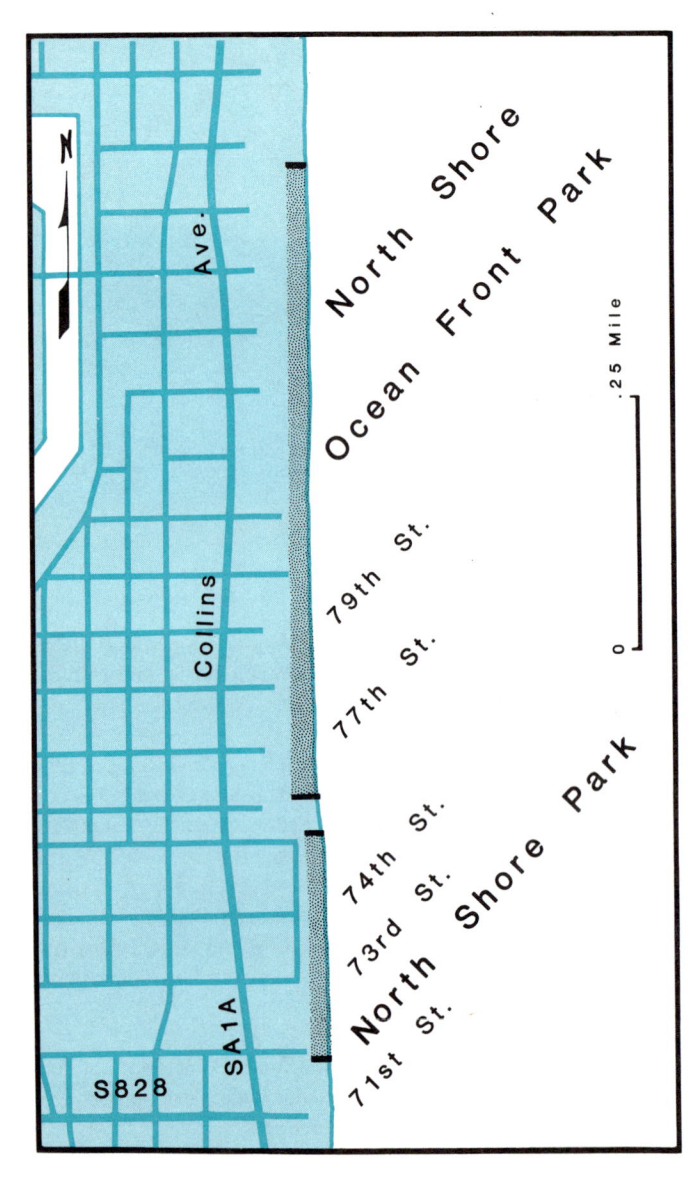

Miami Beach (inset 1)

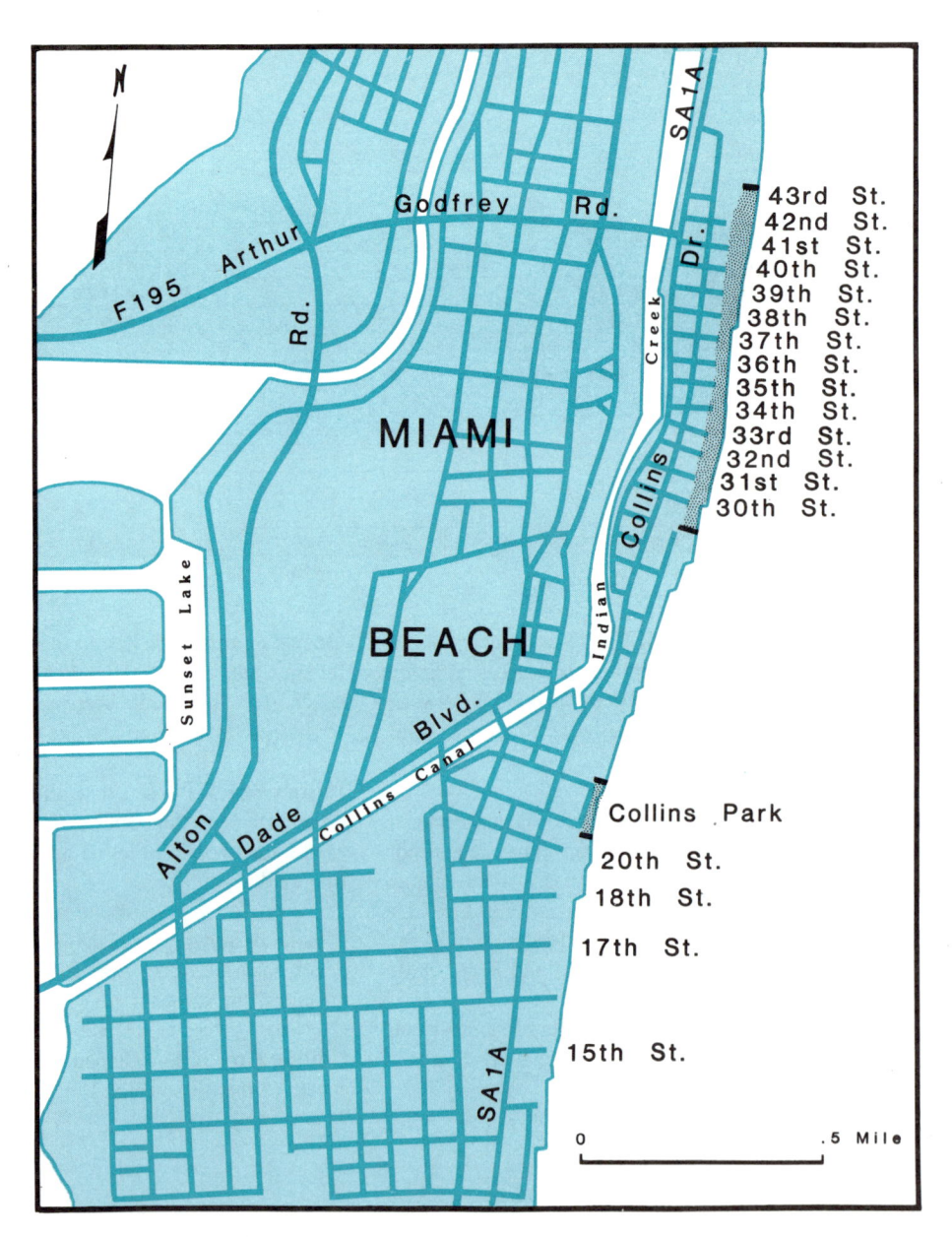

96

Miami Beach (inset 2)

Miami Beach before restoration

After beach restoration

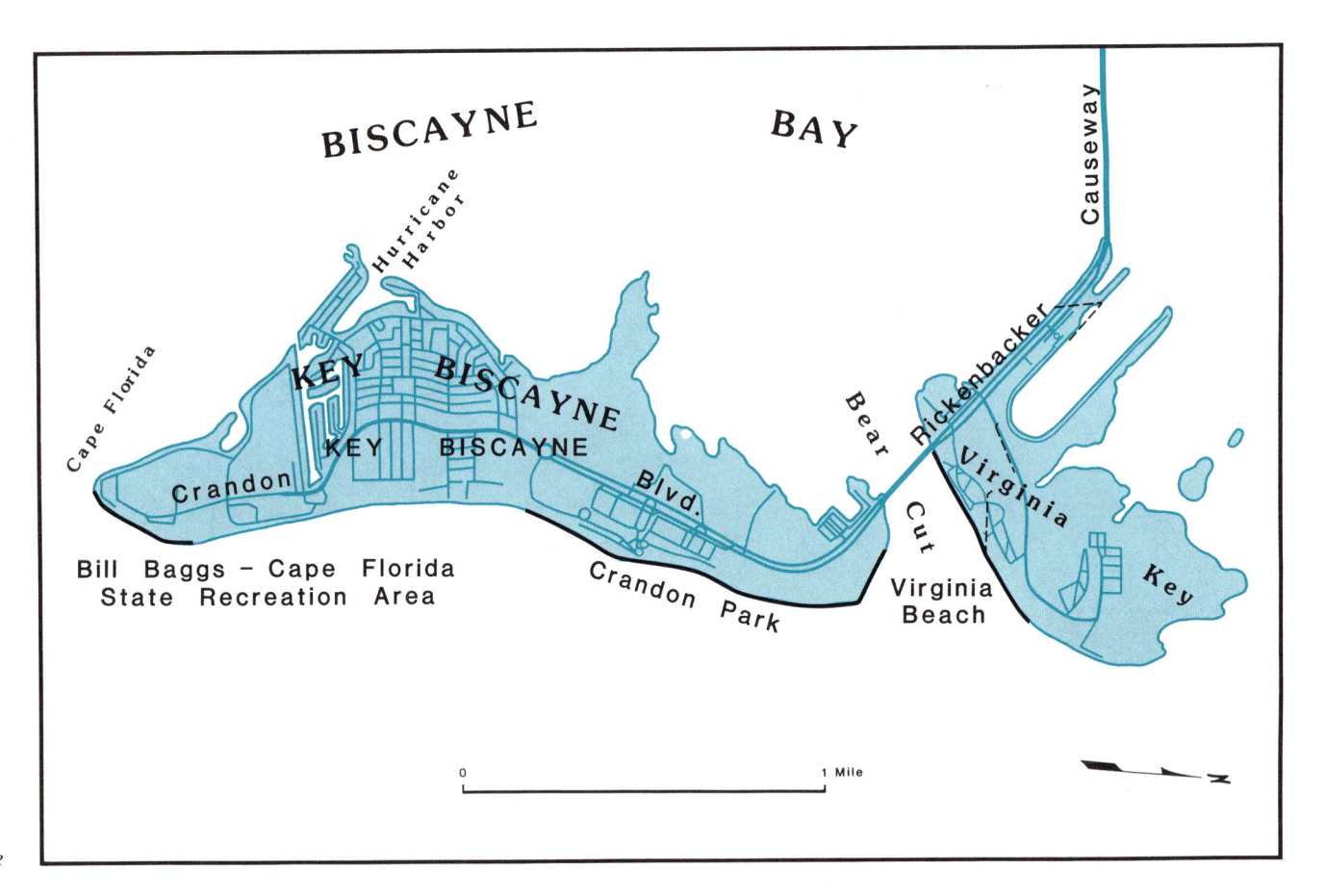

Key Biscayne

Matheson Hammock County Park has 580 acres of park and bay beach. Access to this developed park is off Old Cutler Rd. at 97th St. and Bay St. in South Coral Gables.

Virginia Beach has 145 acres on Virginia Key between the mainland and Key Biscayne. Access is off Rickenbacker Causeway. Park is still in the planning-construction phase in 1985.

Crandon Park has 698 acres of developed ocean beach, north of Key Biscayne, on Crandon Blvd. Golf course is located within park.

Bill Baggs/Cape Florida State Recreation Area lies at the south end of Key Biscayne. The 406-acre park is accessible off Crandon Blvd. No alcohol is allowed.

Dade County

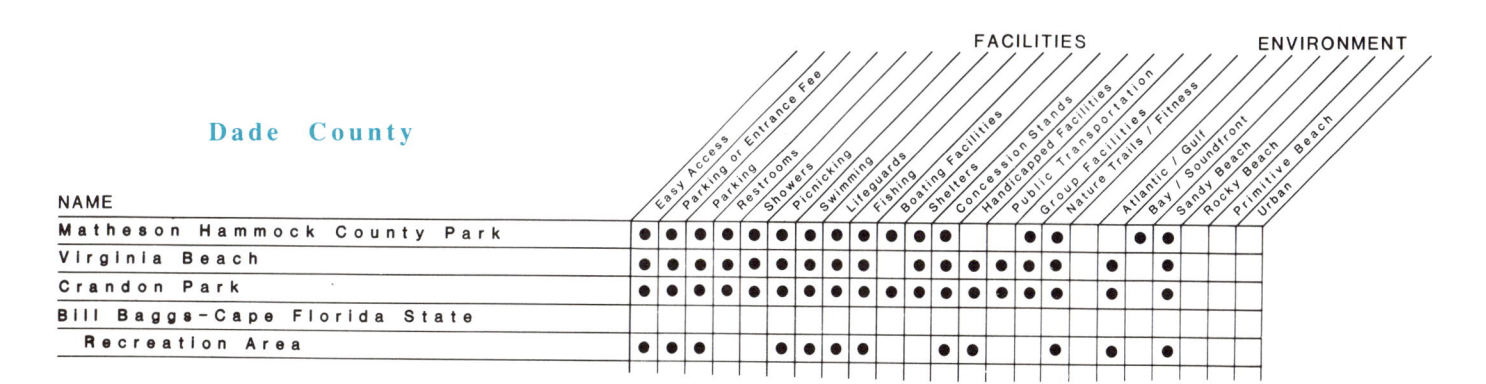

NAME	Easy Access	Parking or Entrance Fee	Parking	Restrooms	Showers	Picnicking	Swimming	Lifeguards	Fishing	Boating Facilities	Shelters	Concession Stands	Handicapped Facilities	Public Transportation	Group Facilities	Nature Trails / Fitness	Atlantic / Gulf	Bay / Soundfront	Sandy Beach	Rocky Beach	Primitive Beach	Urban
Matheson Hammock County Park	●	●	●	●	●	●	●	●	●	●	●	●			●	●			●	●		
Virginia Beach	●	●	●	●	●	●	●		●	●	●	●	●		●			●	●			
Crandon Park	●	●	●	●	●	●	●	●	●	●	●	●	●	●	●		●	●				
Bill Baggs-Cape Florida State Recreation Area	●	●	●		●	●	●		●		●	●		●		●	●					

Crandon Park

99

Cape Florida Lighthouse

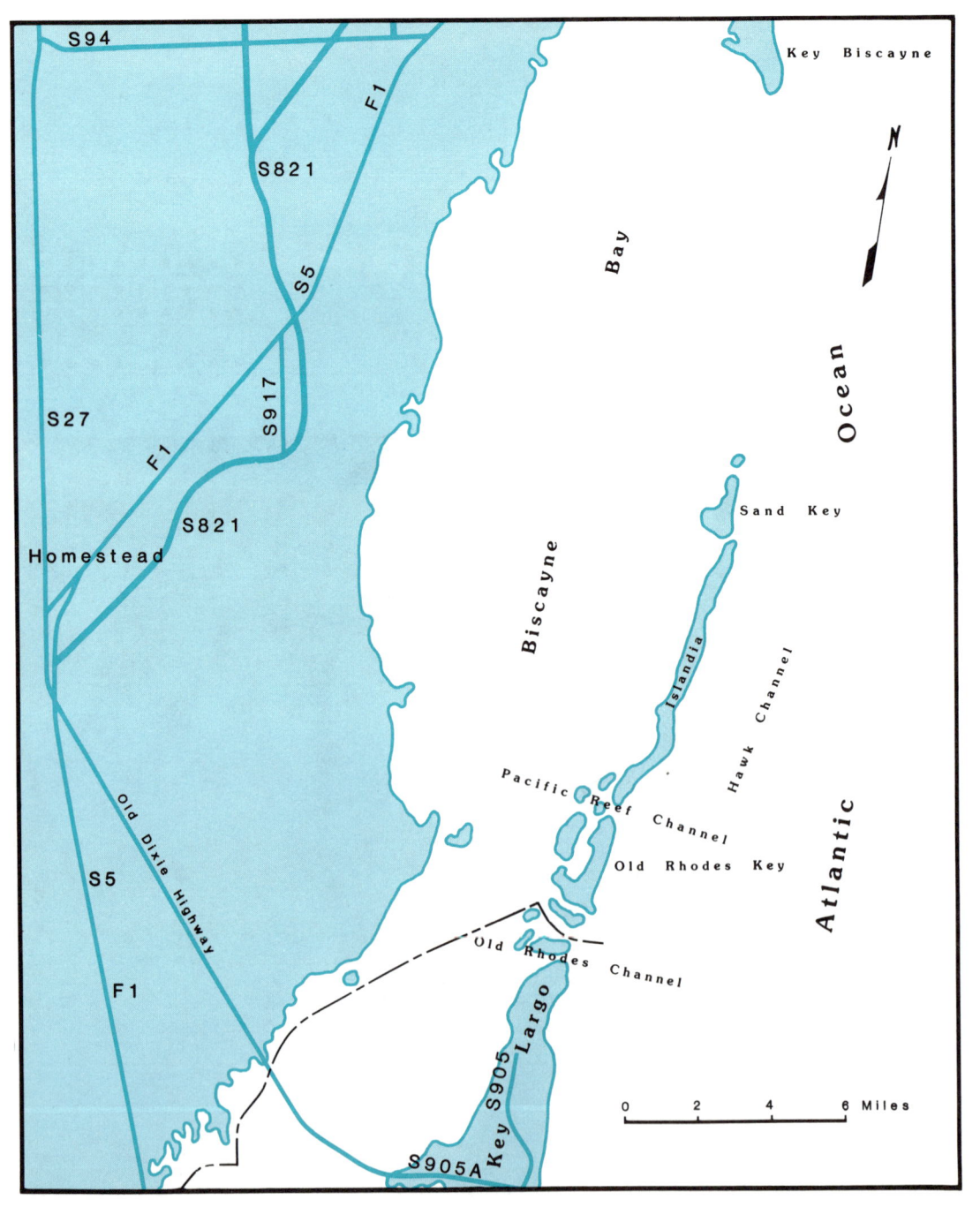

Southern district

Miami Beach Beachfront Park and Promenade

A major new attraction on recently renourished Miami Beach is the construction of a boardwalk between 21st Street and 46th Street in Miami Beach. This project, called the Beachfront Park and Promenade, is 1.8 miles long and has public entrances from 22 access street ends and walkovers from private developments. There will be accesses for handicapped persons at 22d, 36th, and 46th streets. The boardwalk, constructed on top of the storm-surge protective berm, shaped by the U.S. Army Corps of Engineers, is part of the reconstructed beach. The 122-foot-wide boardwalk will have landscaping on both sides to hold the sand in place, provide some shade, and create an attractive tropical setting for strollers. In addition, park benches, lighting, and signs will make it easy for many different groups to use the boardwalk. The City of Miami Beach is accelerating the construction and landscaping phase to accommodate anticipated tourist and resident use. The park is an exciting addition to the beach; it shows imagination, demonstrates agency cooperation, and opens access to the beach for the public.

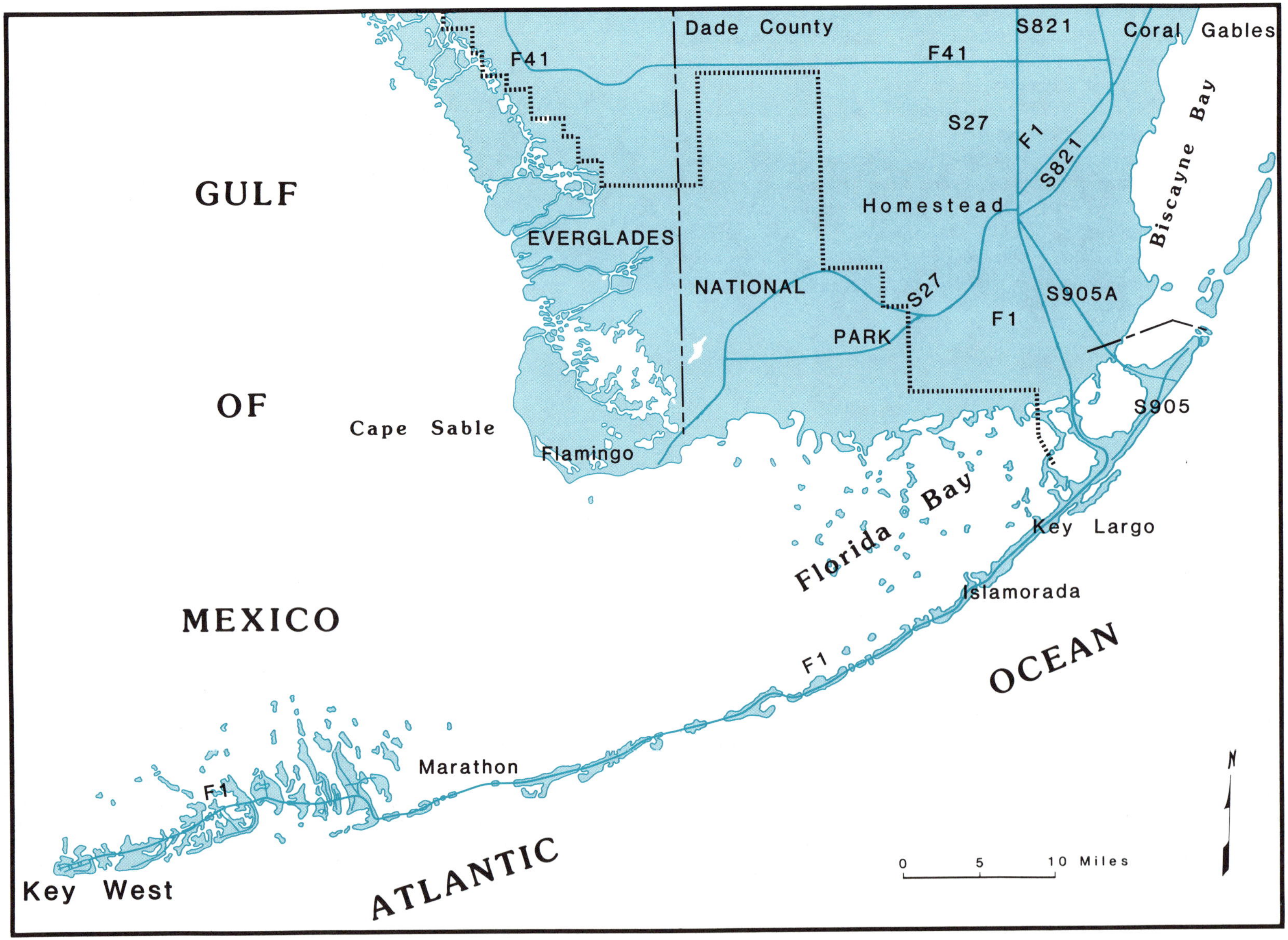
Dade County
S821
Coral Gables
F41
F41
S27
F1
S821
Biscayne Bay
Homestead
GULF
EVERGLADES
NATIONAL
S27
S905A
F1
PARK
OF
S905
Cape Sable
Flamingo
Florida Bay
Key Largo
Islamorada
MEXICO
F1
OCEAN
Marathon
F1
0
5
10 Miles
Key West
ATLANTIC

Monroe County

Monroe County, named for President James Monroe, lies on the southwestern tip of the Florida peninsula. The mainland is undeveloped, and most of it is included in Everglades National Park. Offshore Monroe County, familiarly known as the Florida Keys, extends more than 100 miles into the Straits of Florida, which separate the Atlantic Ocean from the Gulf of Mexico. Most of the county's population resides on these limestone isles, where boating and diving are popular recreational activities.

The keys are connected by the Overseas Highway, the "highway that goes to sea," from Key Largo to Key West, at the southern end of U.S. 1. Originally the Overseas Highway was the roadbed for the old Florida East Coast Railroad (FEC) built by Henry Flagler. This dangerous and expensive venture, often called "Flagler's Folly," took a work force of 5,000 men seven years and nine months to complete. Seven hundred of those men lost their lives working on the job that was completed in 1912 at a cost of $50 million. The railroad operated until a hurricane destroyed 41 miles of track and trestles on Labor Day 1935. Its replacement, the Overseas Highway, was opened in 1944.

Located on Cayo Hueso (Bone Key), literally at the end of U.S. 1, lies the county seat of Monroe County—Key West. The southernmost city in the United States, it has been occupied by pirates, "wreckers" (those who salvage ships wrecked on the reefs), spongers from Greece, and Cuban cigar manufacturers. The descendants of the original settlers are called Conchs, a name derived from their liking for broths, chowders, and fritters made from the muscle of the Queen Conch.

The exotic island city of Key West lures many artists and authors—both the aspiring and the accomplished. In the late 1920s Ernest Hemingway was enchanted by the island. While living there he finished *A Farewell to Arms* and *To Have and Have Not*. Later Tennessee Williams wrote *The Glass Menagerie* and *A Streetcar Named Desire* while he lived in Key West.

About 70 miles west of Key West lie the Dry Tortugas. This cluster of coral reefs was discovered in 1513 by Ponce de Leon, who named them "las tortugas" (the turtles) because of their large breeding population of turtles. The lack of fresh water brought these islands their later name, Dry Tortugas. Strategically important for controlling navigation of the Gulf of Mexico, these reef islands became a major part of the coastal defense system. Fort Jefferson, established in 1846, was dubbed "the Gibraltar of the Gulf." It was here that the "Lincoln conspirators" were imprisoned and the first naval wireless station was built.

Monroe County's environmentally sensitive lands—the Everglades and the Keys—have been recognized as areas of critical state concern. The need for protection of a unique state resource has brought forth regulations from state and county agencies that monitor development. The intention of these regulations is to preserve the natural beauty of the area for residents and visitors.

Mainland

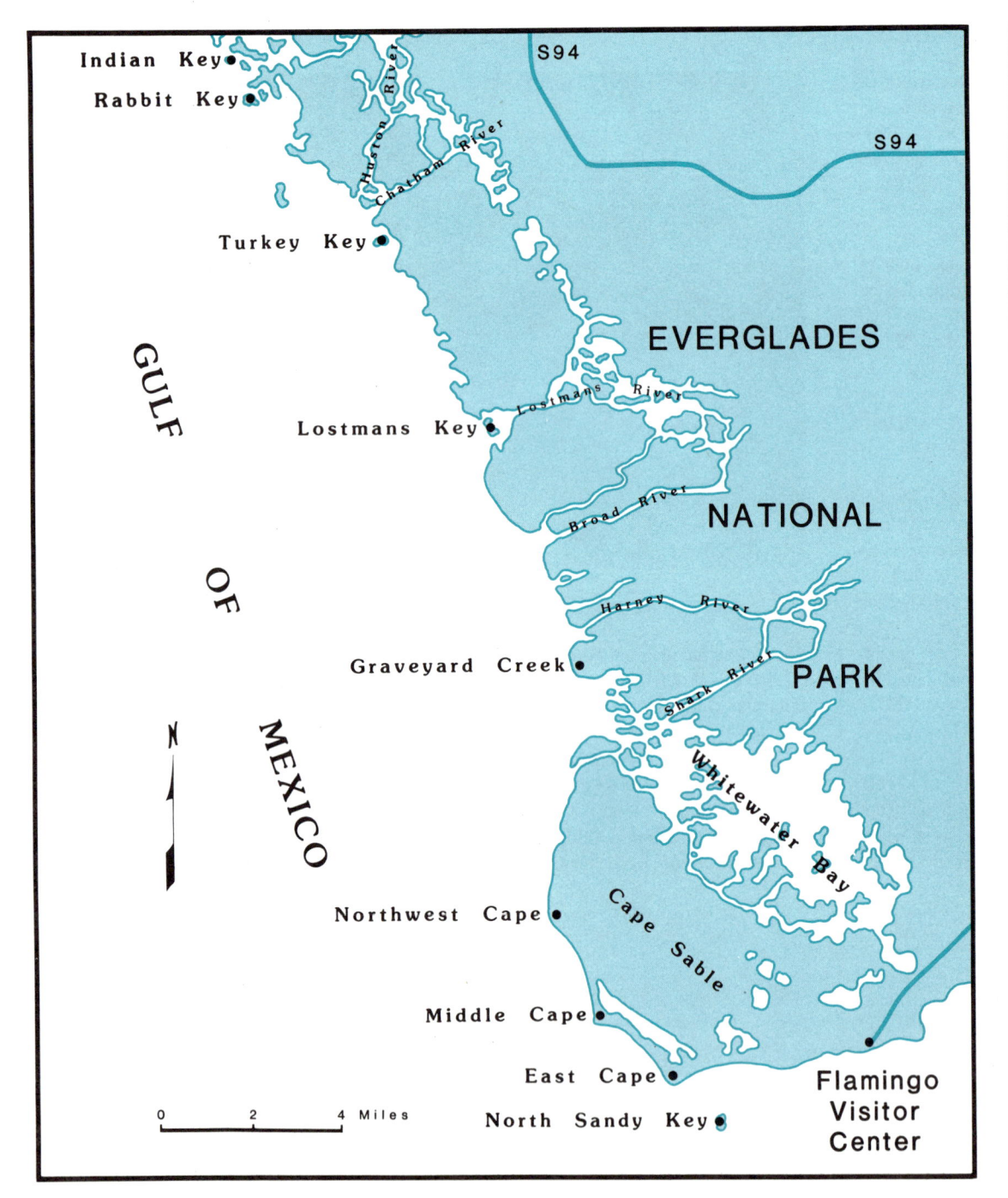

Pelicans

Monroe County

FACILITIES / ENVIRONMENT

NAME	Easy Access	Parking or Entrance Fee	Parking	Restrooms	Showers	Picnicking	Swimming	Lifeguards	Fishing	Boating Facilities	Shelters	Concession Stands	Handicapped Facilities	Public Transportation	Group Facilities	Nature Trails / Fitness	Atlantic / Gulf	Bay / Soundfront	Sandy Beach	Rocky Beach	Primitive Beach	Urban
Everglades National Park	●	●	●	●	●	●	●		●	●	●	●		●	●		●	●		●		
Indian Key	●	●		●		●	●	●	●	●							●		●		●	
Turkey Key	●	●		●			●	●		●	●							●		●		●
Lostmans Key	●	●		●			●	●		●	●							●		●		●
Graveyard Creek	●	●		●			●	●		●	●							●		●		●
Northwest, Middle, and East Cape	●	●		●			●	●		●	●							●		●		●
North Sand Key	●	●		●			●	●		●	●						●	●		●		
Rabbit Key	●	●		●			●	●		●	●							●		●		●
Flamingo Visitors Center	●	●	●	●	●	●	●			●	●	●	●			●		●		●		

Everglades National Park has 1.4 million acres of marshy land covered with tall grasses that support a variety of birdlife, manatee, the Florida panther, and alligator.

The following are located off Monroe County's Gulf coast. They are remote, wilderness beaches accessible only by boat. Major use is by overnight campers.

Indian Key
Rabbit Key
Turkey Key
Lostmans Key
Graveyard Creek
Northwest Cape
Middle Cape (Cape Sable)
East Cape (Cape Sable)
North Sandy Key
North Nest (Eastern Keys Inset)
Rabbit Key (Eastern Keys Inset)

Flamingo Visitors' Center is located at the southern end of the 100-mile Wilderness Waterway through the park's expanse of marsh area to the shallow waters of Florida Bay.

105

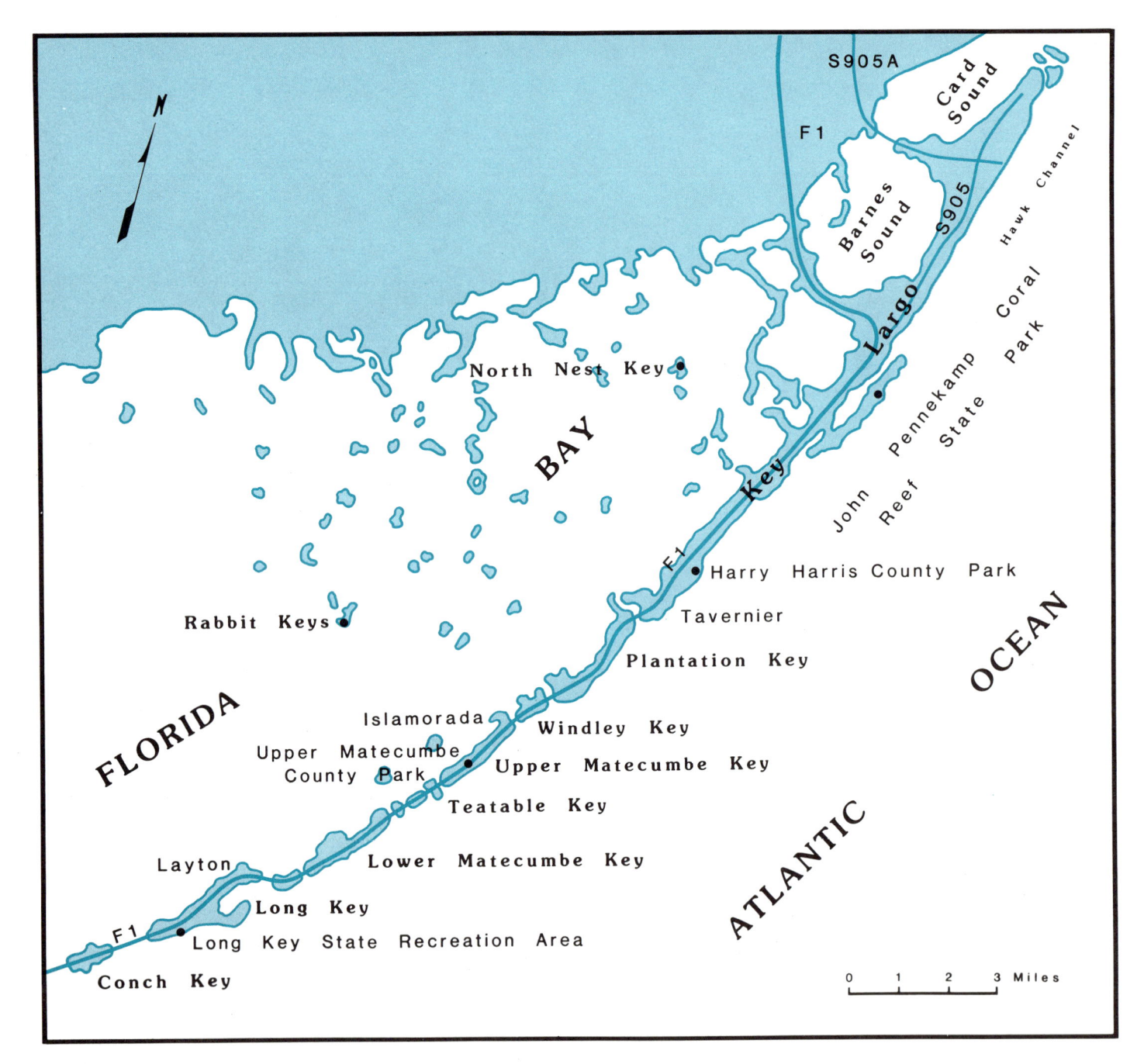

Eastern Keys

Monroe County

NAME	Easy Access	Parking or Entrance Fee	Parking	Restrooms	Showers	Picnicking	Swimming	Lifeguards	Fishing	Boating Facilities	Shelters	Concession Stands	Handicapped Facilities	Public Transportation	Group Facilities	Nature Trails / Fitness	Atlantic / Gulf	Bay / Soundfront	Sandy Beach	Rocky Beach	Primitive Beach	Urban
John Pennekamp Coral Reef State Park	●	●	●	●	●	●	●	●	●	●	●	●	●		●	●		●		●		
Harry Harris County Park	●	●		●		●	●		●	●					●		●		●			
Upper Matecumbe County Park	●	●		●		●	●		●	●							●	●	●			
Long Key State Recreation Area	●		●	●		●	●		●	●	●				●	●	●					

John Pennekamp Coral Reef State Park has 178 nautical square miles of coral reefs, seagrass beds, mangrove swamps, and a 2,640-foot land base with recreational facilities. This underwater state park fronts the Atlantic Ocean on Key Largo, south of Miami on U.S. 1.

Harry Harris County Park is 250 feet of Atlantic beach off Burton Drive in Tavernier, Key Largo.

Upper Matecumbe County Park has 200 feet of beach fronting the Altantic Ocean on Upper Matecumbe Key.

Long Key State Recreation Area has 1.5 miles of passive recreation and camping fronting the Atlantic Ocean. Abundant wading bird populations inhabit this area of mangrove lagoons and tropical hardwood hammocks on an ancient coral reef.

George Smathers Beach has 3,945 feet of developed beach on South Roosevelt Blvd., Key West.

Clarence Higgs Memorial Beach has 300 feet of developed Atlantic beach off Atlantic Blvd., Key West.

South Beach has 200 feet fronting Hawk Channel at the south end of Duval St., Key West.

Fort Jefferson National Monument is an isolated, primitive beach on Garden Key, Dry Tortugas. Access by boat (docking permitted only at designated locations not to exceed 2 hours) or seaplane.

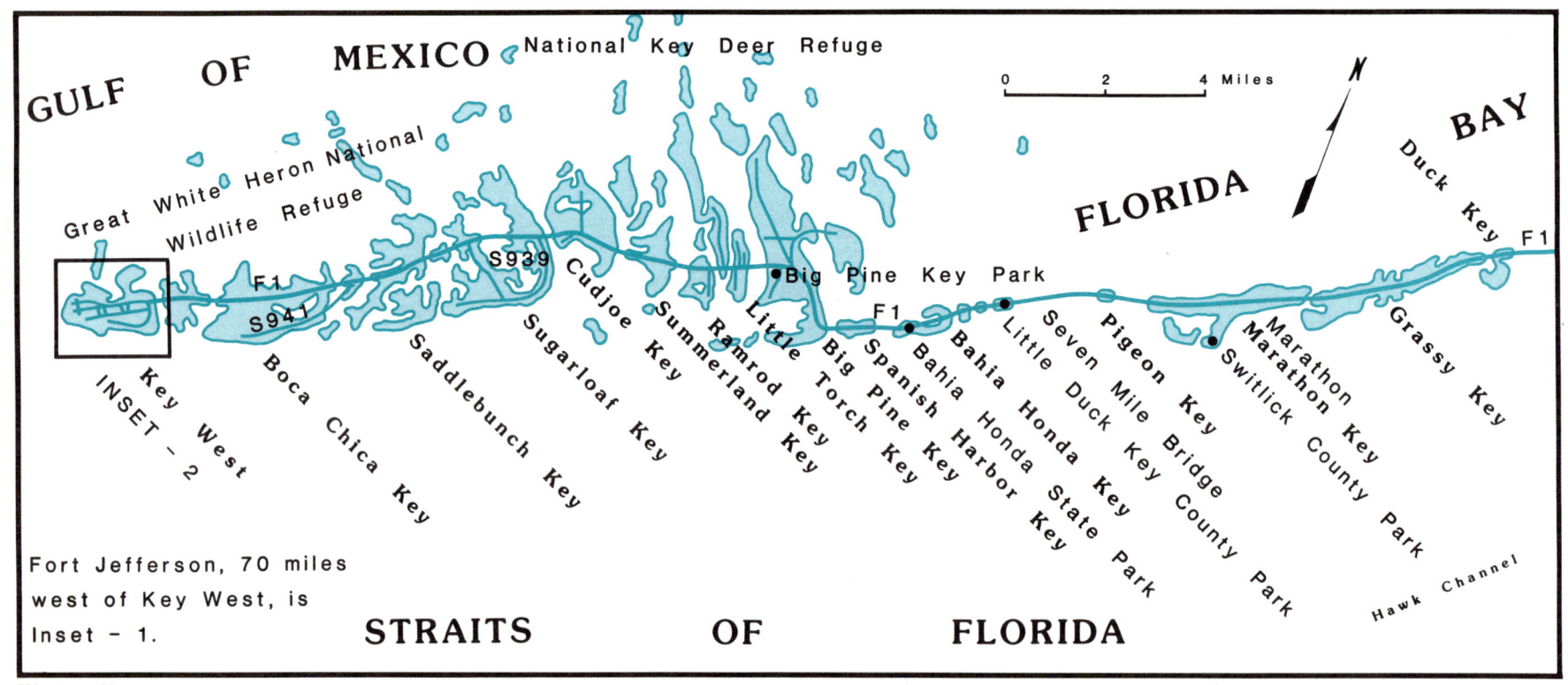

Western Keys

Fort Jefferson

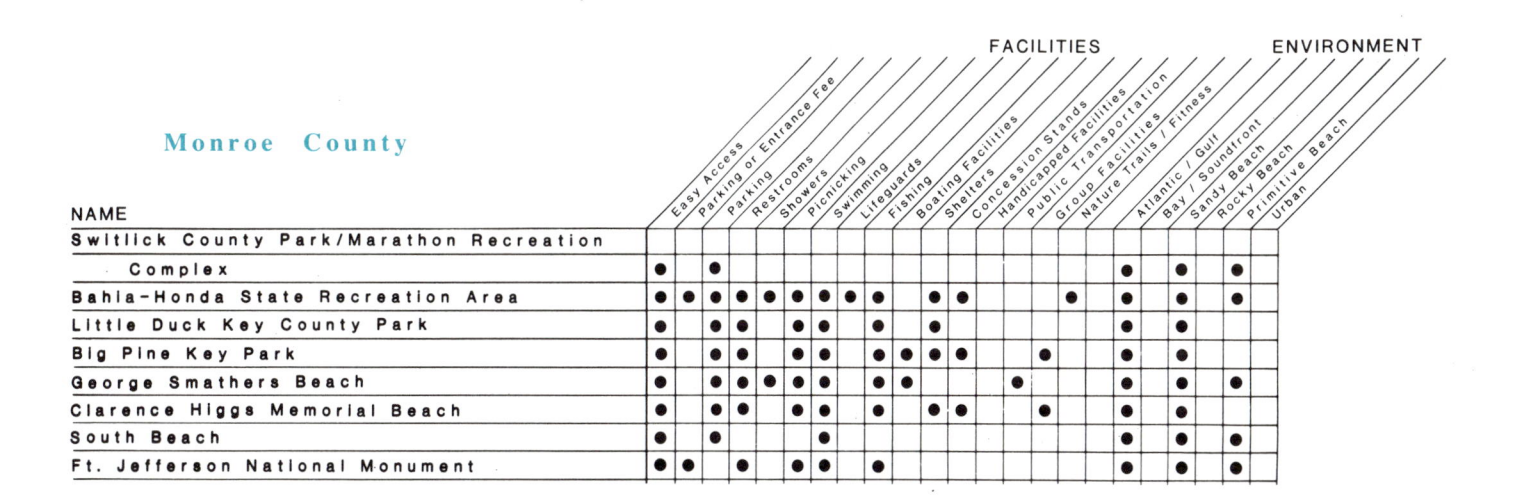

Monroe County

NAME	Easy Access	Parking or Entrance Fee	Parking	Restrooms	Showers	Picnicking	Swimming	Lifeguards	Fishing	Boating Facilities	Shelters	Concession Stands	Handicapped Facilities	Public Transportation	Group Facilities	Nature Trails / Fitness	Atlantic / Gulf	Bay / Soundfront	Sandy Beach	Rocky Beach	Primitive Beach	Urban
Switlick County Park/Marathon Recreation Complex	•	•															•		•			•
Bahia-Honda State Recreation Area	•	•	•	•	•	•	•		•	•			•				•		•			
Little Duck Key County Park	•		•	•		•	•		•								•		•			
Big Pine Key Park	•		•	•		•	•		•	•	•			•			•		•			
George Smathers Beach	•		•	•	•	•	•		•	•			•				•		•			•
Clarence Higgs Memorial Beach	•		•	•		•			•		•	•		•			•		•			
South Beach	•	•				•											•		•			
Ft. Jefferson National Monument	•	•		•		•	•		•								•		•			

Switlick County Park/Marathon Recreation Complex has 1,500 feet of beach located at the end of Sombrero Beach Road on Marathon Key.

Bahia Honda State Recreation Area has 1,548 feet on Bahia Honda Key. This skeleton of an ancient coral reef covered by beach, dunes, and mangroves is the state's southernmost recreation area. Tropical plants and rare species of birds flourish in the warm climate.

Little Duck Key County Park has 1,400 feet fronting the Atlantic Ocean on Little Duck Key.

Big Pine Key Park is an undeveloped county park fronting Florida Bay on Big Pine Key.

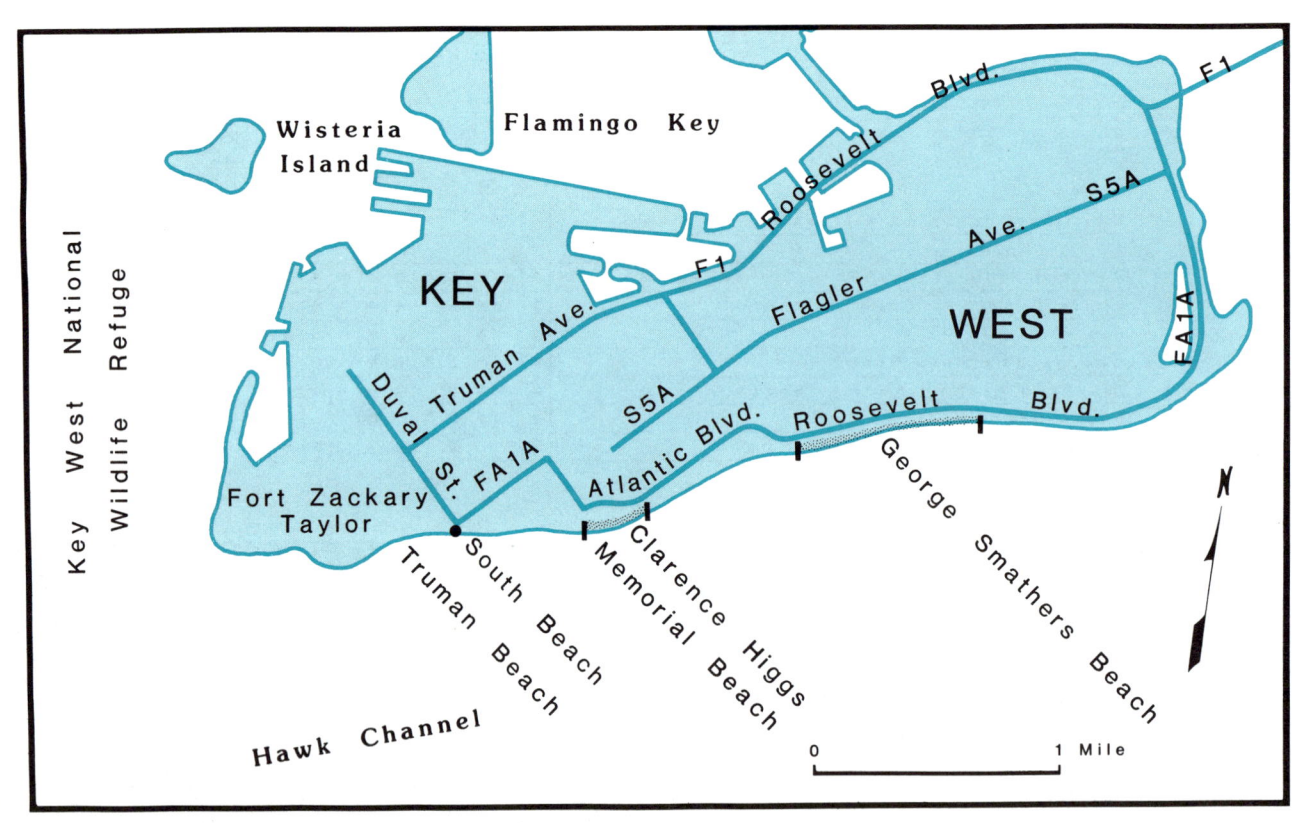

Key West

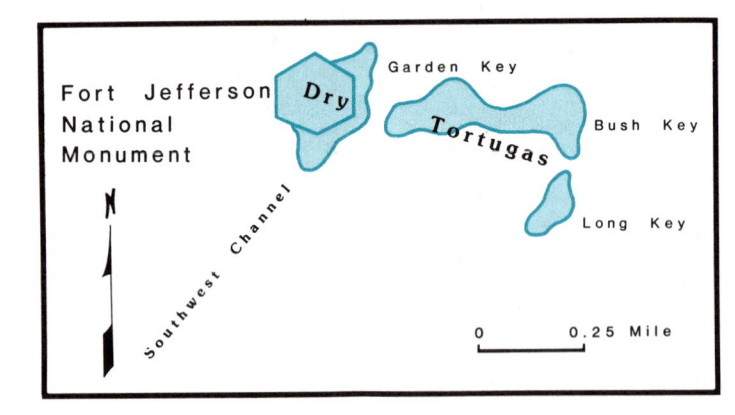

Fort Jefferson (inset 1)

Key West (inset 2)

Coral Reef Parks

The area seaward of Key Largo contains extensive coral reef communities, which flourish in the crystal water. These coral reef areas, the most diverse and productive of all natural marine communities, provide an abundance of vegetation, wildlife, and marine life to intrigue and delight skindivers and naturalists. To help protect this aquatic wonderland, three separate refuges have been established.

Adjacent to Key Largo in the Atlantic Ocean lies John Pennekamp Coral Reef State Park. Extending 21 miles in length and approximately 3 miles into the ocean, it was established as the first underwater state park in the United States in 1961. Directly seaward of the park and stretching its end boundaries out to 8.5 miles is Key Largo Coral Reef Marine Sanctuary, a federally protected area. Since 1975 it has served, in conjunction with the state park, to preserve 236 square miles of mangrove swamps, seagrass beds, and coral reefs. Sharing a common northern border and expanding the protected area by 15 more miles to the north is Biscayne National Monument. It encompasses both the Atlantic side of several keys and the Biscayne Bay area to the mainland. These parks combine to provide a twofold panorama, one above the surface and one below.

Coral reefs are made up of calcium deposited predominantly by the marine coral polyp. These creatures secrete a limey substance which becomes a hard outer skeleton around themselves. As they die, succeeding generations build on this foundation until a reef is established. This process requires clean, warm, circulating seawater with an unchanging salinity. Coral building is enhanced by the sheltering landmass of the keys and generally thrives best in waters less than 40 feet in depth. The many varieties of coral exhibit myriad shapes and colors.

These reefs are inhabited by a wide variety of marine organisms. Seagrasses, sponges, and mangroves provide food and a sheltered habitat along the reefs. Colorful fish such as neon gobies, queen angelfish, and blue chromis can be seen easily in the clear waters; and, for the sportsman, spiny lobster, stone crabs, grouper, and snapper are available in great abundance.

This area also contains culturally significant landmarks and structures. "Christ of the Deep," a nine-foot bronze statue located in 20 feet of water, symbolizes peace for mankind; it can be seen by divers and snorkelers. The Craysfort Reef Lighthouse, built in 1848 by the U.S. Lighthouse Service, still stands 100 feet above the water, warning ships to beware of hidden underwater dangers.

Before lighthouses were established, hundreds of galleons and frigates sank on these reefs in storms or high winds. Many others were sent to the bottom or raided by pirates, who lurked throughout the keys, taking advantage of the reefs with their smaller, more maneuverable sailing vessels. Today many wreck sites in the parks are marked by floating buoys. Their treasures have long since been plundered, and their cannons, now covered with barnacles, rest as monuments to a past fraught with peril and adventure.

Boating, fishing, snorkeling, diving, and swimming are popular activities in the parks. Local shops and marinas provide rental equipment for the experienced diver and sailor. Park stations can provide detailed guide maps and tours of the reefs with explanations of surrounding plant and wildlife. Because of the delicate balance within this marine ecosystem and the 400,000 yearly visitors to it, extra care must be taken by all to obey park regulations in order to preserve this resource.

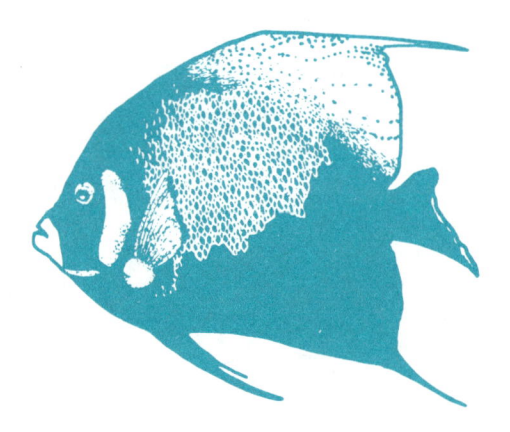

Southwest Coast

Sea oats

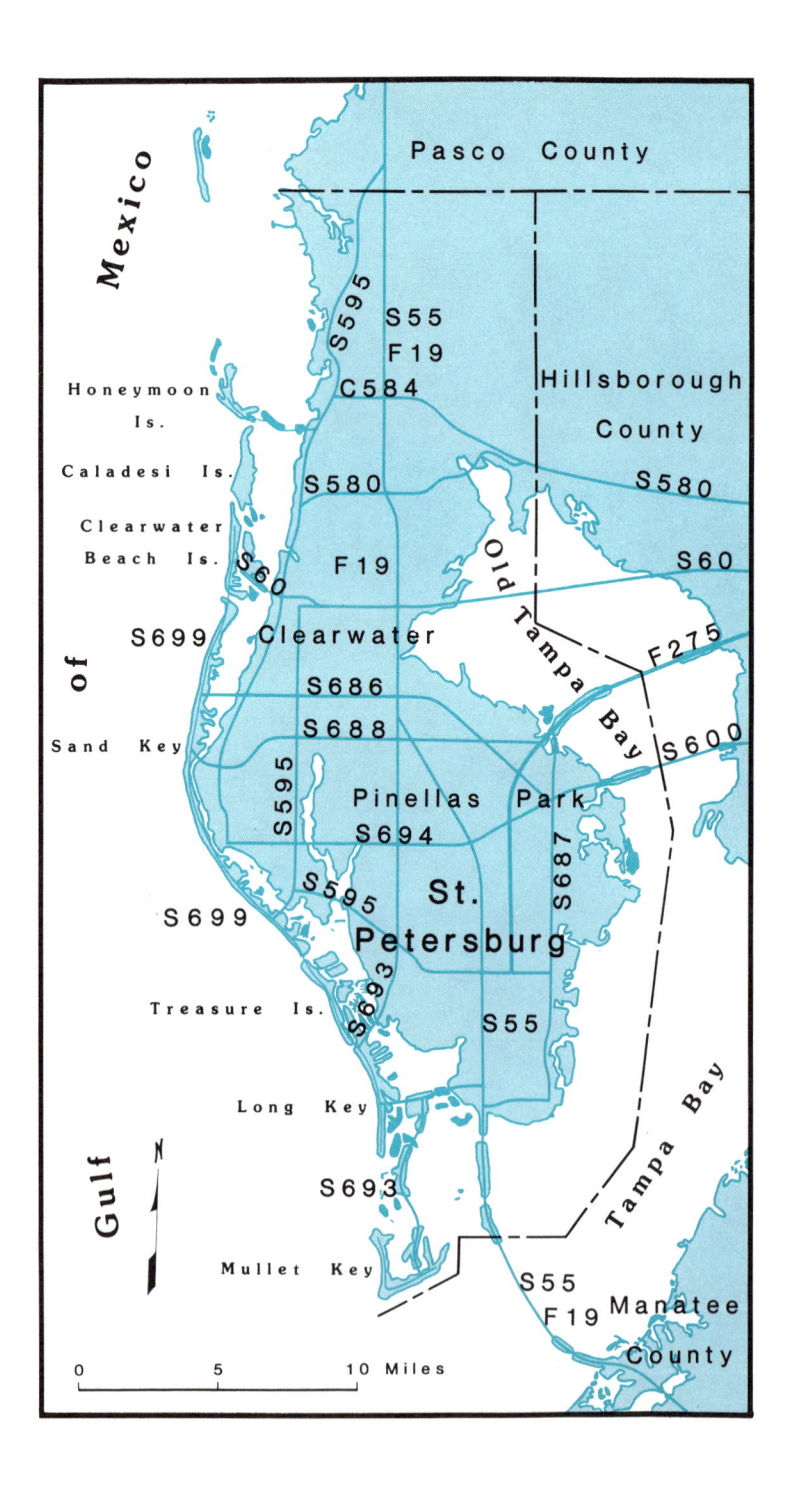

Pinellas County

Pinellas County, the second smallest in area of Florida's counties, lies predominantly on a peninsula in the central Gulf section of the state. This peninsula was known to the Spaniards as "Punta Pinal," which means "point of pines." Its present name was derived from this original title. Despite its small size, Pinellas County ranks third in population in the state and, with over 800,000 residents, has by far the greatest population density of all counties. The addition of 3.6 million tourists a year and 100,000 seasonal residents encourages diversity in entertainment and recreation.

The main attraction of Pinellas County is 28 miles of wide, sandy beach that extends the length of the Gulf of Mexico shoreline including offshore barrier islands, some of which remain in a natural state. Though it was divided into numerous municipalities and heavily developed, proper planning includes scores of state, county, and city parks. Many provide the public with beach access. Gulf Boulevard stretches the length of the islands, while Bayfront Boulevard skirts parts of the peninsula's coast. Countless streets off these roads also provide limited beach access to the visitor as do parks and exhibits on the eastern, bayfront side of the county.

The first recorded explorer of the county was Panfilo de Narvaez in 1528. A long succession of Hispanic explorers and pirates followed. Not until 1823 was the first permanent settlement established. Count Odet Phillipe, formerly a physician in Napoleon's army, is credited not only with founding this settlement but also with introducing the grapefruit tree to Florida. Population growth on the peninsula was fast in comparison with much of the western coast of Florida because of its unlimited water access. By 1892, St. Petersburg had become a flourishing commercial and residential area.

Three of the county's most outstanding coastal features are Caladesi Island, Fort DeSoto, and The Pier at St. Petersburg. Caladesi Island, one of the few remaining undisturbed barrier islands in Florida, lies off the northwestern coast of Pinellas. Accessible only by boat, this pristine state park features nature trails, guided walks by park rangers, and a 60-foot observation tower, which offers a panoramic view of the area. The park opens at 8:00 A.M. daily, and ferry service is available from Honeymoon Island State Park to the north.

At the southern tip of the county lies Fort DeSoto Park, a group of five connected barrier islands. The park has over 900 acres and 7 miles of beachfront. Owned and operated by the county, the area has been dedicated as a bird, animal, and plant sanctuary and is open daily until dark. A past interwoven with historical significance, including Fort DeSoto, gives the area added dimension. Considerable parking, fishing, camping, and picnicking facilities are available.

In Tampa Bay off St. Petersburg lies The Pier. Actually the culmination of several piers built over the last century, the present facility is owned by the City of St. Petersburg. It includes a recreation center and reflects an international flavor featuring ethnic gift shops, crafts, and cafés. Unique architecture and style highlight this hub of the city's two-mile waterfront district.

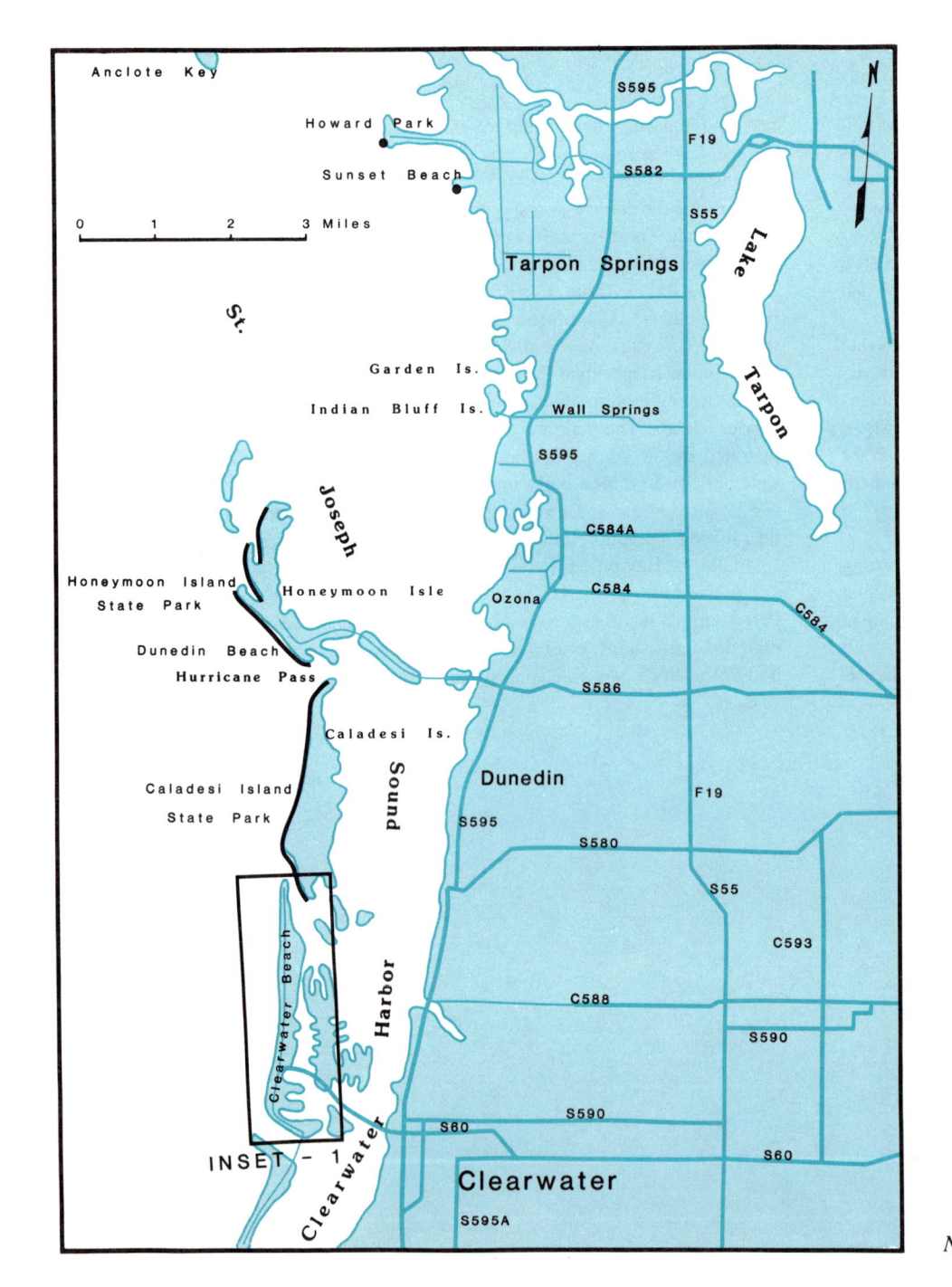

Clearwater Beach

Northern district

Pinellas County

FACILITIES / ENVIRONMENT

NAME	Easy Access	Parking or Entrance Fee	Parking	Restrooms	Showers	Picnicking	Swimming	Lifeguards	Fishing	Boating Facilities	Shelters	Concession Stands	Handicapped Facilities	Public Transportation	Group Facilities	Nature Trails / Fitness	Atlantic / Gulf	Bay / Soundfront	Sandy Beach	Rocky Beach	Primitive Beach	Urban
NORTHERN DISTRICT																						
Anclote Key							●		●								●	●	●		●	
Howard County Park	●		●	●	●	●	●	●	●	●	●			●	●		●	●	●			●
Sunset Beach			●				●	●	●								●	●	●			
Honeymoon Island State Park	●	●	●	●	●	●	●	●	●			●			●	●	●	●	●			
Dunedin Beach							●	●									●	●	●			
Caladesi Island State Park	●	●	●	●	●	●	●	●	●	●	●	●	●	●	●	●	●	●	●			
Clearwater Beach Island	●		●	●	●		●	●	●			●	●	●			●		●			●
Mandalay Park	●		●	●	●		●	●	●			●	●	●			●		●			●
Clearwater Beach Park	●	●	●	●	●		●	●				●	●	●			●		●			●

Anclote Key is an undeveloped state recreation area in the Gulf of Mexico 3 miles off the Tarpon Springs mainland. The 3,000-foot beach is not accessible by land.

Howard County Park has 155 developed acres available for offshore recreation. It is connected to the mainland by a mile-long causeway. Access is a half mile north of Keystone Rd. (S.R. 582) on Florida Ave. N. in Tarpon Springs. No alcohol allowed. Group and bicycle facilities.

Sunset Beach is an undeveloped beach with a municipal pier located in Tarpon Springs at the end of Gulf Rd. W. (S.R. 582).

Honeymoon Island State Park has over 10,000 feet of beachfront with dune walkovers and group facilities. Access is via S.R. 586 west over Dunedin Causeway. No alcohol allowed.

Dunedin Beach is an undeveloped beach on Honeymoon Island reached by taking Curlew Rd. to Causeway Blvd.

Caladesi Island State Park may be reached only by boat. There is ferry service from the pier at Honeymoon Island State Park to the north in Dunedin. It has 2 miles of developed beach with group facilities. No alcohol allowed.

Clearwater Beach Island lies between Clearwater Beach and the Gulf of Mexico and is connected to the Clearwater central business district by Memorial Causeway (S.R. 60). It is a developed beach with lifeguards. Additional access to the gulfside off Mandalay Ave. north on Kendall St., Avalon St., Hellwood St., Glendale St., Idlewild St., and Cambria St.; gulfside off Eldorado Ave. on Bohemia Cir. S., Mango St., Bohemia Cir., Gardenia St., Aurel St., and Juniper St.; and channelside on Mandalay Ave., Somerset St., and Aster St.

Mandalay Park has 500 feet of city-owned, developed beachfront. Access is gulfside off Mandalay Ave. a half mile north of S.R. 60 (Memorial Causeway) on Clearwater Beach Island.

Clearwater Beach Park is a developed beach (approximately a half mile long) with lifeguards, located on the south end of Clearwater Beach Island off Gulfview Blvd. No alcohol allowed. Fishing pier and bicycle facilities.

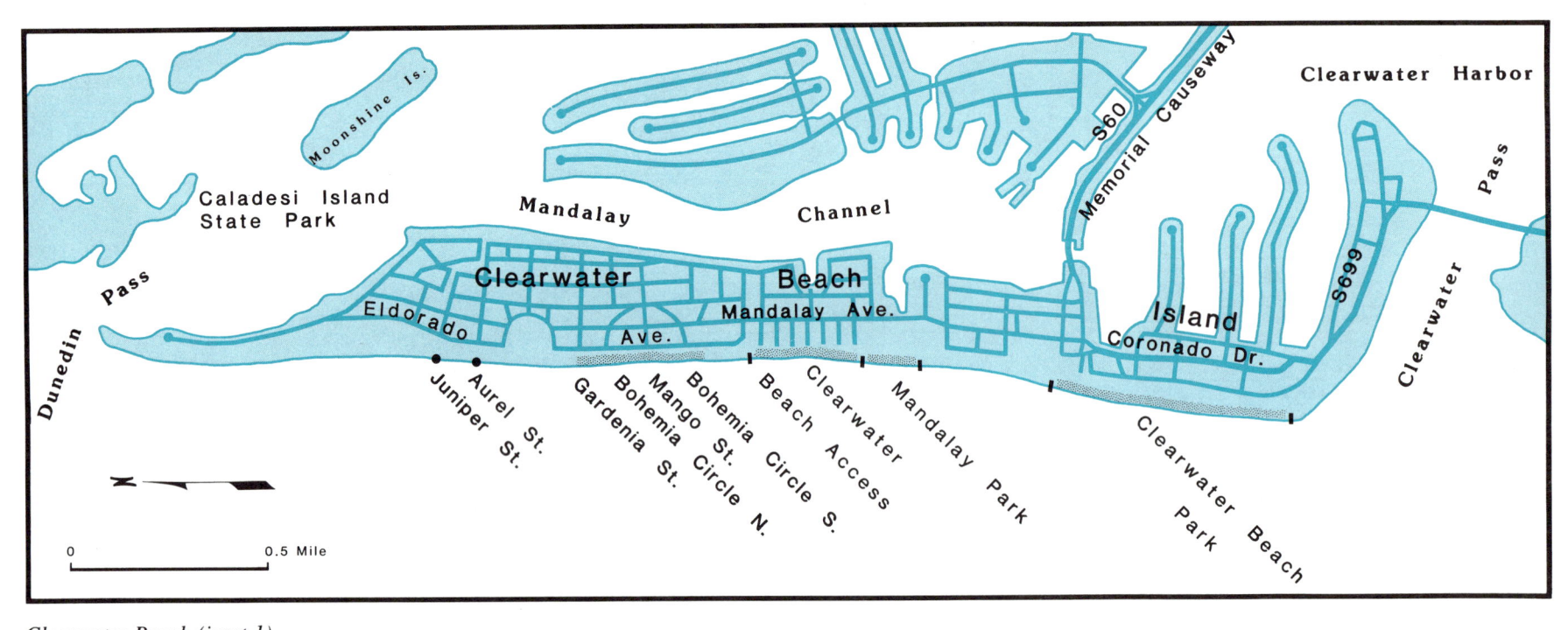

Clearwater Beach (inset 1)

Clearwater Beach

Pinellas County

NAME	Easy Access	Parking or Entrance Fee	Parking	Restrooms	Showers	Picnicking	Swimming	Lifeguards	Fishing	Boating Facilities	Shelters	Concession Stands	Handicapped Facilities	Public Transportation	Group Facilities	Nature Trails / Fitness	Atlantic / Gulf	Bay / Soundfront	Sandy Beach	Rocky Beach	Primitive Beach	Urban
NORTH CENTRAL DISTRICT																						
Sand Key County Park (not open)																						
Belleair Shores Beach Access				●	●												●		●			
Indian Rocks Beach	●	●		●	●				●								●		●			
Indian Shores Beach Access	●	●		●	●												●		●			

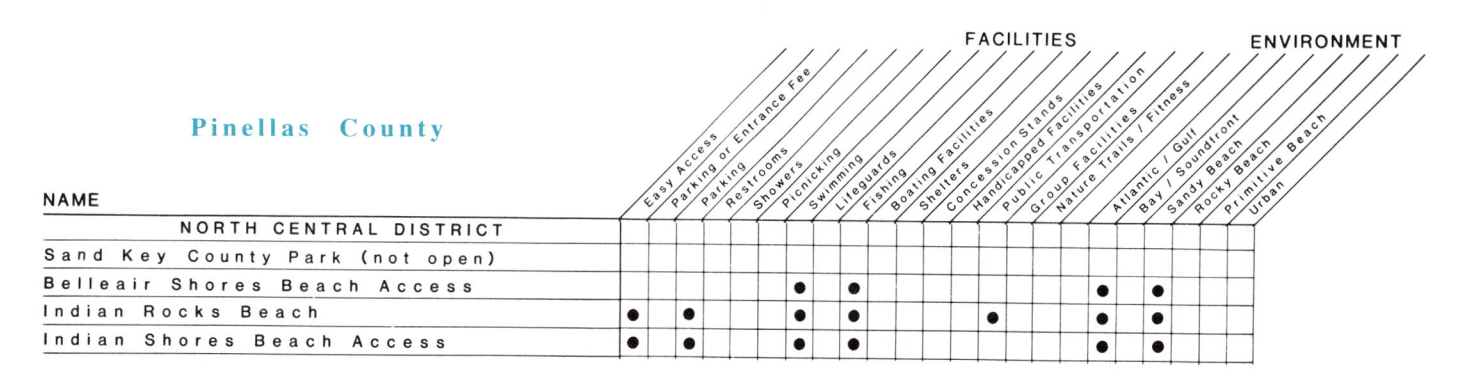

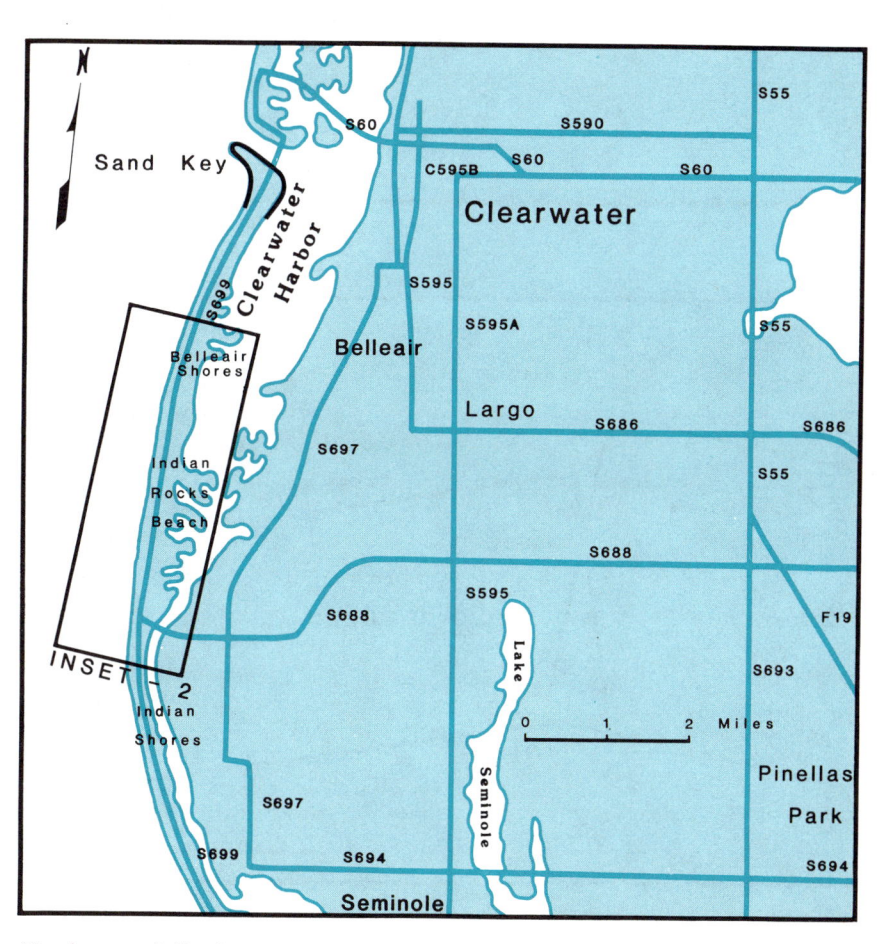

Sand Key County Park has 2,100 feet of beachfront temporarily closed (1984) for development. Access on Clearwater Beach Island is south on S.R. 699 about 1.5 miles from Memorial Causeway (S.R. 60). Local parking is very limited.

Belleair Shores Beach Access off N. Gulf Shore Blvd. (S.R. 699) on 7th St., 13th St., and 19th St. Local parking is unavailable.

Indian Rocks Beach has 3 miles of developed beach and is along N. Gulf Shore Blvd. (S.R. 699) at 27th Ave. through 15th Ave., 12th, 10th, and 8th through 2d avenues.

Indian Shores Beach Access is an undeveloped beach along S.R. 699 (Gulf Blvd.) at 186th, 188th, 189th, 190th, 193d, 195th, 197th, 198th, 199th, and 200th streets.

North-central district

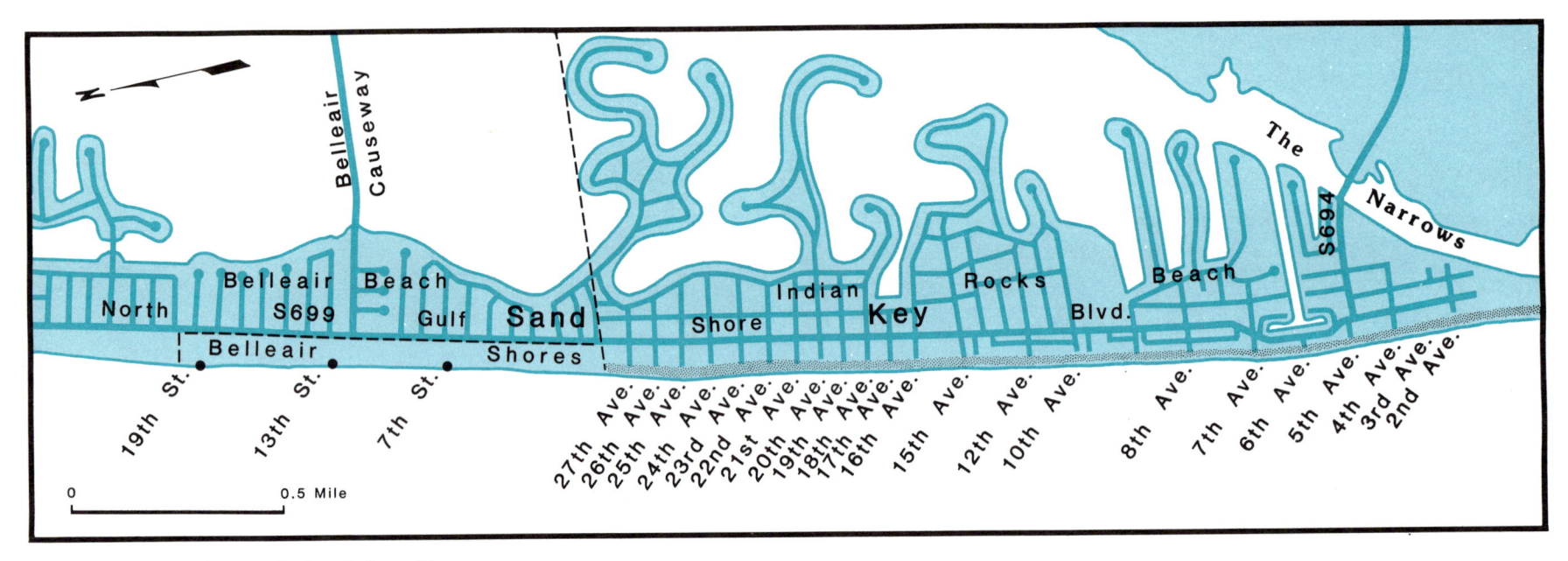

Belleair Beach/Indian Rocks Beach (inset 2)

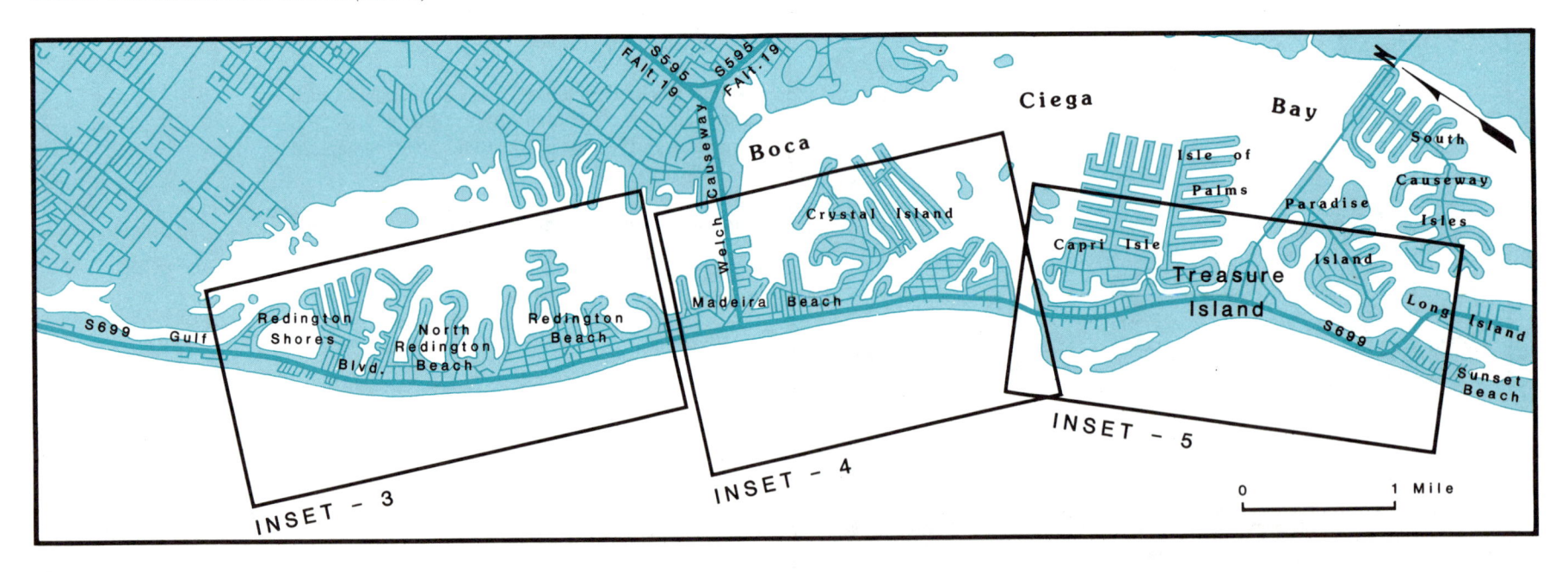

Central district

Pinellas County

FACILITIES / ENVIRONMENT

NAME	Easy Access	Parking or Entrance Fee	Parking	Restrooms	Showers	Picnicking	Swimming	Lifeguards	Fishing	Boating Facilities	Shelters	Concession Stands	Handicapped Facilities	Public Transportation	Group Facilities	Nature Trails / Fitness	Atlantic / Gulf	Bay / Soundfront	Sandy Beach	Rocky Beach	Primitive Beach	Urban
CENTRAL DISTRICT																						
Redington Shores County Park	●		●	●	●		●					●	●				●		●			●
North Redington Beach Access	●		●			●	●										●		●			
Redington Shores Beach						●	●										●		●			
Archibald Memorial Beach	●	●	●	●	●	●	●		●	●	●	●		●	●		●		●			●
Madeira Beach Access						●	●										●		●			
Madeira Beach County Park	●		●			●					●	●					●		●			
Johns Pass Beach and Park	●	●	●	●	●	●	●		●			●	●	●			●		●			●
Treasure Island Beach Access			●			●	●		●				●	●			●	●	●	●		
Treasure Island Beach	●	●		●		●	●	●	●			●	●				●		●			

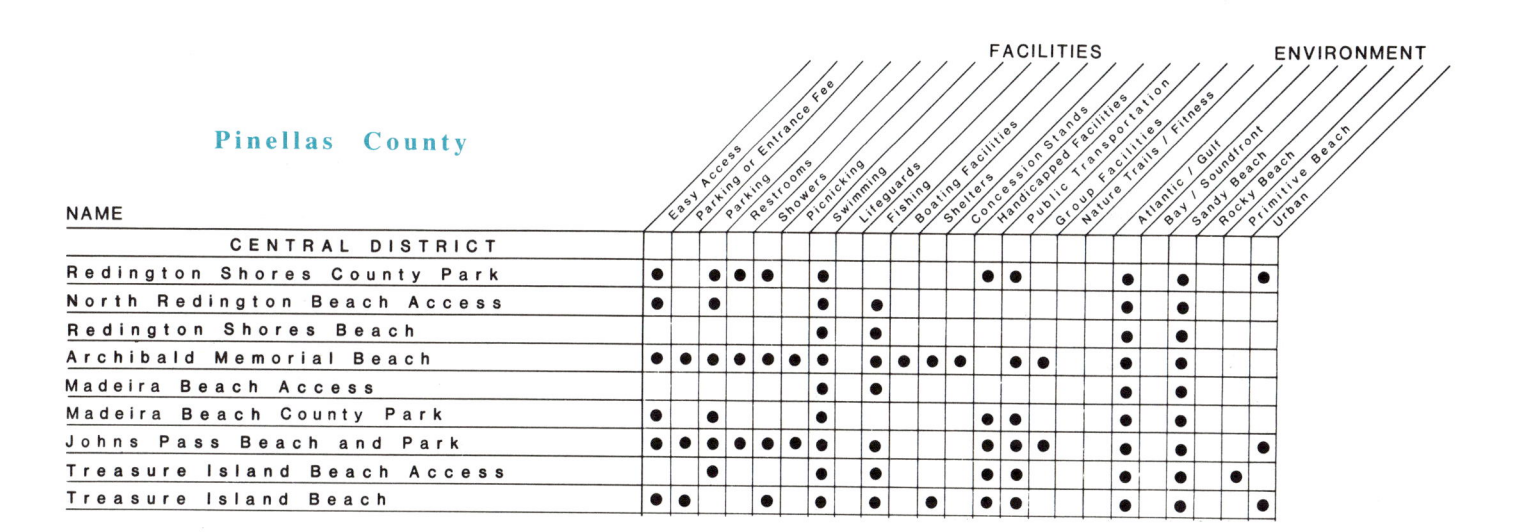

Redington Shores County Park is on Gulf Blvd. (S.R. 699) in Redington Shores at 182d Ave. It has about 50 feet of undeveloped beach. No alcohol allowed.

North Redington Beach Access is south of Redington Shores on S.R. 699. Public walkways to the gulf are at 173d Ave., between lots 3 and 4, at 171st Ave., at 170th Ave., and at Bath Club Cir.

Redington Shores Beach is a 350-foot-long developed beach on the Gulf of Mexico. Access is via Gulf Blvd. (S.R. 699) on 183d Tr. N., 183d St., Coral St., Beach St., Atoll St., and 178th St. through 174th Tr. No alcohol allowed.

Redington Shores Beach Access is off S.R. 699 at 155th Ave. in Redington Beach. No alcohol allowed.

Archibald Memorial Beach has 5.5 acres of developed recreation area on the Gulf. Group facilities. Marked access is off S.R. 699 about 0.2 mile north of Welch Causeway.

Madeira Beach Access is 1.5 miles of undeveloped beachfront on Gulf Blvd. in Madeira Beach (S.R. 699) at 148th St., 142d St., 141st St., and 137th St. through 129th St.

Madeira Beach County Park is a half mile south of Welch Causeway on S.R. 699. This park has 300 feet of developed beach frontage.

Johns Pass Beach and Park lies 1.5 miles south of Madeira Beach on S.R. 699. It is developed with dune walkovers to 500 feet of beachfront and a fishing pier. No alcohol allowed.

Treasure Island Beach Access lies north of St. Petersburg Beach on S.R. 699. There are 45 dune walkovers and street ends giving access to the gulfside beach between 77th Ave. and 127th Ave. and 20 street ends provide bayside beach access. Three (123d Ave., 100th Ave., and 94th Ave.) have boat ramps. Night access prohibited (1–5 A.M.).

Treasure Island Beach has 0.7 mile on the Gulf of Mexico developed for recreation and has lifeguards. There is access along Gulf Blvd. (S.R. 699) in Treasure Island.

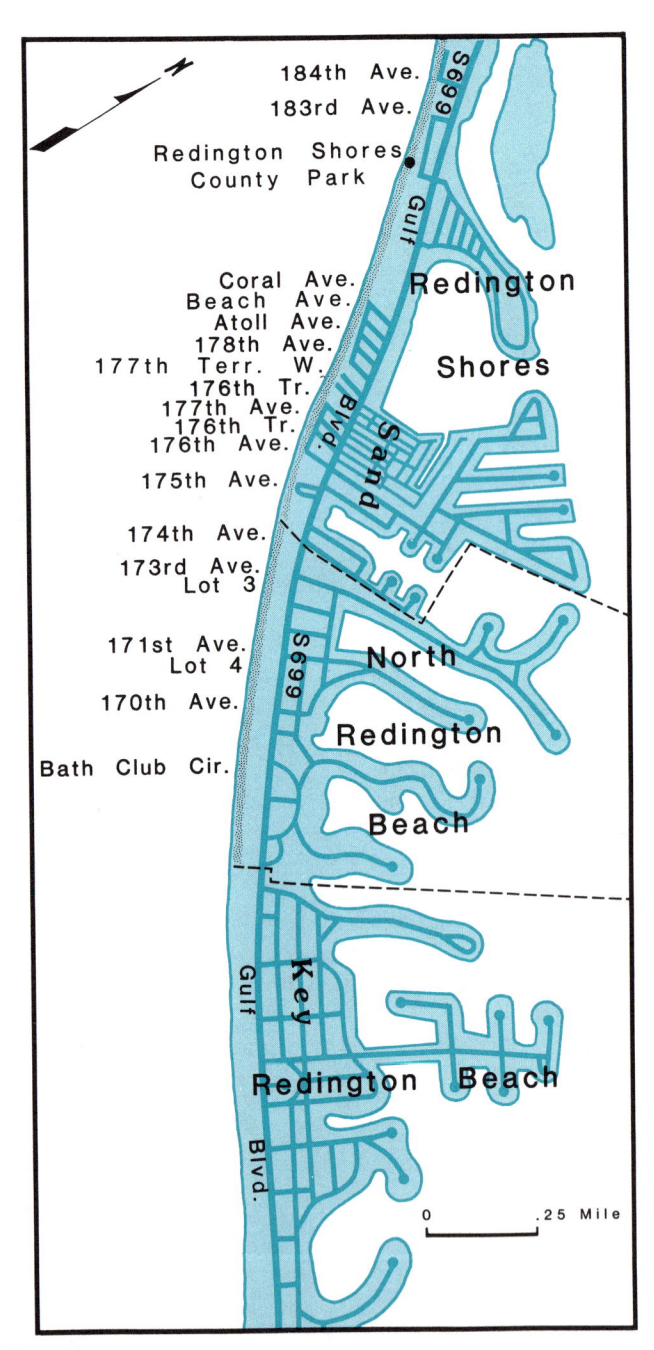

Redington Beach (inset 3)

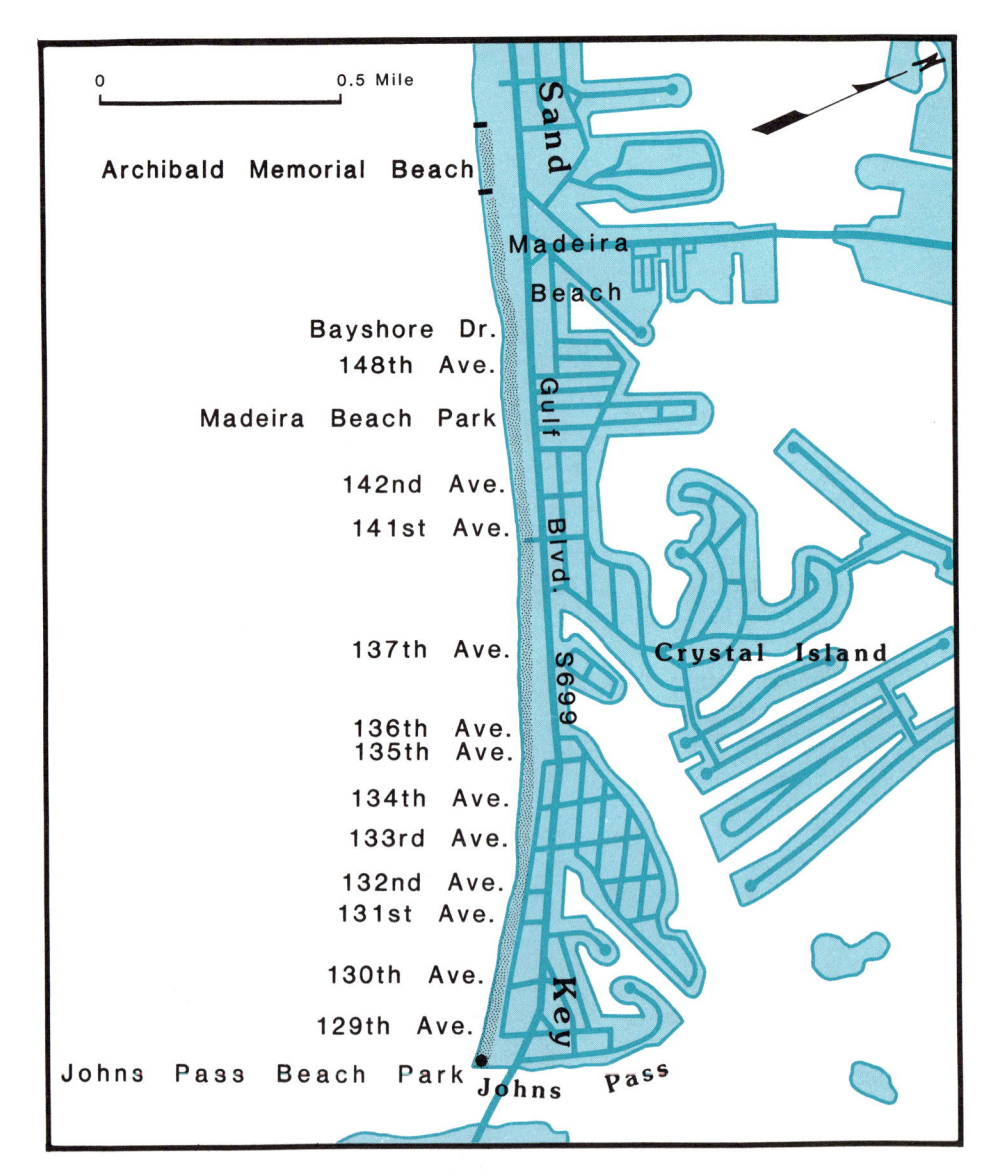

Madeira Beach (inset 4)

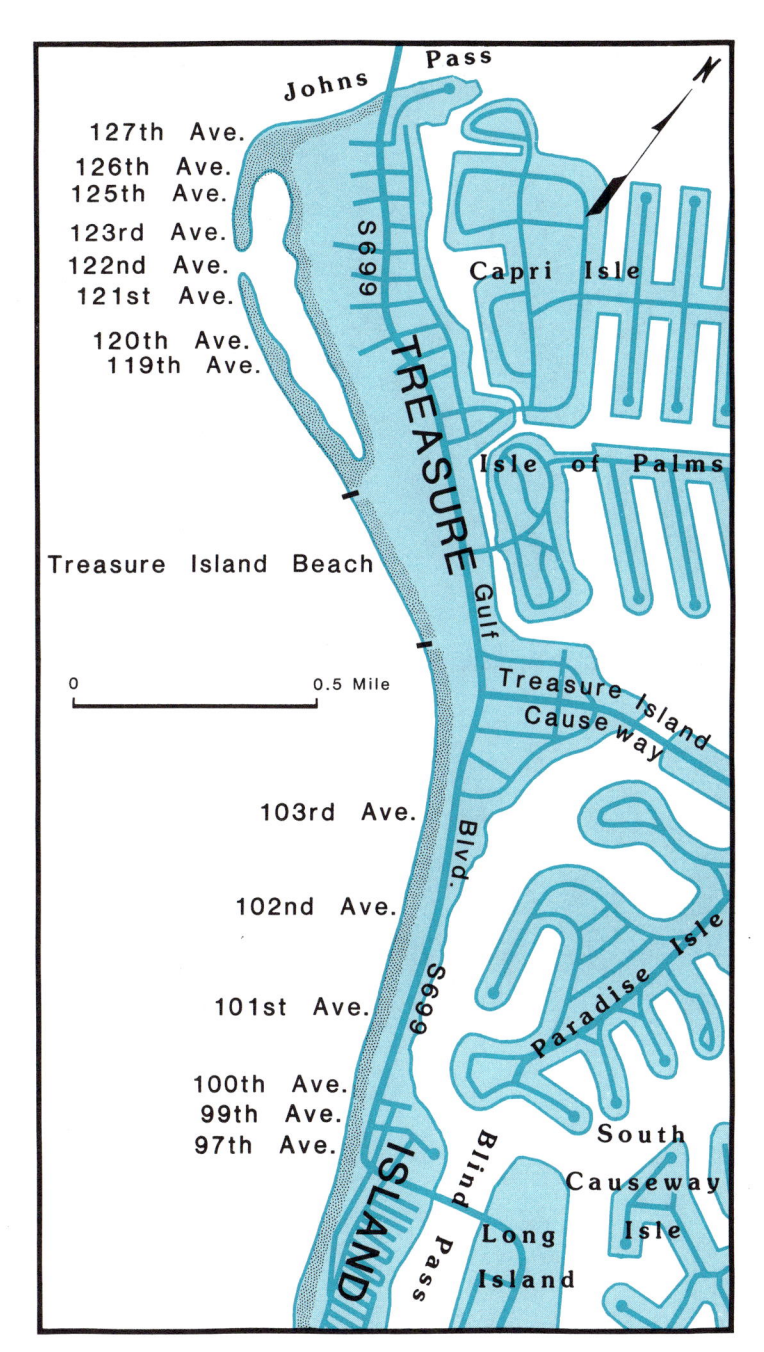

Treasure Island (inset 5)

Wind surfing

123

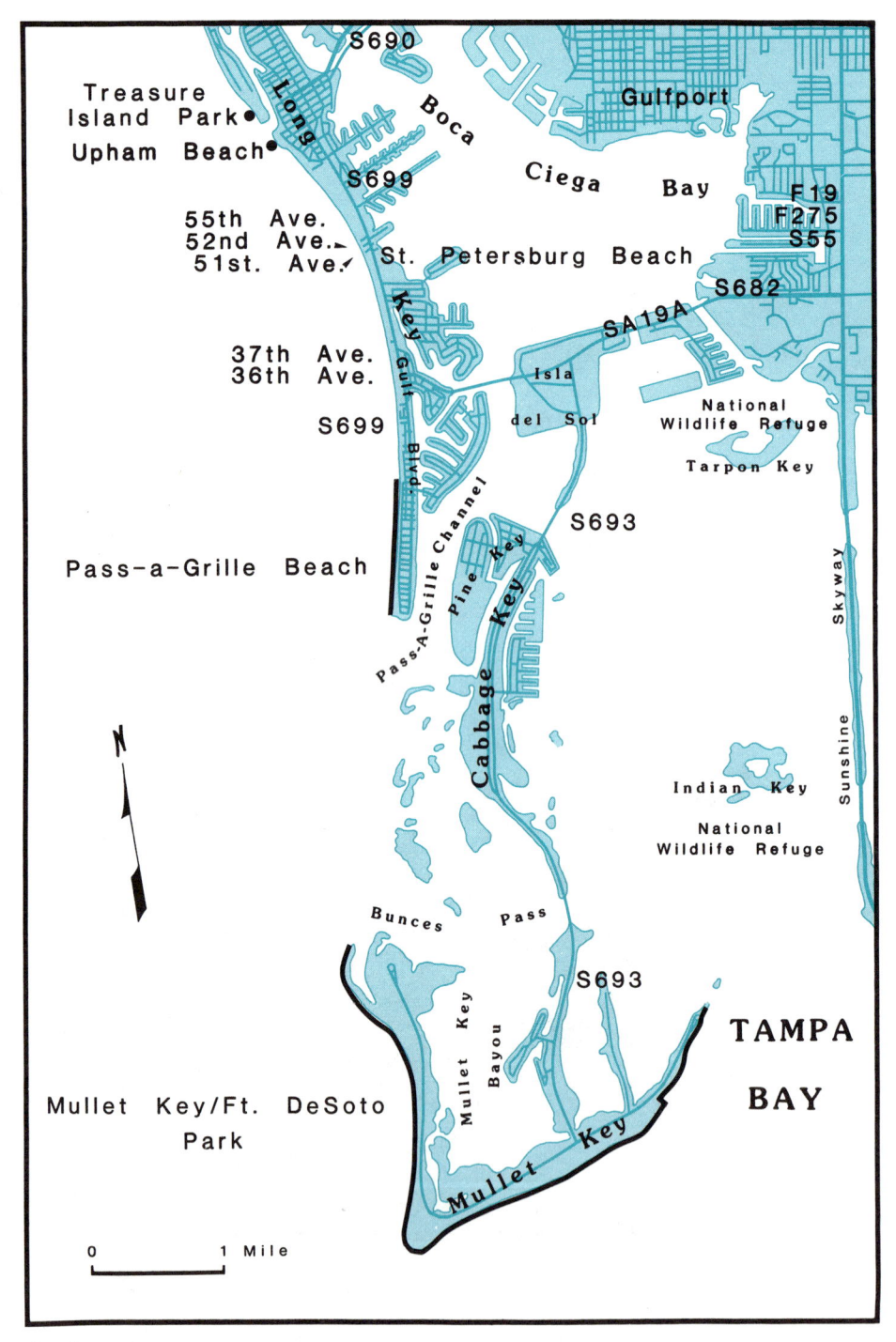

Southern district

Pinellas County

NAME	Easy Access	Parking or Entrance Fee	Parking	Restrooms	Showers	Picnicking	Swimming	Lifeguards	Fishing	Boating Facilities	Shelters	Concession Stands	Handicapped Facilities	Public Transportation	Group Facilities	Nature Trails / Fitness	Atlantic / Gulf	Bay / Soundfront	Sandy Beach	Rocky Beach	Primitive Beach	Urban
SOUTHERN DISTRICT																						
Treasure Island Park		•				•	•								•							
Upham Beach	•	•				•					•					•	•		•			
St. Petersburge Municipal Beach	•		•	•	•	•	•				•	•	•			•	•	•				•
Pass-a-Grille Beach Park		•	•			•					•	•					•					
Mullet Key/Ft. DeSoto County Park	•	•	•	•		•	•	•	•	•	•	•		•	•	•	•	•				

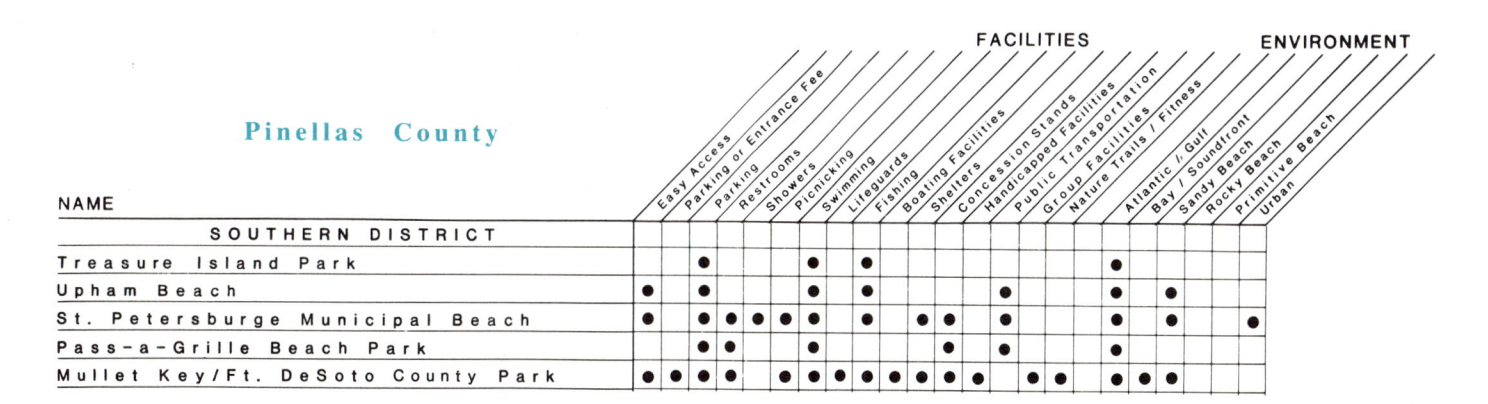

Upham Beach is 9 acres of developed beach on Long Key. Across the channel lies Treasure Island Park. However, access is through St. Petersburg Beach S.R. A19A to 71st Ave. W. No alcohol allowed.

St. Petersburg Municipal Beach is developed with street access from 80th St. to 24th St. off Gulf Blvd. and S.R. A19A.

Pass-a-Grille Beach Park has 5,200 feet of gulf-front with unlimited access.

Mullet Key/Ft. DeSoto Park is comprised of five islands, in all about 900 acres with 7 miles of waterfront. Access is via the Pinellas Bayway (S.R. 679). This is a well-developed recreation area, as well as a bird, plant, and animal sanctuary and a historic area. Overnight camping, group and bicycle facilities, and a fishing pier. No alcohol allowed.

Clearwater Beach

Manatee

The West Indian Manatee, *Trichechus manatus*, is one of the state's most endangered species. A distant cousin to the elephant, these marine mammals can measure 15 feet in length and weigh over a ton. Large and gentle, they have no natural enemies other than man.

The manatee is a mammal: it breathes air, bears its young alive, has hair, and is warm-blooded. It feeds on vegetation in shallow waters near the coast, helping to clear weeds in congested areas. It does not bite but eats by raking plants into its mouth with bristles on its toothless lip pads. Once inside, food is ground up on rear molars and swallowed.

During the winter months when water temperatures drop below the level manatees can tolerate, they congregate around warm-water springs and warm-water discharge of power plants along the coast. Such areas are carefully marked with signs describing them as manatee refuges and urging boats to proceed with caution.

At one time herds of manatees roamed the waters of the southeastern United States. Today their population has dwindled to about a thousand. In 1982 over 120 of the threatened "sea cows" were killed or died of natural causes. A leading cause of death is collision with motor boats. Most live manatees bear scars on their backs from propeller blades. These deaths, coupled with a slow reproductive rate because of a year pregnancy period, have hampered efforts to enlarge the population. They have been protected by the State of Florida since 1893. Today, manatees are protected by the Marine Mammal Protection Act of 1972 and the Endangered Species Act of 1973.

Currently, efforts are under way to learn more about the life cycle and behavior of the manatee. The U.S. Fish and Wildlife Service is establishing a research center for the continued study of the species. Its effort, along with significant private support, may herald a reversal of the manatees' fate.

Manatee County

Named after the large "sea cows" that frequent its waters, Manatee County is located on the western Gulf coast halfway along the Florida peninsula. The majority of its population of 150,000 live in the Bradenton area or on the highly developed barrier islands fronting the Gulf of Mexico. Public access to the islands is somewhat limited: only two roads to the mainland and one road connecting Longboat and Anna Maria keys.

The first visitor to Manatee County was perhaps the Spanish explorer Panfilo de Narvaez in 1528. But the one best remembered is Hernando De Soto, who landed on Terra Cecia Island in 1539. From here his band of soldiers began their four-year trek northward through Georgia, the Carolinas, Tennessee, and Arkansas in an ill-fated search for gold. De Soto's expedition, which included over 700 men, was the first major European penetration into the southern United States.

Although De Soto and most of his men eventually perished on the epic march, the information they supplied about the land and people they encountered stimulated further Spanish and English exploration. To commemorate the landing and expedition of De Soto, a national memorial was established. Located on Tampa Bay off S.R. 64, the park provides insights into sixteenth-century life and the hardships of the conquistadors by a nature trail, live demonstrations, and visual aids.

After the Spanish left the area, hundreds of years passed with little traffic other than pirates and fishing boats. In 1841 the area was opened for homesteading, and settlers soon colonized the coast. Bradenton grew quickly; the county itself was legally formed in 1855. Many early settlers established large plantations for growing and processing sugarcane for export to the north. The John Gamble Mansion, the home of one of the first settlers in Manatee County, is the oldest building on the west coast of Florida and the only Confederate shrine in the state. It is open from 9:00 A.M. to 5:00 P.M. daily.

Today Manatee County's economy is based on a variety of industries. Agriculture and cattle remain important, and commercial fishing also contributes to economic growth. The development of deep-water Port Manatee on the north coast is planned to attract future commerce to this growing area.

All Manatee County beaches may be reached from Interstate 75 or U.S. 41 by taking either Manatee Ave. (S.R. 64) or Cortez Road (S.R. 684). Both terminate on Gulf Drive, which runs north-south along the length of the islands. Beach access is provided by numerous street ends, but parking is limited. More public access can be found at Bayfront Park on the northern tip of Anna Maria Key, Manatee County Beach off Manatee Ave., and Coquina Public Beach at the southern end of the island.

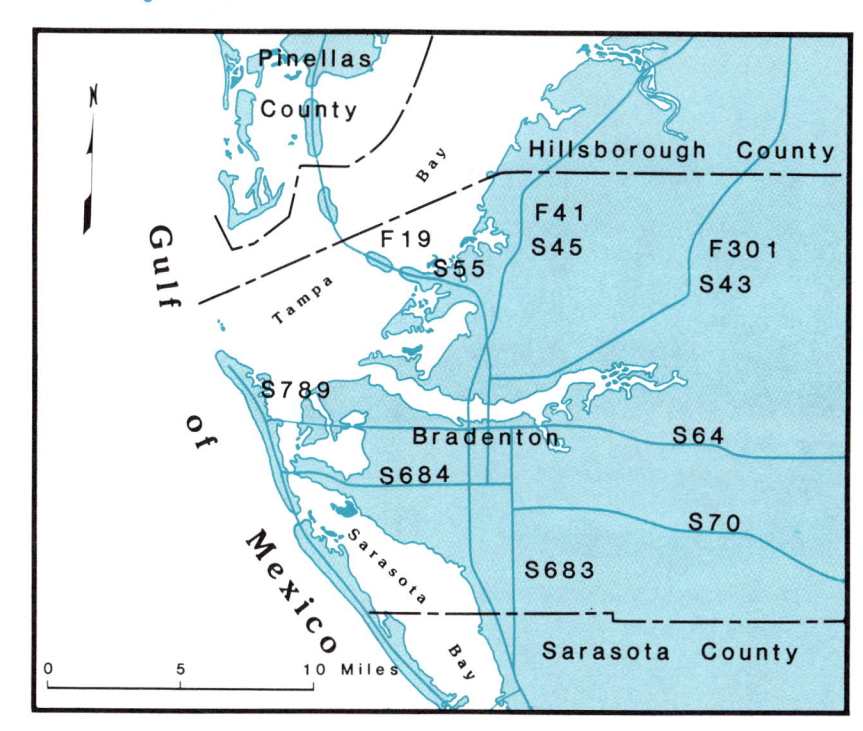

Northern district

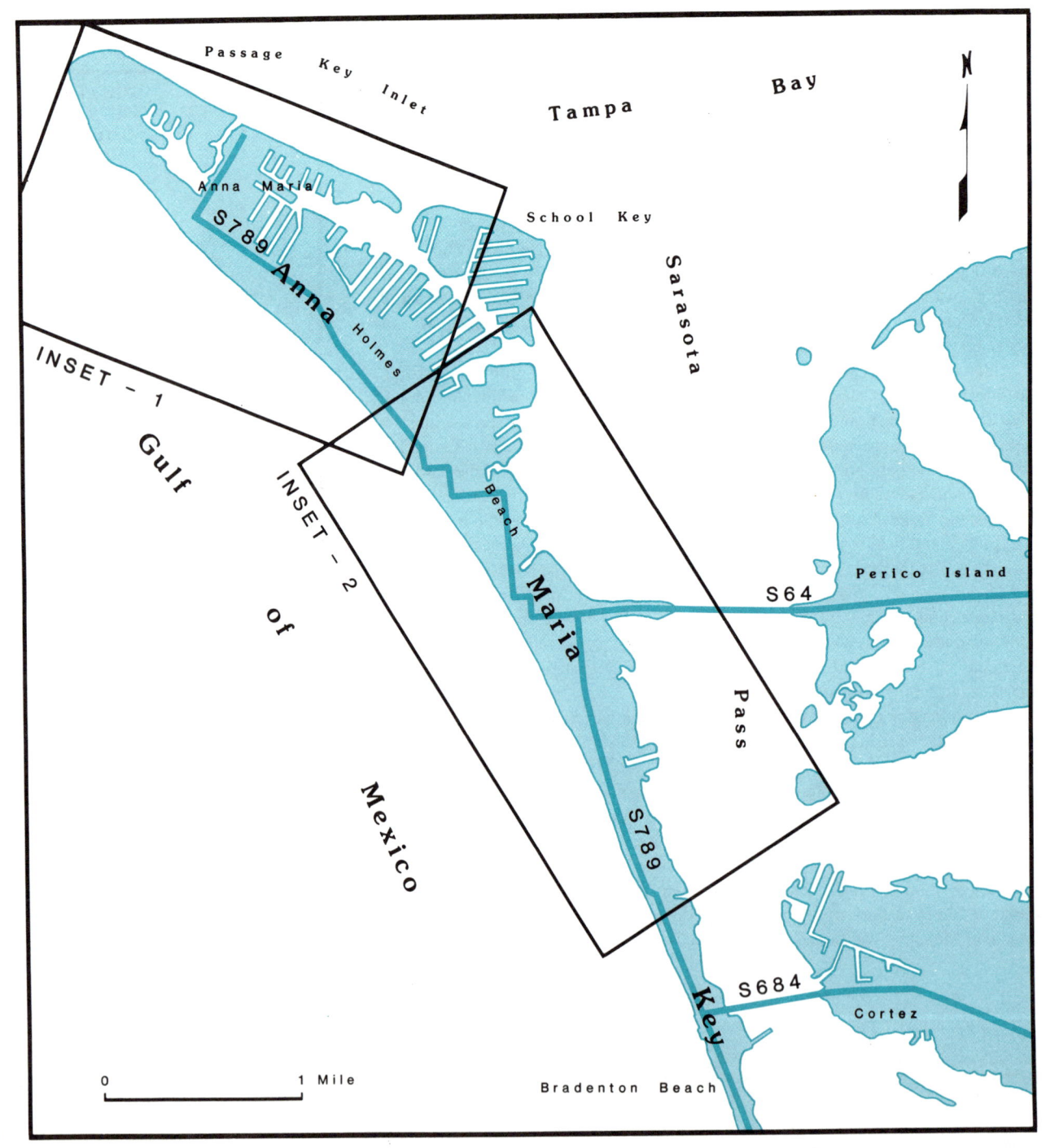

Manatee County

NAME	Easy Access	Parking or Entrance Fee	Parking	Restrooms	Showers	Picnicking	Swimming	Lifeguards	Fishing	Boating Facilities	Shelters	Concession Stands	Handicapped Facilities	Public Transportation	Group Facilities	Nature Trails / Fitness	Atlantic / Gulf	Bay / Soundfront	Sandy Beach	Rocky Beach	Primitive Beach	Urban
NORTHERN DISTRICT																						
Anna Maria Beach		●				●	●			●				●			●		●			
Bayfront Park	●	●	●		●	●		●		●	●	●					●	●	●			
Holmes Beach	●	●				●		●	●			●					●		●			
CENTRAL DISTRICT																						
Manatee County Beach	●	●	●	●	●	●	●	●	●		●	●	●	●	●		●		●			
Palma Sola Causeway	●	●	●		●	●		●	●	●			●	●			●	●	●			
SOUTHERN DISTRICT																						
Cortez Beach	●	●	●	●	●	●	●	●	●		●	●	●	●			●		●			
Coquina Beach	●	●	●	●	●	●	●	●	●	●	●	●	●	●			●		●			
Beer Can Island		●				●	●				●						●	●	●		●	
Longboat Key Beach Access		●				●	●										●	●	●		●	

Anna Maria Beach, on the north point of Anna Maria Island, has 3 acres of undeveloped beach with access from private residential street ends facing the Gulf. Bayside fishing piers are on Alamanda and Pine Ave. No alcohol allowed.

Bayfront Park has 1,000 feet of beach along Tampa Bay. Located in the city of Anna Maria, this developed park with 150 parking places can be reached off Bay Blvd.

Holmes Beach runs 14,400 feet along the Gulf of Mexico. It is reached off Gulf Dr. (S.R. 789) from Beach Ave. to 27th St. Soundside boat ramp is located at Anna Maria Bridge (S.R. 641/Manatee Ave.).

Manatee Co. Beach has 900 feet of developed beach fronting the Gulf of Mexico. Marked access is at the intersection of Gulf Dr. (S.R. 789) and 40th St. (S.R. 64) in the city of Holmes Beach. It has 120 parking places.

Palma Sola Causeway, the Manatee Ave. (S.R. 64) crossing of Palma Sola Bay, has access to 3,000 feet of developed beach on both sides and at each end of the causeway.

Cortez Beach has 140 feet of undeveloped beach reached off Gulf Dr. (S.R. 789) between 5th and 13th streets, fronting the Gulf of Mexico in Bradenton Beach.

Coquina Beach has over 5,000 feet of developed beach on the Gulf of Mexico with a boat ramp to Sarasota Bay. Access is off Gulf Dr. (S.R. 789) at the southern end of Anna Maria Island, about one mile south of S.R. 684 (5th St. N.) in the City of Bradenton Beach. A catamaran launch site and fishing groin are off to the right before the bridge over Longboat Pass is crossed. Parking for 1,350 vehicles.

Beer Can Island has 2,000 feet of undeveloped beach on the northern end of Longboat Key. Access to the hooked-spit beach is by walking north from the street end of N. Shore Rd. off the Gulf of Mexico Dr.

Longboat Key Beach Access extends along the barrier island south from N. Shore Rd. to the county line. Access to the undeveloped beach is from residential street ends: Jay, Coral, Palmetto, Seabreeze, Broadway, and Gulfside Rd.

Anna Maria Beach/Holmes Beach (insets 1 and 2)

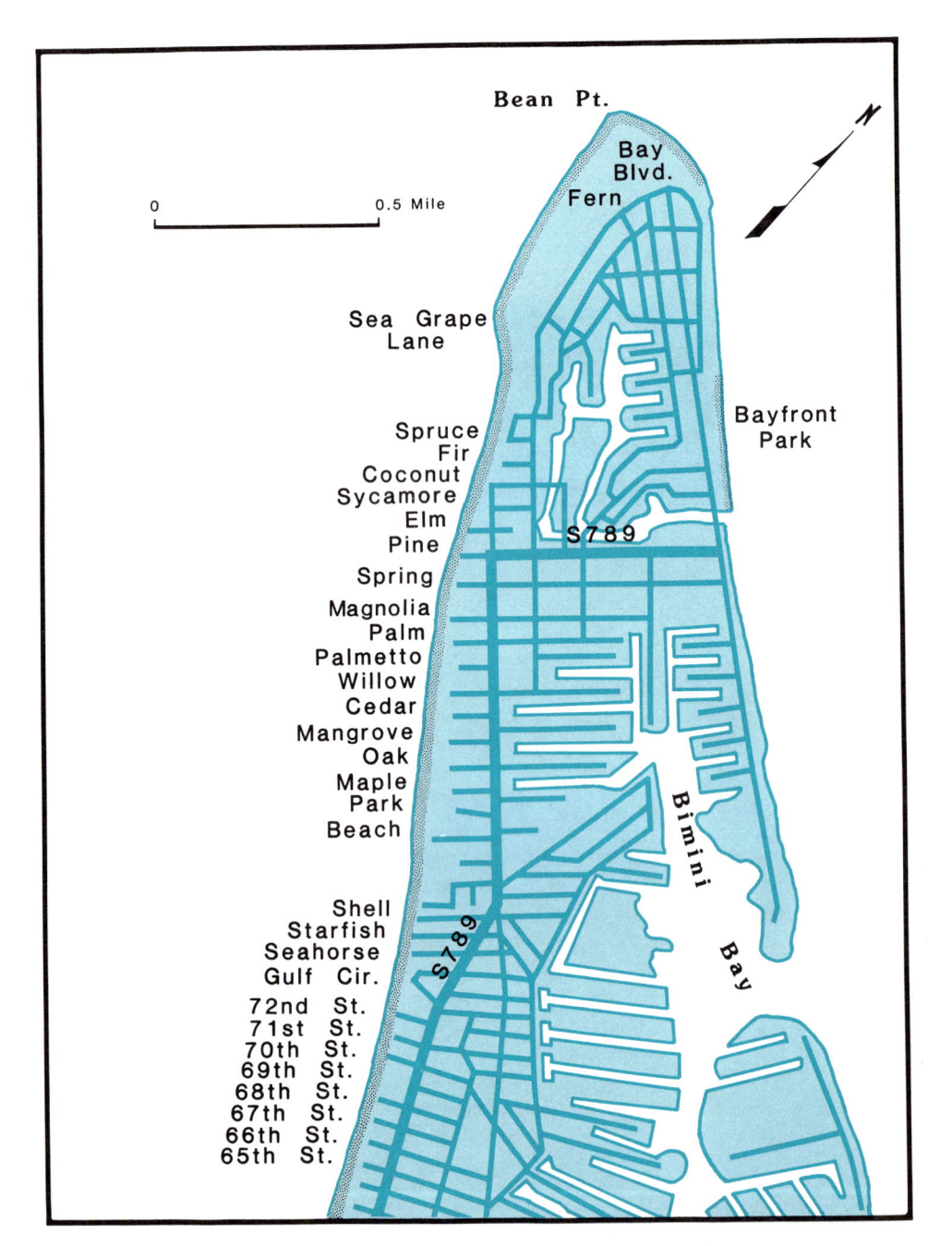

Para-sailing

130

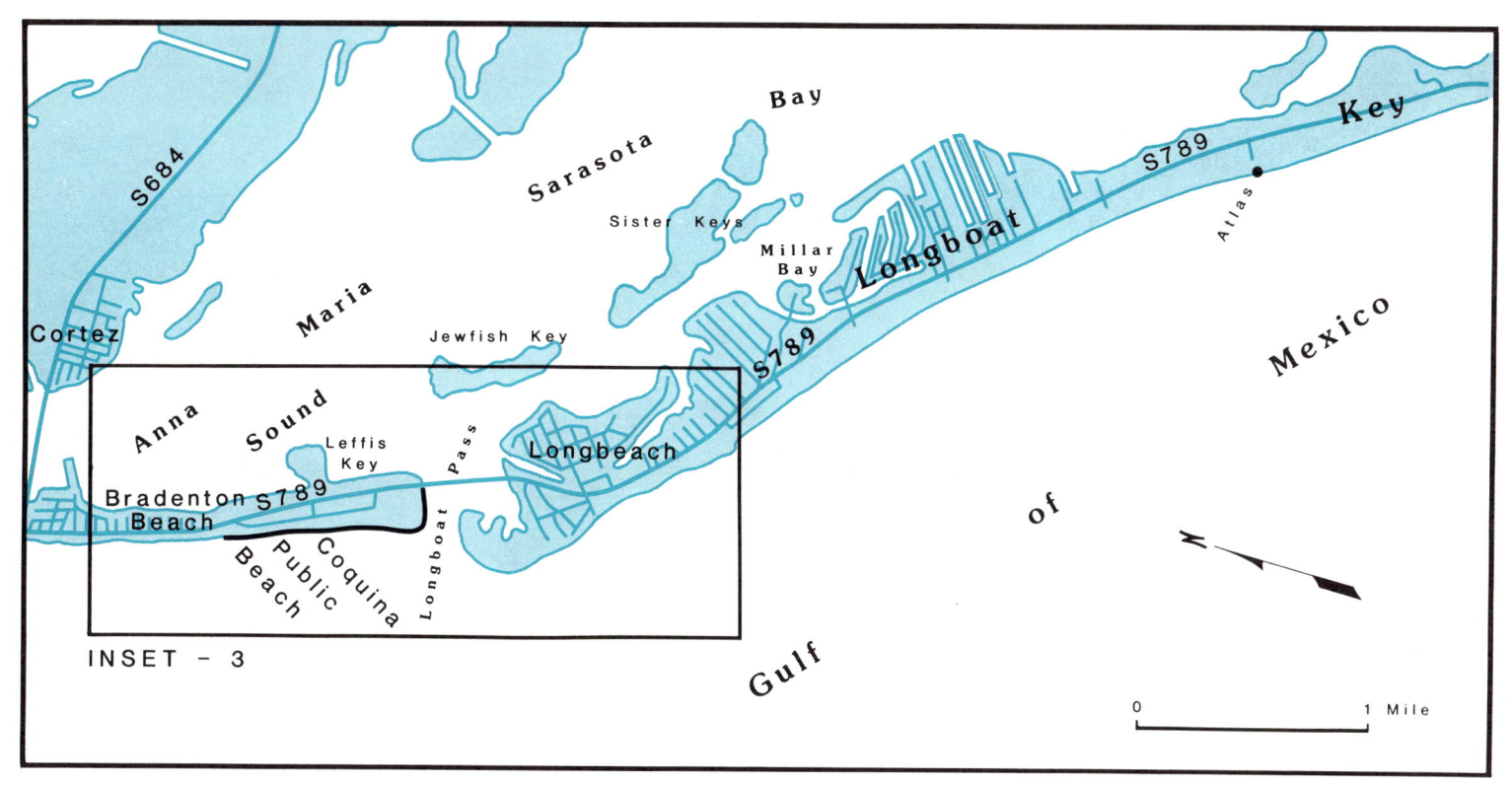

Southern district

Common bay Scallop

131

Bradenton Beach/Long Beach (inset 3)

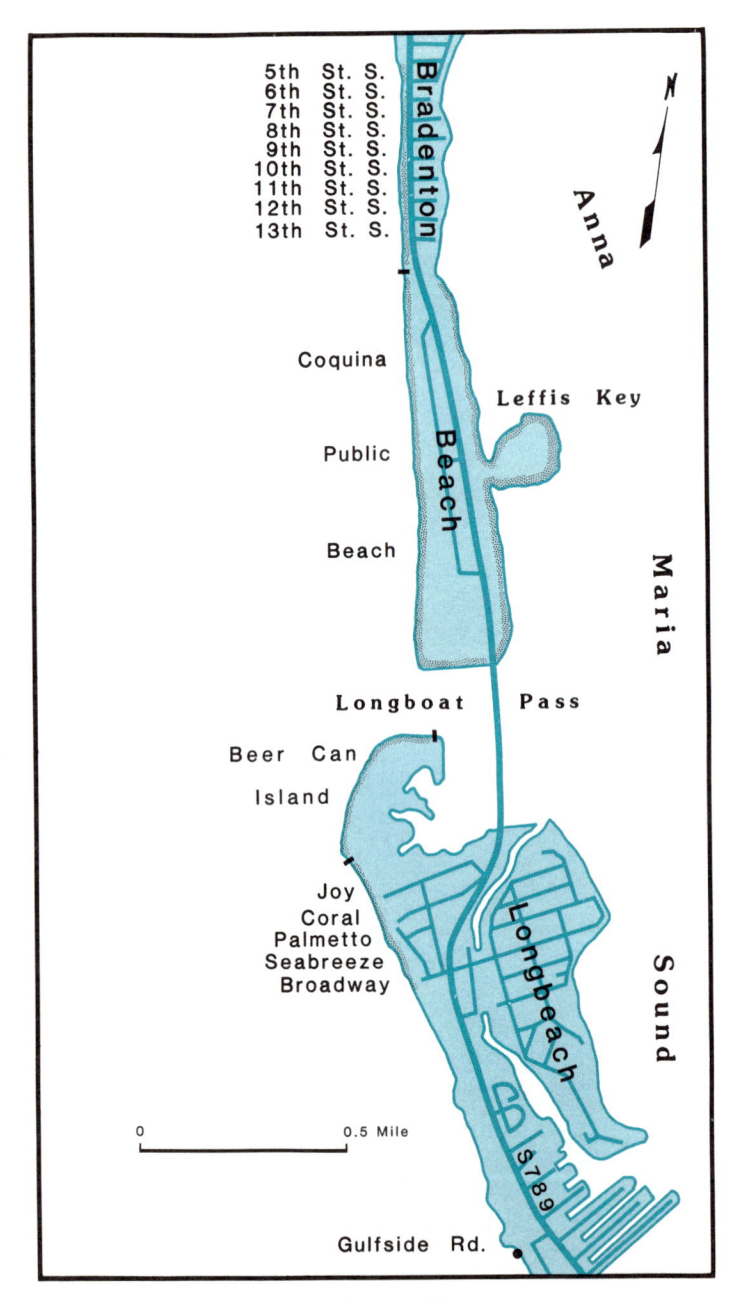

Longboat Key

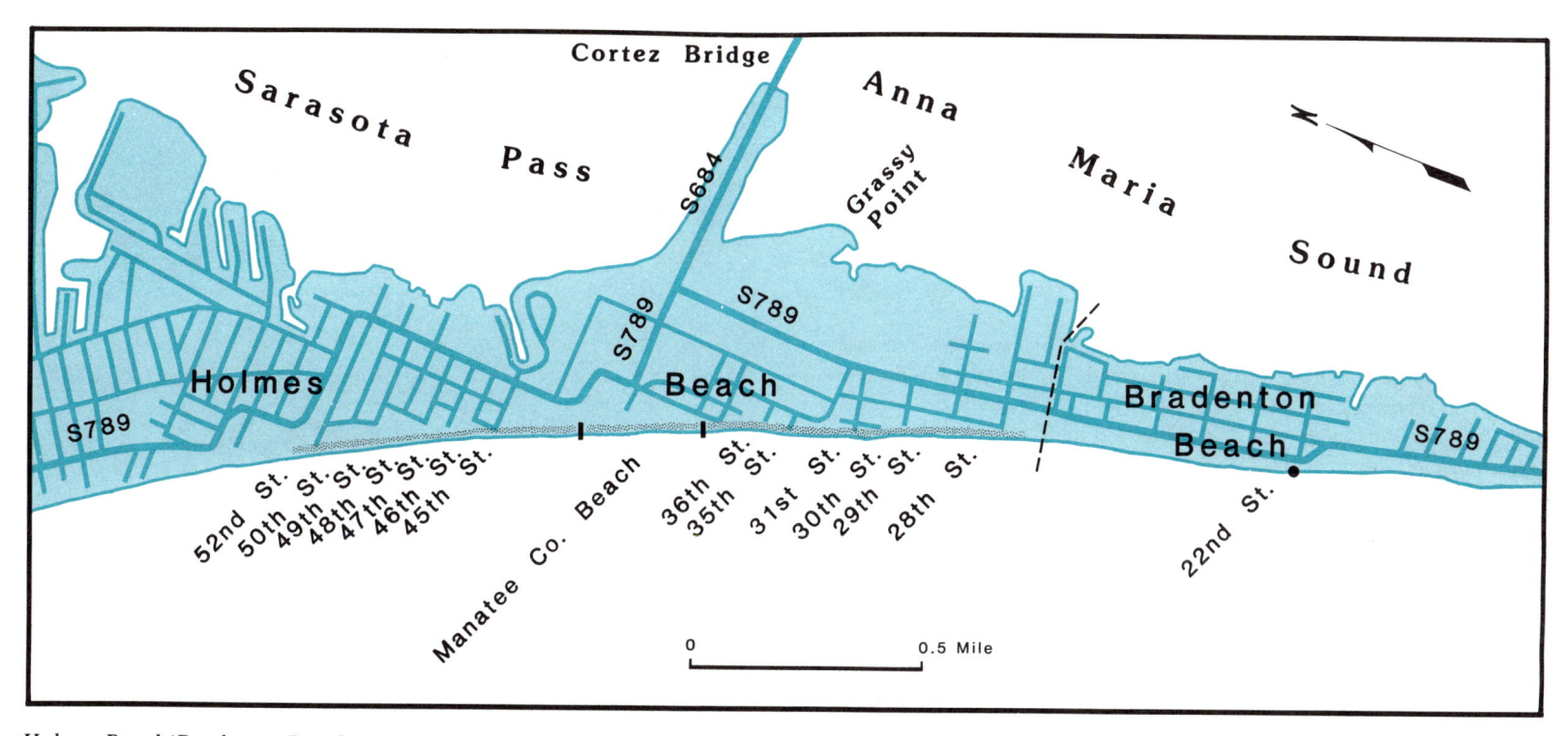

Holmes Beach/Bradenton Beach

Marine Mammal Protection Act of 1972

Under the provisions of the Marine Mammal Protection Act, there is a moratorium, with certain limited exceptions, on the taking and importation of marine mammals and marine mammal parts and products. These animals include whales, dolphins, porpoises, seals, sea lions, walruses, and polar bears and any parts or products made from these animals such as skins, furs, teeth, bones, and oil. Such items will be subject to seizure by agents of the U.S. Government.

Complete information may be obtained from Director, National Marine Fisheries Service, Washington, DC 20235, and from Regional Director, Southeast Region, Duval Building, 9450 Gandy Boulevard, St. Petersburg, FL 33702.

Artificial Reefs

The coastal waters of Florida have long been renowned for the abundance and variety of sport fish. Recognizing this, the state, in cooperation with federal and county authorities and local universities, has developed a comprehensive system of artificial reefs that is unparalleled in the United States. Over 150 artificial reefs have been established along the Atlantic and Gulf coasts to supplement natural reef areas and boost fish populations.

The value of artificial reefs as an attraction for fish has long been appreciated. The first artificial reefs were built in the 1700s by commercial fishermen in Japan. Constructed of bamboo lashed into a frame, they were weighted and sunk in 120 feet of water. The experiment was so successful it has been duplicated repeatedly since that time.

In South Carolina during the 1860s artificial reefs were first introduced in America. Oak and pine logs were used there; concrete modules were employed in New York in 1916 and abandoned vessels sunk off Cape May, New Jersey, in 1935.

As Florida's reef program expanded, research was initiated on reef behavior in order to maximize potential benefits. Much knowledge has been obtained, thus improving site selection and providing a more attractive habitat to many species. This information has led to the proliferation of many effective artificial reefs along the state's coasts.

These reefs are constructed of a variety of inexpensive, discarded materials: abandoned ships, baled tires, building rubble, concrete culverts, clay pipe, quarry rock, and abandoned oil rigs have all been used. These materials, when spread over several acres of normally sandy, flat sea bottom, provide several advantages. First, they supply a base on which barnacles, algae, and many tiny aquatic plants can grow. In turn, the small fish that forage on these specimens are afforded protection by the irregular topography of the reef structure. Finally, the larger sport fish that anglers pursue are attracted by their bountiful smaller counterparts, and an aquatic food chain is established. Studies show these reefs are not only more effective producers than the regular sea bottom but are even more productive than most natural reefs.

Besides having superior fish-attracting abilities, artificial reefs are generally located within a few miles of shore and are thus accessible to small boats and available for wide public use. They are marked by large buoys painted in lateral 6-inch-wide white and international orange stripes; local marinas and bait shops can provide directions to the reefs in the area one wishes to fish. Though the Pinellas County area leads the state with its active reef program, anywhere there is coastline a productive reef will be found close by.

Sarasota County

Sarasota County borders the Gulf of Mexico with a chain of peninsulas and barrier islands. Along these waters the first county inhabitants lived off the Gulf's bounty. Scattered oyster-shell mounds remind us of their early presence.

After the Spaniards had driven the Indians inland in the late 1500s, Sarasota remained uninhabited. During the eighteenth century Cuban and Spanish fish camps dotted the islands. The first permanent American settlers came from Georgia in 1842, and in 1885 they were joined by Scottish colonists who helped found Sarasota.

The area was popularized in the early twentieth century by Mrs. Potter (Bertha) Palmer, a wealthy Chicagoan. Taken with the vista and gently lapping bays, she accumulated thousands of acres between Venice and Sarasota. She also brought many friends to the area who built vacation homes and invested in cultural luxuries. Growth was slow and steady after the depression, with a resulting tranquil atmosphere accompanied by cultural refinements usually found only in larger cities.

The county's coastline begins in the north at Manatee County, with which it shares the City of Longboat Key. Beaches are typified by soft, white sands, and in some areas seawalls and groins have been constructed. Many private homes, clubs, and condominiums on the beach minimize public access. Longboat Key is connected to Lido Key by New Pass Bridge, which lies directly west of Sarasota and has several public beaches. Lido Key may also be reached from the mainland via John Ringling Causeway. To the south, separated by Big Sarasota Pass, lies Siesta Key, a large barrier island. Numerous public beaches, most with supporting facilities, dot its shoreline. Midnight Pass, the only "wild" or unmaintained pass along the coast, separates Siesta Key from its long, thin neighbor to the south, Casey Key. This, the most remote of Sarasota's keys, is reached at Blackburn Point or Nokomis Beach. Like all beaches in the county, it is well known for shell and fossilized-shark-tooth collecting.

Directly south across Venice Inlet is the lovely City of Venice, patterned after its namesake in Italy. It has a large, well-maintained public beach and a long, concrete fishing pier.

Further south, extending to Charlotte County, the Manasota Peninsula boasts three public beaches and may be entered at Manasota Beach. Between the peninsula and mainland is Lemon Bay which, along with Sarasota and Little Sarasota bays to the north, provides a calm, estuarine habitat. The numerous grassbeds are home to many aquatic species integral to the Gulf's food chain. Throughout these bays are many tiny mangrove islands and shallows.

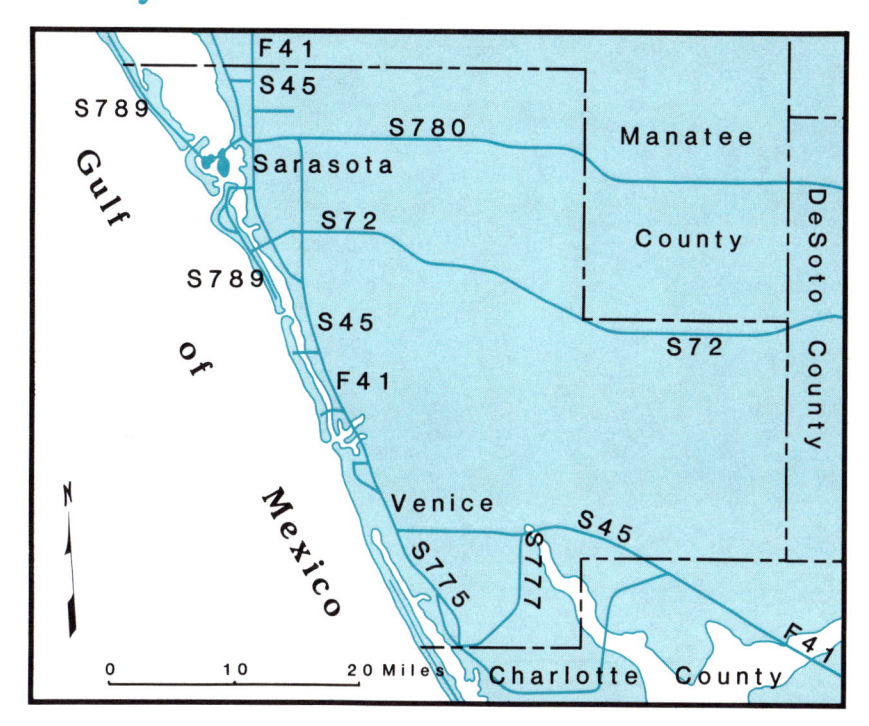

Besides the year-round semitropical climate and miles of beautiful beaches, Sarasota has much to offer the vacationing visitor, including the world-famous Ringling Museum and the Circus Hall of Fame. Bellm Cars and Music of Yesterday feature more than 70 classic old cars and the world's greatest collection of music boxes. The Performing Arts Center, designed by Taliesin architects of the Frank Lloyd Wright School, is noted for its beauty.

Northern district

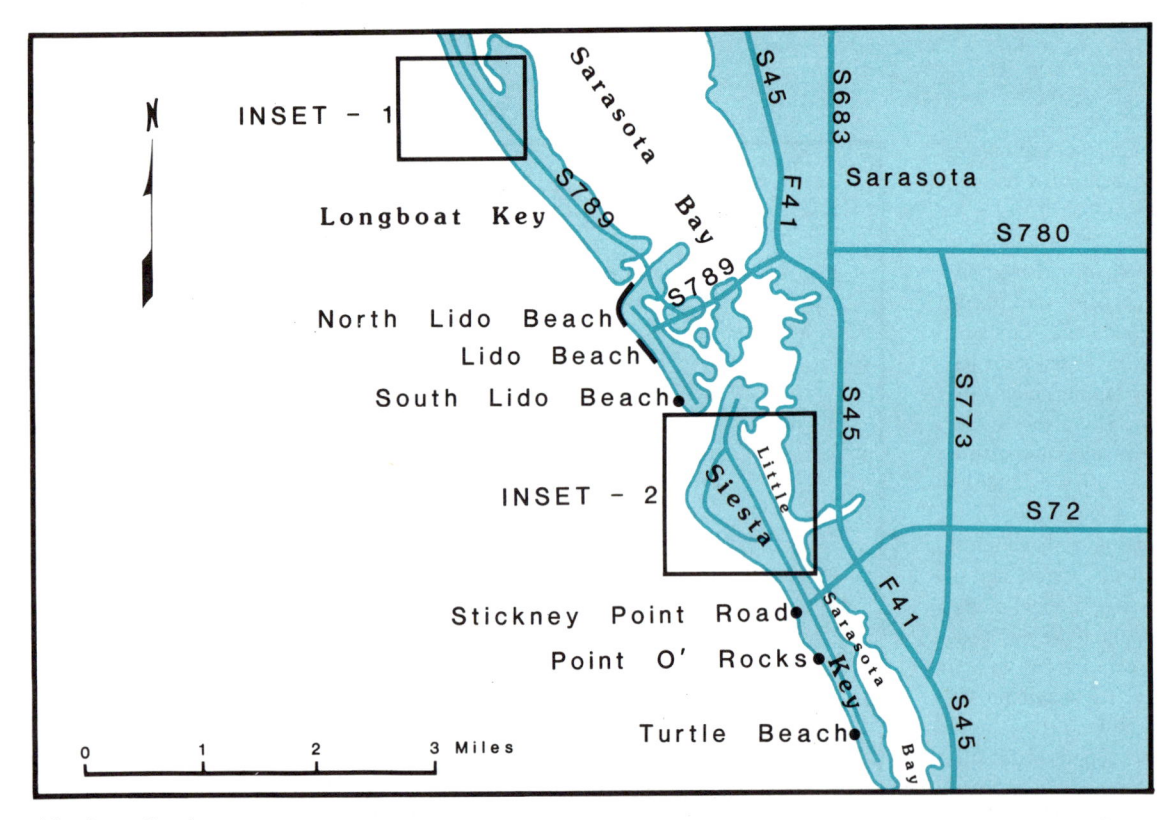

Longboat Key (inset 1)

Dolphin

Sarasota County

NAME	Easy Access	Parking or Entrance Fee	Parking	Restrooms	Showers	Picnicking	Swimming	Lifeguards	Fishing	Boating Facilities	Shelters	Concession Stands	Handicapped Facilities	Public Transportation	Group Facilities	Nature Trails / Fitness	Atlantic / Gulf	Bay / Soundfront	Sandy Beach	Rocky Beach	Primitive Beach	Urban
Longboat Key Beaches						●											●		●		●	
North Lido Beach			●			●					●						●		●		●	
Lido Beach			●	●	●		●	●		●		●					●		●			●
South Lido Beach	●		●	●	●	●	●	●	●			●		●			●	●	●			●
Siesta Key Beach			●			●											●		●	●		
Siesta Key Beach Park	●		●	●	●	●	●	●	●		●	●	●	●			●		●			●
Turtle Beach	●		●	●	●	●	●	●	●	●		●	●	●			●	●	●			●

Longboat Key Beaches are 3.75 miles northwest of Lido Key along S.R. 789 (Gulf of Mexico Dr.). Access is via Triton and Mayfield and the south ends of Bay Isles Rd., Neptune Ave., and Buttonwood Dr. Alcohol is not permitted at these accesses. Parking is very limited.

North Lido Beach is not highly developed. It is more than 3,000 feet long on the west side of West Way Dr., a quarter mile west of John Ringling at the north end of Lido Key. Other access available from southwest side of Harding Cir.: go southwest 2 blocks to N. Polk Dr., turn right for 3 blocks, then left on Emerson Dr. North Lido Beach is connected to Lido Beach. Parking is very limited.

Lido Beach is 3,000 feet long; it has lifeguards 9:00 A.M. to 5:00 P.M. and a 25-meter pool and is developed. The beach is 2.5 blocks southwest of Harding Cir.: turn left along Benjamin Franklin.

South Lido Beach is 1,000 feet long and is developed. It has lifeguards Memorial Day to Labor Day. It is 1.5 miles southeast of Harding Cir.: go southeast on South Blvd. to Taft Dr., right one block on Taft Dr., left on Benjamin Franklin to end of road. Night access prohibited.

Siesta Key Beach accesses, with one exception, are street endings, and most are between Shell Rd. on the north and Calle de Invierno on the south. Shell Rd. is 1 block north of intersection of Midnight Pass Rd. (S.R. 789) and Higel Ave. Go north on Higel one block to Shell Rd. and turn left. Night access prohibited. Stickney Point Rd. access is at Gulf end of Old Stickney Point Rd., half a block south of new road end. Point of Rocks Rd. access is along Gulf side of both that road and Point of Rocks Cir.: go south 6 blocks from Stickney Point Rd. on Midnight Pass Rd. (S.R. 789) and turn right on Point of Rocks Rd.

Siesta Key Beach Park, almost ½ mile long, is developed and has lifeguards. It is along south side of Beach Rd. west of Midnight Pass Rd. intersection.

Turtle Beach, slightly less than a quarter mile long, is developed and has seasonal lifeguards (Memorial Day to Labor Day). It is at the south end of Siesta Key, 2.5 miles south of Stickney Point Rd. on Midnight Pass Rd. (S.R. 789). Turn right on Turtle Beach Rd. Public transportation available.

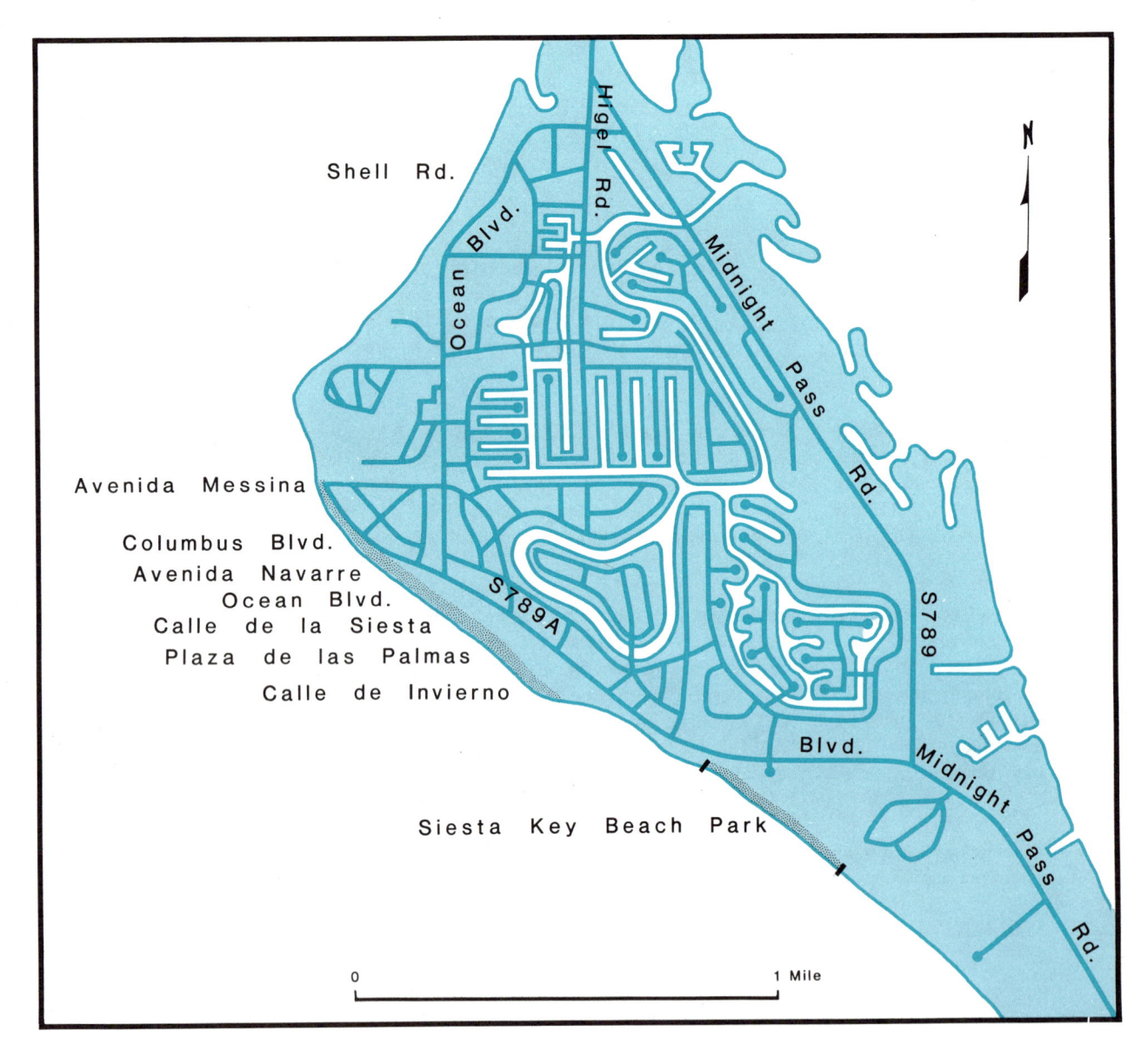

Siesta Key (inset 2)

Sarasota and Lido Key

Southern district

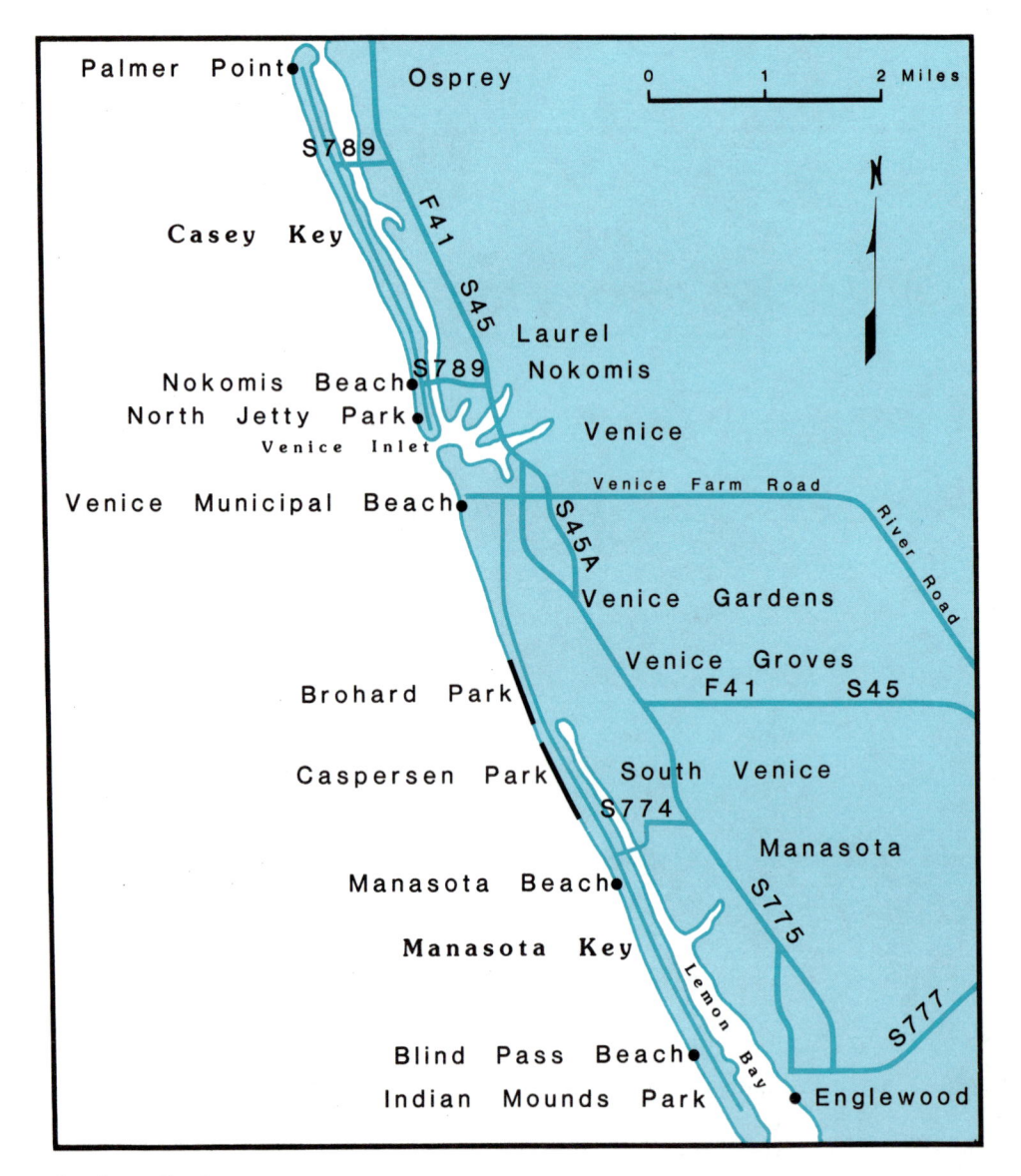

Treasure hunting

140

Sarasota County

NAME	Easy Access	Parking or Entrance Fee	Parking	Restrooms	Showers	Picnicking	Swimming	Lifeguards	Fishing	Boating Facilities	Shelters	Concession Stands	Handicapped Facilities	Public Transportation	Group Facilities	Nature Trails / Fitness	Atlantic / Gulf	Bay / Soundfront	Sandy Beach	Rocky Beach	Primitive Beach	Urban
Palmer Point						•	•										•	•	•		•	
Nokomis Beach	•		•	•	•	•	•	•	•		•	•					•	•	•			•
North Jetty Park	•		•	•	•	•	•	•	•		•	•	•				•	•	•			•
Venice Municipal Beach	•		•	•	•	•	•	•	•		•	•	•	•			•	•	•			•
Brohard Park Beach	•		•	•	•	•	•					•					•	•	•			•
Caspersen Park Beach	•		•	•	•	•	•					•					•	•	•			•
Manasota Beach	•		•	•	•	•	•	•				•					•	•	•			
Blind Pass Beach	•		•	•	•	•	•	•	•								•	•	•			
Indian Mounds Park			•	•		•	•		•	•				•			•	•	•		•	

Palmer Point is located at the north end of Casey Key; take S.R. 789 west from S.R. 45, north on beach road.

Nokomis Beach is a developed, lifeguarded beach 0.2 mile long. It includes beaches on the Gulf and is one mile west of the Tamiami Trail (U.S. 41) on S.R. 789 at intersection of Casey Key Rd.

North Jetty Park, almost 0.2 miles long, is developed and has lifeguards. It also fronts both Gulf and bay and is 0.6 mile south of S.R. 789 on Casey Key Rd. at sound end of key. One of the Gulf's best surfing beaches.

Venice Municipal Beach, 600 feet long, is developed and has lifeguards. It is 0.9 mile west of U.S. 41 on Venice Ave. W. at Gulf.

Brohard and Caspersen Park Beaches are developed and occupy 3 miles between Venice Municipal Airport and the Intracoastal Waterway and Gulf. The north end of these connected beaches is 1.5 miles south of Venice Ave. W. on Harbor Dr. S. Several unpaved roads and Harbor Dr. S. provide access to both. Approximately 225 parking spaces.

Manasota Beach, a quarter mile long, is developed and has lifeguards year-round. It is on Manasota Key at west end of Manasota Bridge.

Blind Pass Beach, 2,000 feet long, is undeveloped. It is 3.6 miles south of Manasota Bridge and fronts both Lemon Bay and the Gulf.

Indian Mounds Park, named for a historic Indian mound, has 1,100 feet on Lemon Bay and is developed. It is east of Winson St. and a third of a mile south of S.R. 775A. Follow park signs.

Brown Pelican (*Pelecanus occidentalis*)

The brown pelican is a well-known coastal bird in Florida. It is found in saltwater areas and, on very rare occasions, in freshwater regions. Large groups of the birds can be found nesting on small coastal islands and sitting on top of docks and pilings and near fishing boats.

Easily identified by the pouch under the bill, which gave rise to the saying "its beak can hold more than its belly," the bird's brown color, large size, and webbed feet are also distinguishing features. A close relative, the white pelican, is larger in size. It resides in Florida during the fall, winter, and spring and travels elsewhere to nest.

The brown pelican's primary nesting season is spring and early summer in most of Florida. In the Florida Keys it will nest anytime. Pelicans select one mate and remain together during their lives. Usually three eggs are laid in one season.

Saltwater fish found at the surface or in shallow water provide its primary food source. Fish remains thrown overboard by commercial fishing boats and handouts by tourists and sport fishermen are quickly scooped up in the birds' pouches. Their eagerness often entangles pelicans in fishing lines and hooks. Fishermen should take precautions to prevent the entanglement of this large bird.

Brown pelican populations in North America have been reduced significantly in the past ten years because of DDT contamination and other factors. When the birds have high levels of DDT compounds the eggs have shells too thin to hatch. This reproductive rate reduction could eventually lead to the extinction of the species. Fortunately in Florida, pelican eggs have not become too thin, and reproduction rates appear to be normal at this time. Florida's pelican population is estimated to be 20,000. The Florida Game Management Division has been studying the species for five years in the hope that this distinctive coastal bird will continue to share the shoreline.

Charlotte County

Charlotte County is named for Charlotte Harbor, which intrudes into the county mainland and is the largest indentation in the southwest Florida coast. The name is said to honor the wife of King George III, Charlotte Sophia. Ponce de Leon returned to the New World (La Florida) in 1521 as a governor appointed by the King of Spain. Accompanied by two ships of soldiers, de Leon selected Charlotte Harbor as his seat of government to provide a base from which he could continue his search for the elusive Fountain of Youth. Though this outpost lasted only five months it opened the door for future settlements in the New World.

In 1539, Hernando De Soto came to La Florida as Spain's new governor. He set forth from Seville with 10 ships and 700 men and the expectation of conquering the hidden wealth of the new land, as he had in his conquest of Peru. Early maps of La Florida drawn by de Lisle (ca. 1781) show that De Soto landed at Charlotte Harbor, although some historians believe the site was Fort Myers. A federal committee in the 1930s traced the expedition to a point west of Bradenton. On this site, at the mouth of the Manatee River on Tampa Bay, the De Soto National Memorial has been designated with a monument bearing the inscription that "Near here, Hernando De Soto and His Men landed May 30, 1539, and Began the March West to the Mississippi River"—a trip that took four years.

Charlotte County was established in 1921 as Florida's fifty-third county. Although only 17 miles north to south, it has over 120 miles of coastline. The county seat, Punta Gorda, lies at the mouth of the Peace River which empties into Charlotte Harbor.

Punta Gorda, meaning "Broad Point" in Spanish, was the early terminus of the railroad serving the southwest Gulf coast. Directly on the sparkling blue waters of Charlotte Harbor, this subtropical city with gently waving palms and exotic foliage has an enchanting feel of peace and security, but it is also a growing and progressive city. Towering palms line its streets, and the waterfront district combines shopping, boating, and park complexes to serve many needs. Due to the prime location on the harbor, surrounded by miles of rivers, creeks, and canals, fishing is one of the most popular recreational activities in the area. The best known sport fish is the Silver King Tarpon (May to September). Other game fish are snook and tarpon.

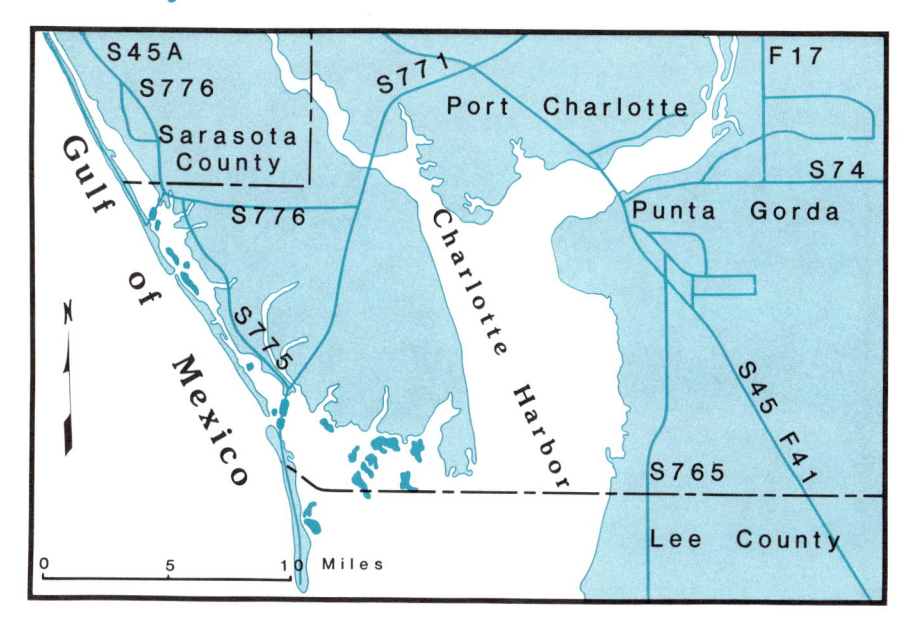

Port Charlotte, two miles north of Punta Gorda, is one of the largest retirement city complexes in Florida. Even though the county has five miles of barrier islands situated between the Gulf of Mexico and Lemon Bay, beach access is very limited. Englewood Beach and Port Charlotte Beach State Park at the south end of Manasota Key are the only county public coastal beaches. The remaining barrier islands are undeveloped and inaccessible by land.

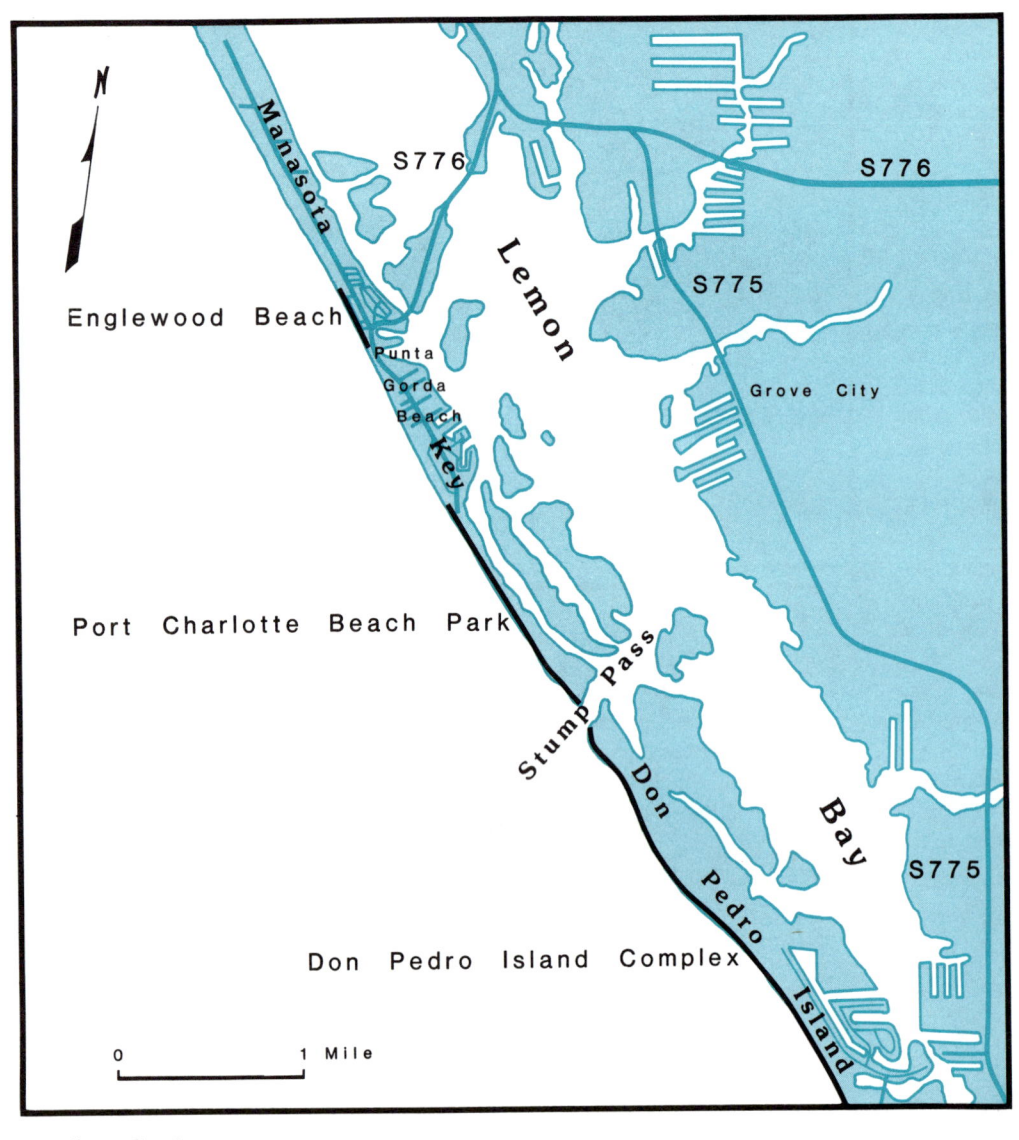

Englewood Beach is developed with 600 feet of coastline, situated between Lemon Bay and the Gulf of Mexico on Manasota Key. Access is via S.R. 776 from the mainland. No alcohol allowed.

Port Charlotte Beach Park has 25 acres south of Englewood Beach at the southern end of Manasota Key. The only access to this undeveloped beach is S.R. 776.

Don Pedro Island complex is an undeveloped beach on the barrier island that can be reached only by boat.

Northern district

Southern district

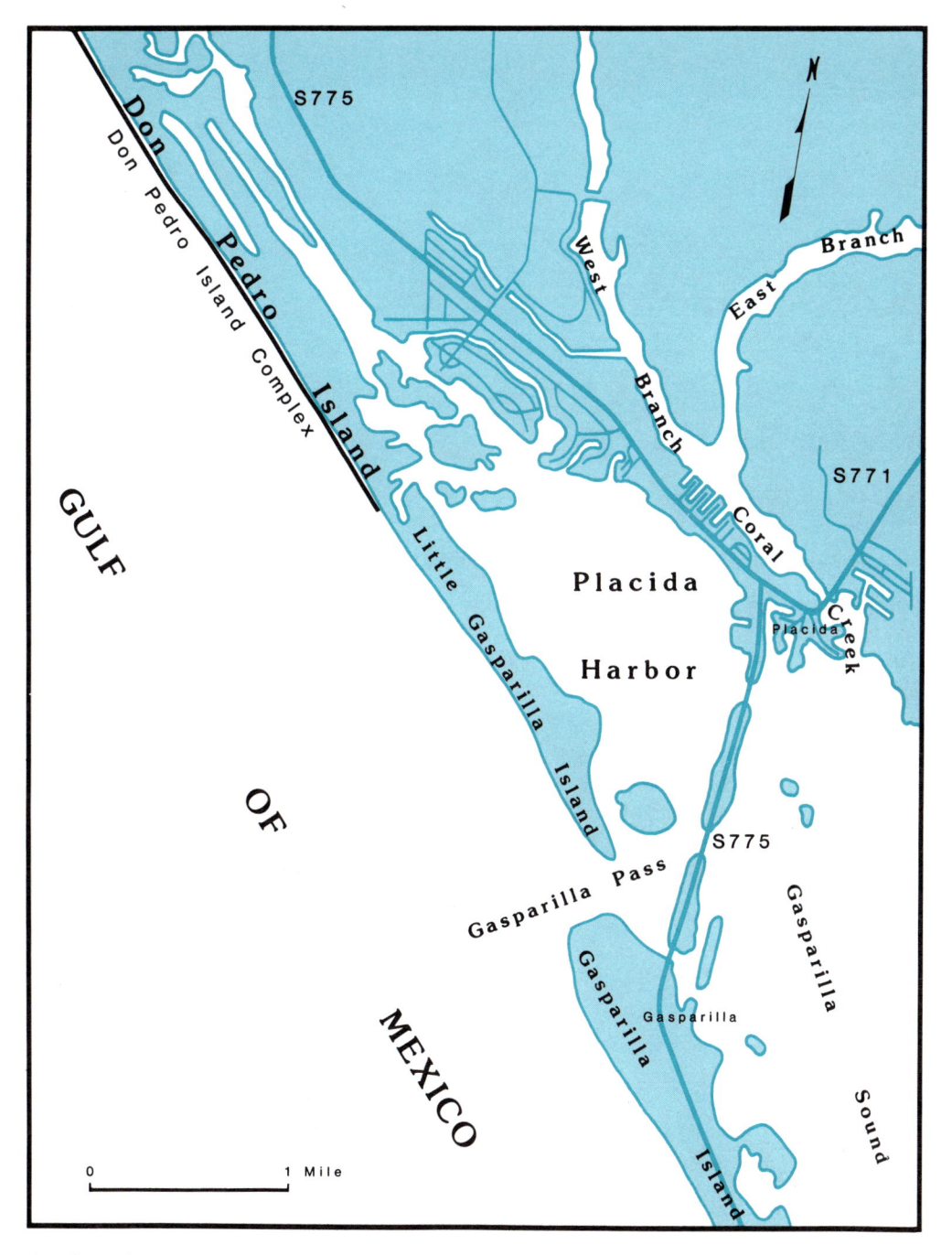

Charlotte County

FACILITIES / ENVIRONMENT

NAME	Easy Access	Parking or Entrance Fee	Parking	Restrooms	Showers	Picnicking	Swimming	Lifeguards	Fishing	Boating Facilities	Shelters	Concession Stands	Handicapped Facilities	Public Transportation	Group Facilities	Nature Trails / Fitness	Atlantic / Gulf	Bay / Soundfront	Sandy Beach	Rocky Beach	Primitive Beach	Urban
Englewood Beach	●		●	●	●	●	●	●	●		●	●	●		●		●		●			●
Port Charlotte Beach Park							●		●								●		●		●	
Don Pedro Island Complex	●		●	●	●	●	●	●		●	●	●	●				●	●	●			●

Florida Bicycle Laws

Vehicles with a seat height of 25 inches or more from the ground are legally included in the definition of bicycle. This excludes toy sidewalk bikes and includes adult three-wheelers.

Every bicycle in use between sunset and sunrise shall be equipped with lights or reflectors.

Any person operating a bicycle upon a one-way highway with two or more marked traffic lanes may ride as near the left-hand curb or edge of such roadway as practicable.

A person propelling a vehicle by human power upon and along a sidewalk, or across a roadway upon and along a crosswalk, shall have all the rights and duties applicable to a pedestrian under the same circumstances.

No bicycle shall be used to carry more persons at one time than the number for which it is designed, except an adult rider may carry a child securely attached to his or her person in a backpack or sling.

Bicyclists will have the full legal responsibilities and protection of vehicle operators. This includes obedience to traffic laws.

Bicyclists are required to ride with the flow of traffic on the right portion of the roadway. This permits the bicyclist to leave the right edge and enter the traffic lane to avoid hazards.

Bicyclists may not impede normal traffic flow and they are restricted to a single lane on multiple-laned highways. The law removes the requirement that bicyclists use all bike paths adjacent to roadways.

Bicyclists are permitted a new left-turn option of riding along the right portion of the roadway to the far side of the intersection, stopping, and then proceeding with the new traffic when the traffic signal changes, similar to turns made by pedestrians.

West Coast Intracoastal Waterway

Beginning at the mouth of the Caloosahatchee River in Lee County, the West Coast Intracoastal Waterway follows a winding course through channels, bays, harbors, and sounds northward toward the Anclote River west of Tarpon Springs. The waterway is federally maintained to a depth of 9 feet and a width of 100 feet. A number of inlets and passes along the waterway are sufficiently stable and well lighted to be safely entered at night on a rising tide. However, the best time is when the sun is high, lighting shoal areas and shallow waters.

The waterway is relatively well protected from high winds and rough water; the only open water is Charlotte Harbor and San Carlos Bay. A rich and varied birdlife flourishes along the channel waters—brown and white pelicans, cormorants, herons, roseate spoonbills, osprey, and egrets. These waters are also abundant in game fish—trout, snook, mangrove snapper, mackerel, and tarpon.

Cruising the waterway offers the opportunity to venture up tributary rivers or explore islands accessible only by boat, escaping the outside world to lose oneself in the tranquility of nature and Florida's subtropical beauty.

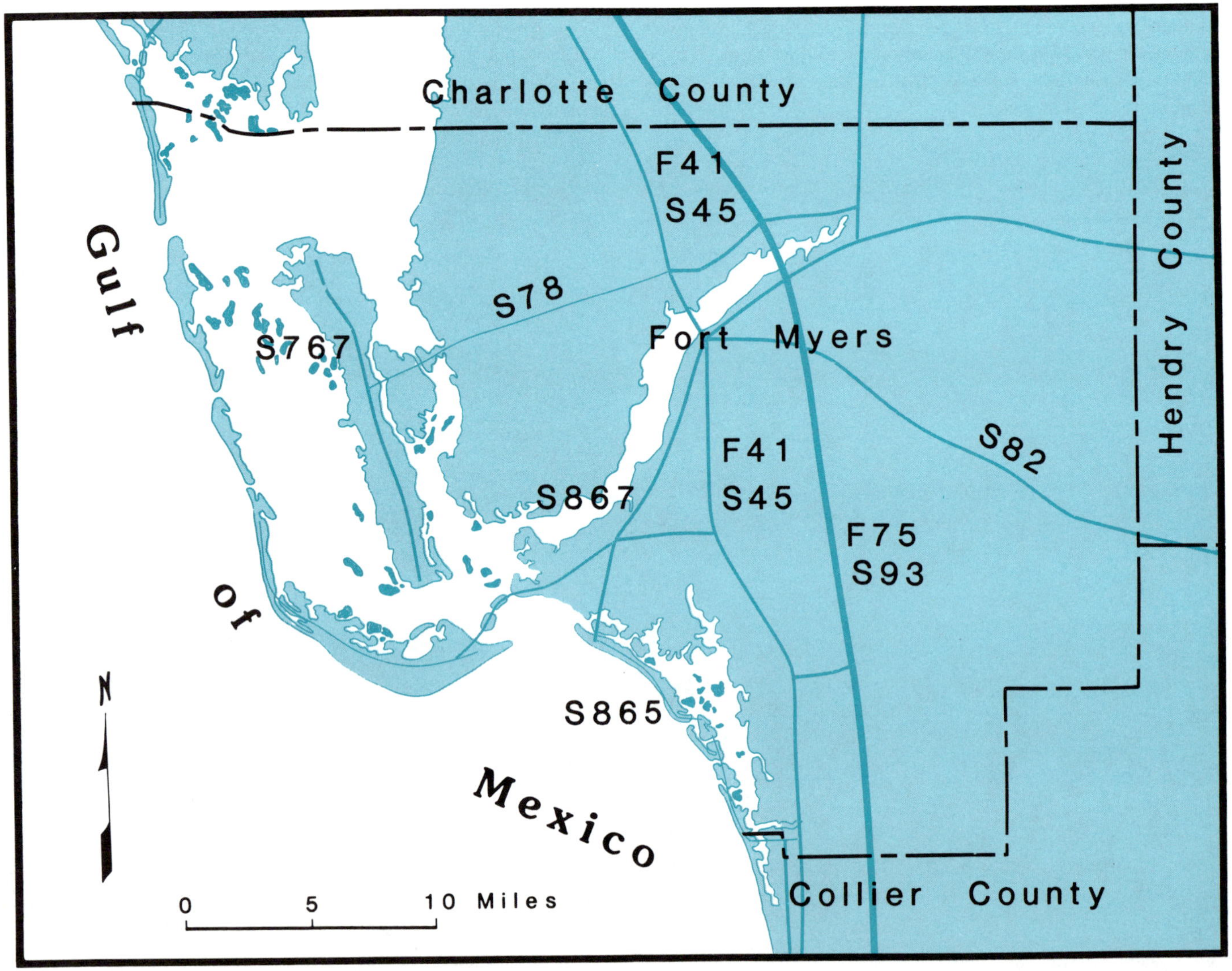
Charlotte County
F41
S45
S78
S767
Fort Myers
F41
S45
S82
S867
F75
S93
S865
Hendry County
Collier County
Gulf of Mexico
N
0
5
10 Miles

Lee County

Lee County lies on the southwest coast of the Florida peninsula. Named by the pioneering Florida cattleman and politician Captain Francis Asbury Hendry in honor of General Robert E. Lee, it became Florida's forty-first county on May 13, 1887. Present-day Lee County includes 51.5 miles of sandy beach, mostly located on the Gulf side of the many barrier islands stretching the length of the county. Interior islands and shallow estuaries between the barrier islands and the mainland combine with extensive mangrove swamps to provide the county with an especially abundant wildlife.

Because of the dynamic nature of the barrier islands, erosion and deposition create constant changes in the islands' topography. Storms and hurricanes can cause dramatic differences overnight. Inlets and passes become shallow over time, while new ones may be created by island breaching. These coastal changes have occurred repeatedly in Lee County, often accelerated where real estate development has changed the coastal topography.

The northernmost island, Gasparilla, is about 7 miles long and averages about one-third mile in width. Boca Grande Pass, which separates Gasparilla from Cayo Costa Island, is especially deep and allowed for the building of a deep-water port on the southern end of the island. The port led to minor development, but the island remains relatively unspoiled despite an erosion problem. Gasparilla is accessible only from Charlotte County to the north, making it inconvenient to residents.

Cayo Costa and Upper Captiva, the next two barrier islands, are not accessible by automobile. Together, the county and state own most of these islands and intend to preserve them in their natural state. Their size and pristine condition make them excellent workshops for the study of barrier island systems.

Sanibel Island is connected to the mainland by the Sanibel Causeway (S.R. 867) and is world famous as a model of thoughtful environmental planning and controlled development. The island has many public and private nature preserves, including the J. N. "Ding" Darling National Wildlife Refuge. Most beach access is near Point Ybel on the island's southern tip. Captiva Island is reached by road from Sanibel.

Across San Carlos Bay from Point Ybel lies Estero Island, frequently referred to as Fort Myers Beach. The island is heavily developed and is the most popular recreation area in the county. Reached by S.R. 865, which runs the length of the island, Estero also features the 80-acre Mantanzas Pass Nature Preserve.

Besides the barrier island beaches, the county has significant amounts of shoreline on the Pine Islands and on its inland coast, much of it covered with mangroves. Fort Myers, noted for its hospitality and quaintness, was the winter home of inventor Thomas Edison. His house and workshop are a museum where daily tours are given.

Captiva Island

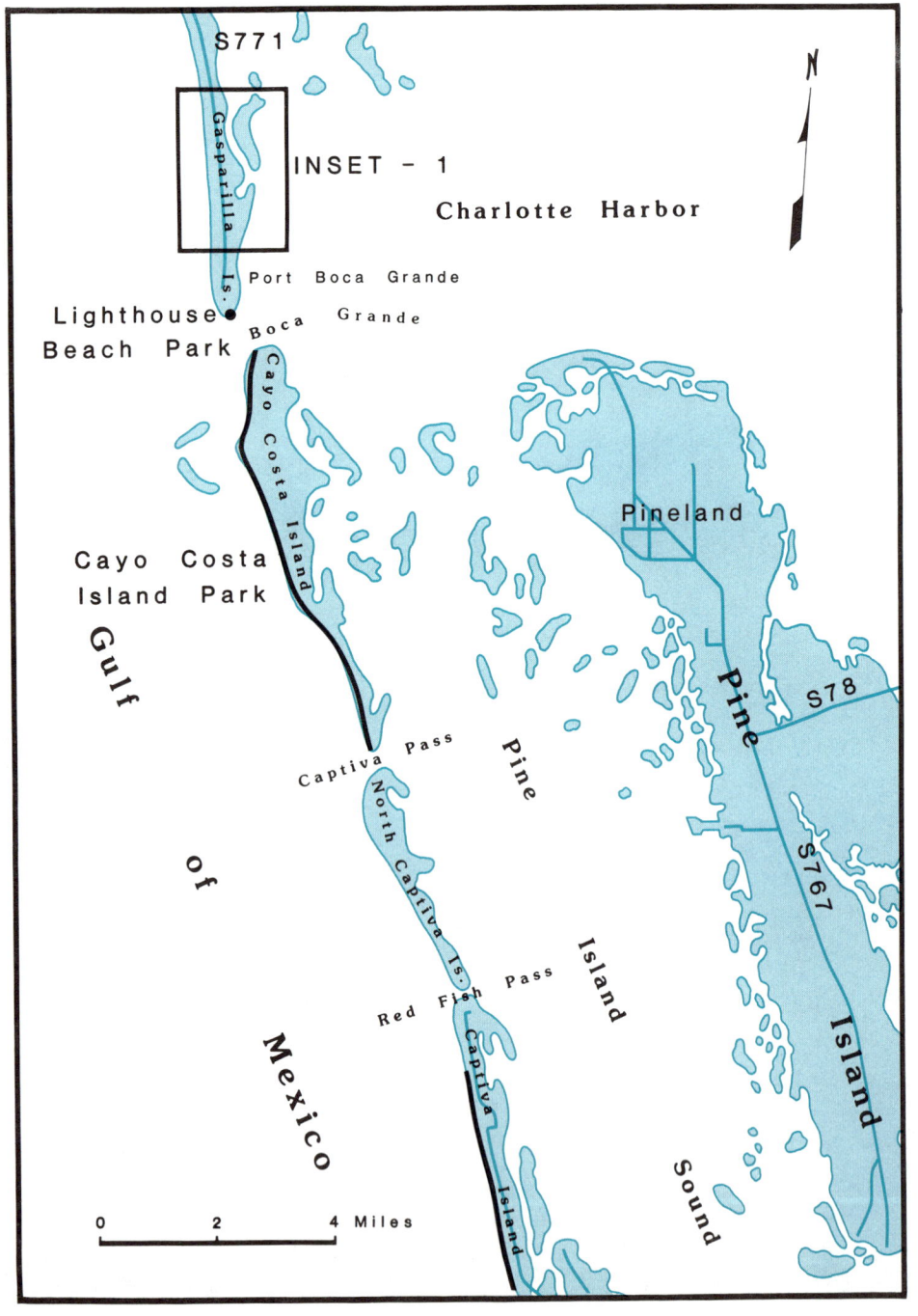

Boca Grande accessways are located on Gasparilla Island, a barrier island between the Gulf of Mexico and Charlotte Harbor. Access to 200 feet of undeveloped city beach is off Gulf Blvd. (S.R. 771 South) street ends: 19th, 17th, 13th, 12th, 11th, 10th, 7th, 5th, and 1st streets.

Boca Grande Beach has 300 feet of developed Gulf beach off S.R. 771, west of the landing field.

Lighthouse Beach Park occupies the southern end of Gasparilla Island and has 400 feet of undeveloped beach. Access is at the southern street end of Gulf Blvd. (S.R. 771). Fishing pier.

Cayo Costa Island Park has 625 acres of isolated and undeveloped beach between the Gulf of Mexico and Pine Island Sound. The only access is by boat. Alcohol is not permitted.

Captiva accessways are located on 5 miles of undeveloped beach accessible from Sanibel-Captiva Rd. (S.R. 867) west from Fort Myers across Sanibel Island to Captiva Island. Street-end access roads are Laika Ln., South Seas, Wightman Ln., Andy Rosse Ln., Cemetery, and Post Office Corner.

Northern district

Lee County

FACILITIES | ENVIRONMENT

NAME	Easy Access	Parking or Entrance Fee	Parking	Restrooms	Showers	Picnicking	Swimming	Lifeguards	Fishing	Boating Facilities	Shelters	Concession Stands	Handicapped Facilities	Public Transportation	Group Facilities	Nature Trails / Fitness	Atlantic / Gulf	Bay / Soundfront	Sandy Beach	Rocky Beach	Primitive Beach	Urban
Boca Grande	●		●			●	●									●		●	●			●
Boca Grande Beach	●		●	●		●	●		●		●						●		●			●
Lighthouse Beach Park	●		●	●			●						●				●		●		●	
Cayo Costa Island Park						●	●						●				●		●		●	
Captiva	●		●			●	●									●		●		●	●	

Captiva Island

151

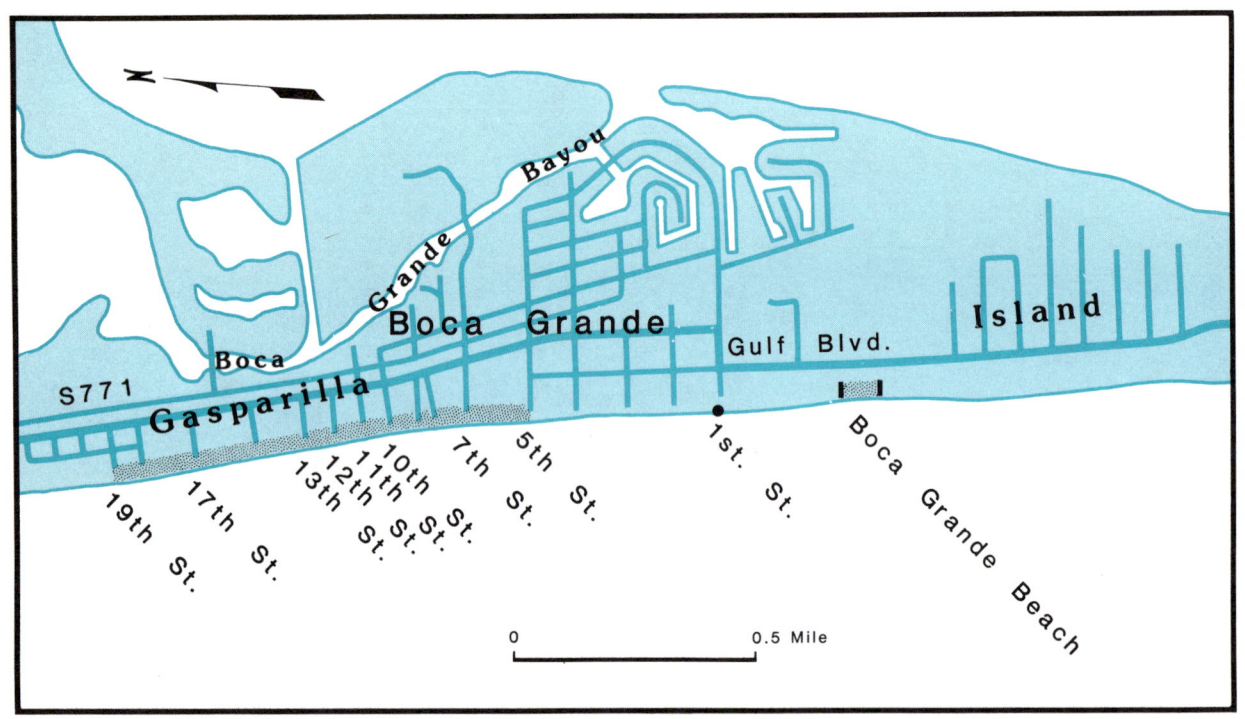

Gasparilla Island (inset 1)

Egrets

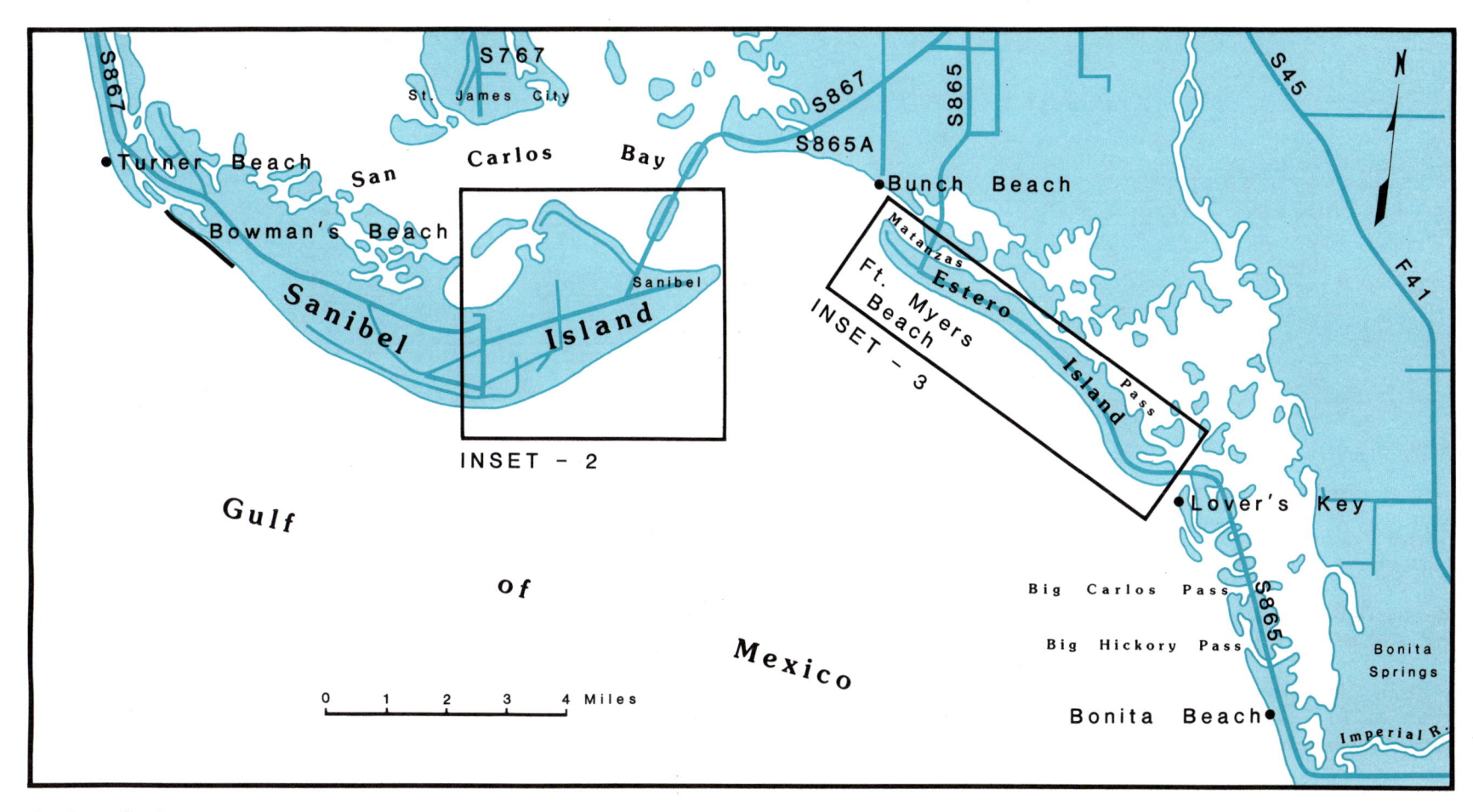

Southern district

Lee County

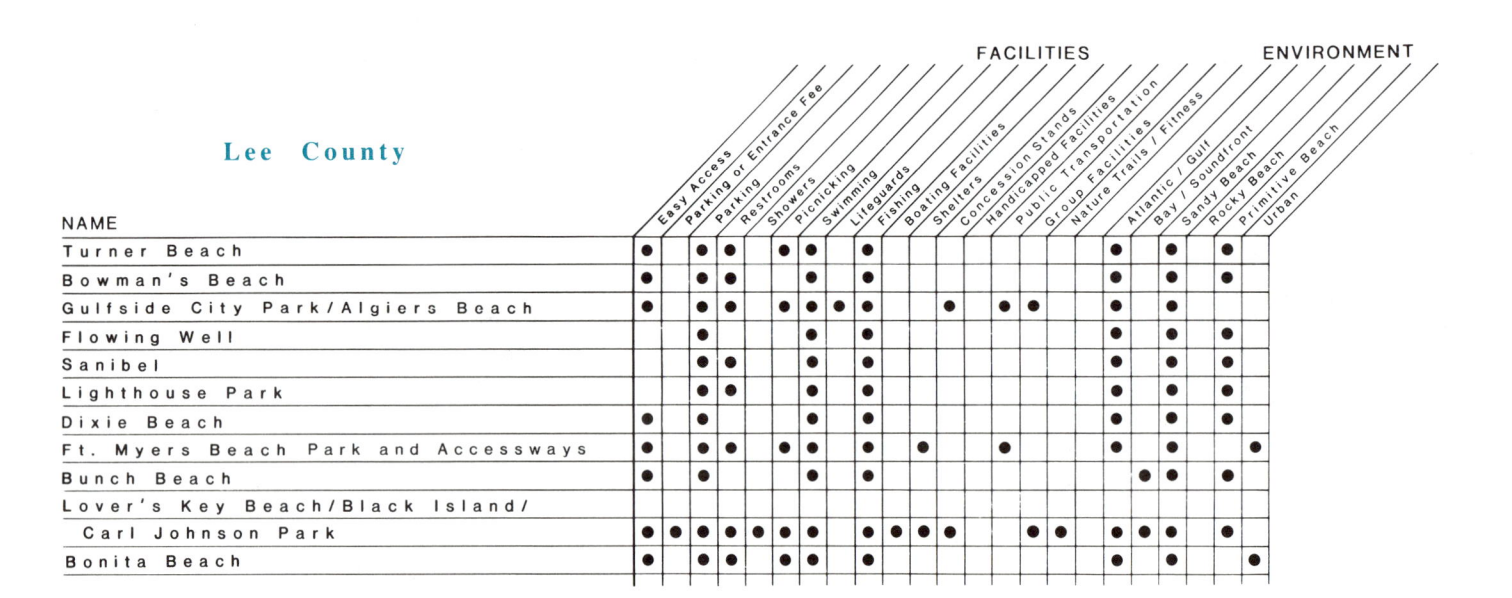

NAME	Easy Access	Parking or Entrance Fee	Parking	Restrooms	Showers	Picnicking	Swimming	Lifeguards	Fishing	Boating Facilities	Shelters	Concession Stands	Handicapped Facilities	Public Transportation	Group Facilities	Nature Trails / Fitness	Atlantic / Gulf	Bay / Soundfront	Sandy Beach	Rocky Beach	Primitive Beach	Urban
Turner Beach	●		●	●		●	●		●								●		●		●	
Bowman's Beach	●		●	●			●										●		●		●	
Gulfside City Park/Algiers Beach	●		●	●	●	●	●				●		●	●			●		●			
Flowing Well				●			●										●		●			
Sanibel			●	●			●										●		●			
Lighthouse Park			●	●			●										●		●			
Dixie Beach	●		●				●										●		●			
Ft. Myers Beach Park and Accessways	●		●	●		●	●		●			●					●		●			●
Bunch Beach	●		●				●											●	●			
Lover's Key Beach/Black Island/ Carl Johnson Park	●	●	●	●	●	●	●		●	●	●			●	●		●	●	●		●	
Bonita Beach	●		●	●		●	●		●								●		●			●

Turner Beach has 3 acres of developed beach at the south end of Captiva Island. Access is off S.R. 867.

Bowman's Beach is an emerged sandbar on the north end of Sanibel Island. Marked access is from S.R. 867.

Gulfside City Park/Algiers Beach is a developed beach with 30 acres on the Gulf of Mexico in Ybel. Access is at the beach end of Southwinds Dr. E.

Flowing Well has access to more than a half mile of undeveloped beach at Falgur St., Donax St., Nerita St., and Beach Rd.

Sanibel beach access (both Gulf and bay) is at Buttonwood and at Seagrape.

Lighthouse Park is 5 acres of undeveloped beach at Point Ybel. Access is at the end of Periwinkle Way in Sanibel.

Dixie Beach faces San Carlos Bay on the east end of Sanibel Island and adjoins J. N. "Ding" Darling National Wildlife Refuge. Access is marked on Periwinkle Way when entering Sanibel.

Ft. Myers Beach access to undeveloped beach is off S.R. 865 on streets marked on map on next page, from Ave. A through Flamingo Ave. Fishing pier.

Bunch Beach is a mainland beach on San Carlos Bay across from Fort Myers Beach (Estero Island). Access is S.R. 867 west from Fort Myers south to John Morris Rd.

Lover's Key Beach/Black Island/Carl Johnson Park are currently under acquisition by the state with plans for future development. Carl Johnson Park is a developed park with many facilities.

Bonita Beach lies on the southernmost border of Lee County. Access is marked off S.R. 865. It has 600 feet of beachfront.

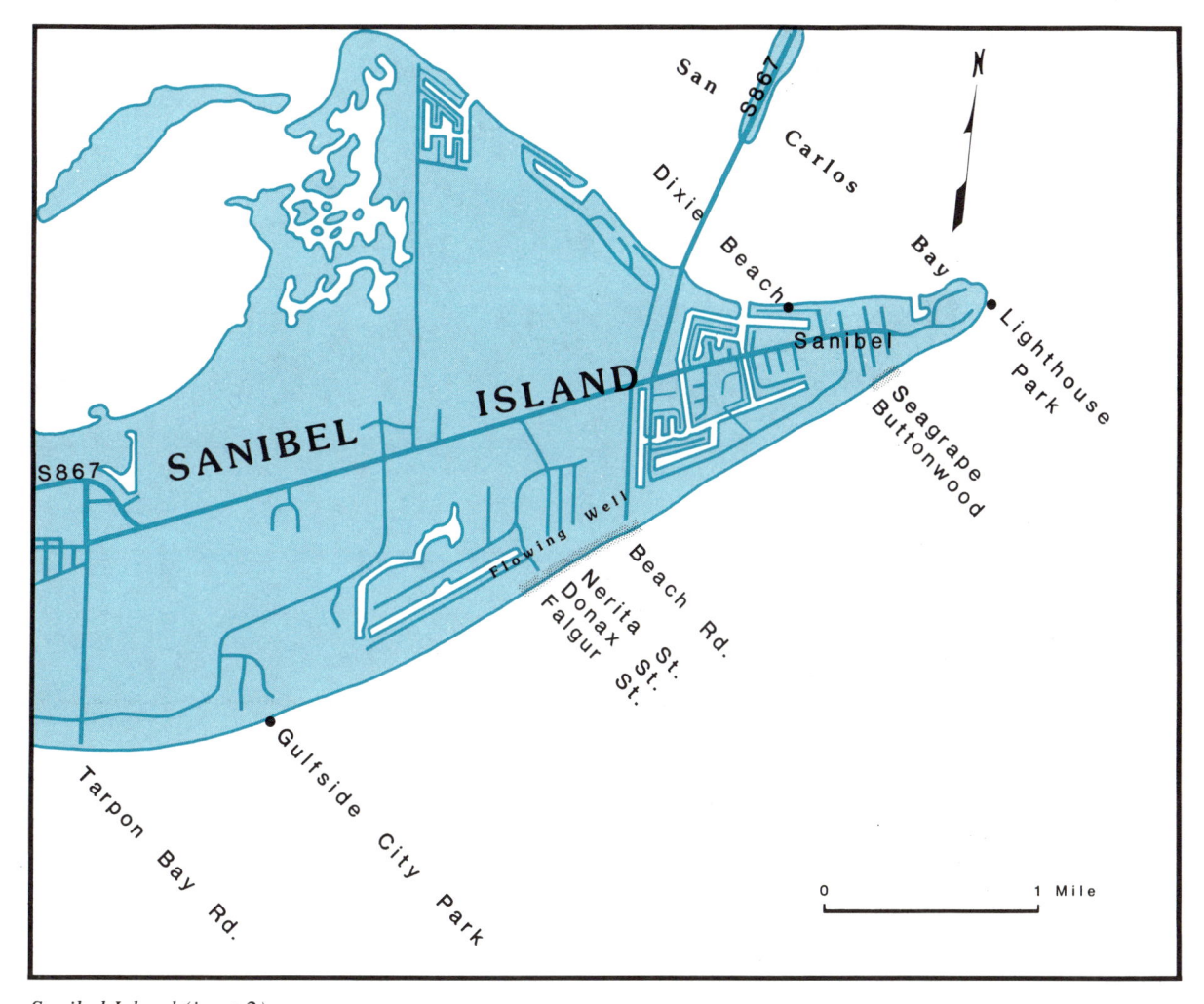

Sanibel Island (inset 2)

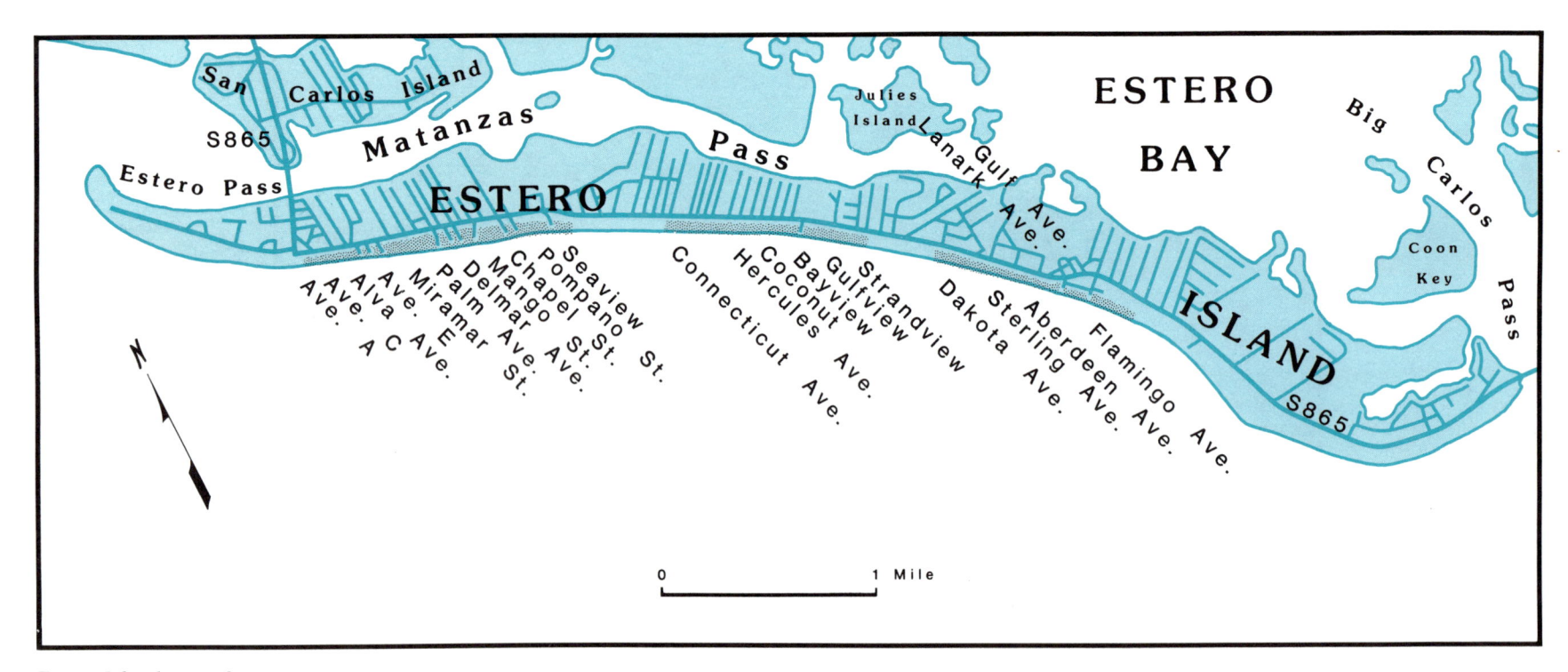

Estero Island (inset 3)

Southwest Shell Collecting

Collecting seashells and sharks' teeth on the beaches of southwest Florida has been a favorite pastime of both visitors and residents for years. Because of the gradually sloping seabed in this area, the gentle, rolling swell of the Gulf washes up a steady supply for the beachcomber. Specimens are especially plentiful on the beach immediately after a storm.

A large variety of shells, from the large brown cockle, up to 5 inches across, to the tiny, colorful coquina, is commonly available to the casual collector. Conch, auger, and olive shells provide a wide diversity of shapes and sizes as further reward for the exploring eye of the attentive searcher. But the fossilized sharks' teeth, perhaps more abundant here than anywhere else worldwide, are the most intriguing shoreline item.

The teeth are all that remain of sharks that inhabited these waters millions of years ago. Colored black, gray, or brown, they have assumed the color of the sediment in which they have been buried for eons. Many came from species long since extinct and vary from 1/8 inch to 3 inches in size.

To simplify the gathering of shells and teeth, many local shops and bookstores sell guides to assist in locating the best hunting grounds and to identify the samples discovered. Many guides supply extensive detail on the teeth, including illustrations, descriptions, and history.

What starts out to be an adventuresome day of collecting on the beach might turn into a lifetime hobby.

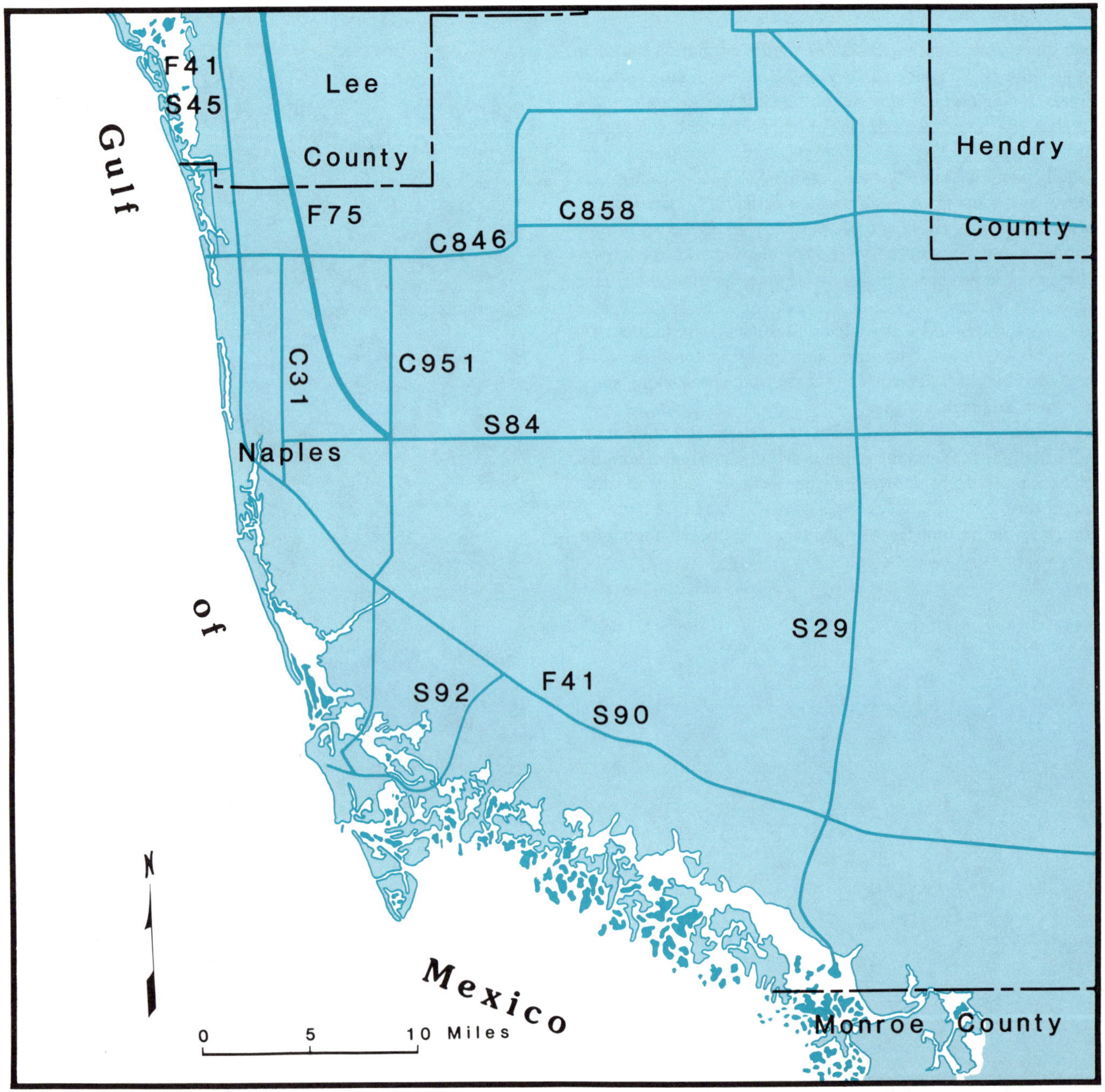
F41
S45
Lee
County
Hendry
County
Gulf
of
Mexico
F75
C858
C846
C31
C951
S84
Naples
S29
S92
F41
S90
N
0
5
10 Miles
Monroe
County

Collier County

Florida's second largest county in land area is Collier County, covering approximately 2,119 square miles. It is also one of the fastest growing areas in the country, having had a population increase of 125 percent from 1970 to 1980. Resident population in 1982 was 101,000, up from 38,040 in 1970.

The county is named for Barron G. Collier who purchased large tracts of land here in the early 1920s. Located between the Gulf and the Everglades, it enjoys moderate temperatures and predictable sunshine. The beautiful white beaches are quartz sand mixed with shell fragments that have been carried southward by rivers and shore currents from Alabama through Georgia and northern Florida. The small tidal range of 3 feet minimizes beach erosion from the adverse effects of tidal actions.

Naples is the largest incorporated municipality in the county and has been the county seat since 1962. In the 1880s the owner and publisher of the *Louisville Courier-Journal*, Walter N. Haldeman, "discovered" and founded this exotic city on the Gulf. At that time the only transportation to Naples was by boat. Other pioneers saw the magnificence and drama of this natural environment as the setting for a resort community. To this day the City of Naples and Collier County are actively involved in protecting and preserving the natural state, yet providing access for visitors and residents to desirable areas.

Those who enjoy a natural habitat where wildlife thrives undisturbed by development will be happy to find over 70 percent of Collier County under preservation. Corkscrew Sanctuary, the oldest and most visited, has a 1.75-mile boardwalk through the last sizable stand of big cypress in Florida. Twenty miles south of Corkscrew and east of Naples on Everglades Parkway (Alligator Alley or S.R. 84) is the 600,000-acre Big Cypress National Preserve.

The unique feature of this federal preserve is that in addition to providing a home for birds, alligator, deer, and the endangered Florida panther, the Park Service allows hunting, oil exploration, and mining. Here the philosophy of preserve management is to promote balanced ecological use of South Florida's environment.

South of Big Cypress is the famed Everglades National Park that extends from Everglades City to Florida Bay along the sensitive coastal estuaries of southwest Florida.

Collier-Seminole State Park, 16 miles east of Naples, has a boardwalk that winds through the mangroves, salt marshes, and cypress trees. Wildlife includes brown pelicans, bald eagles, and manatees.

Between Naples and the Isle of Capri lies a quiet retreat known as the Rookery Bay National Estuarine Sanctuary. This 6,000-acre tidal mangrove system has been set aside as a living laboratory for the study and research of sensitive estuaries.

Other activities important in Collier County include production and retailing of natural gas. Recent years have seen dramatic increases in citrus fruit and winter vegetable production.

Sound planning and a strong determination to safeguard the unique environment in Collier County ensure that this area will be protected and enjoyed by those who visit and live here.

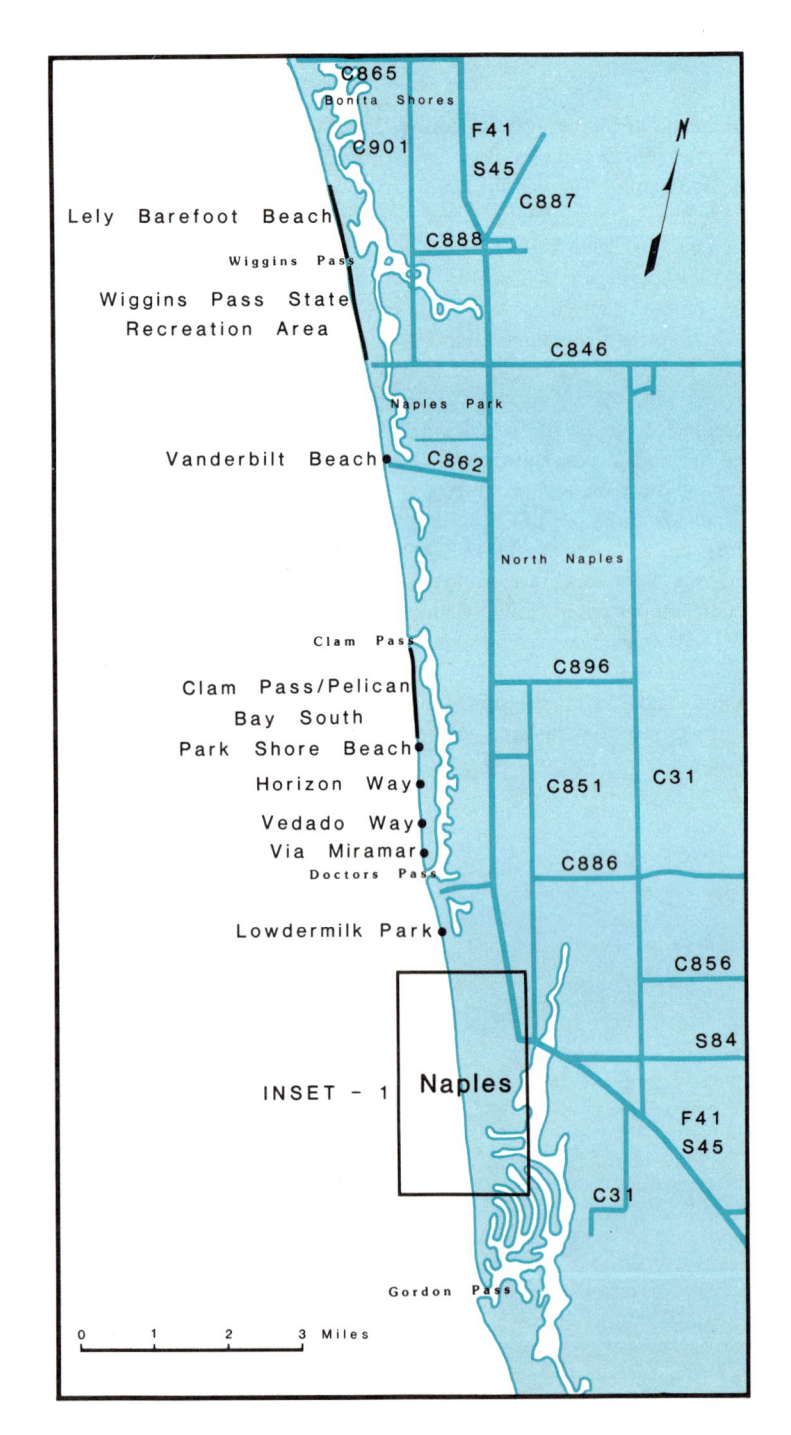

Naples fishing pier

Northern district

Collier County

FACILITIES / ENVIRONMENT

NAME	Easy Access	Parking or Entrance Fee	Parking	Restrooms	Showers	Picnicking	Swimming	Lifeguards	Fishing	Boating Facilities	Shelters	Concession Stands	Handicapped Facilities	Public Transportation	Group Facilities	Nature Trails / Fitness	Atlantic / Gulf	Bay / Soundfront	Sandy Beach	Rocky Beach	Primitive Beach	Urban
NORTHERN DISTRICT																						
Lely Barefoot Beach	●		●	●	●	●		●		●	●						●		●			
Wiggins Pass State Recreation Area	●	●	●	●	●	●	●	●	●	●	●		●		●		●		●			●
Vanderbilt Beach	●		●				●		●				●				●		●			
Clam Pass/Pelican Bay South							●		●								●		●	●		
Park Shore Beach							●		●								●		●	●		
Naples Municipal Beach	●		●	●	●		●		●			●	●	●	●		●		●			●
Lowdermilk Park	●		●	●	●	●	●					●	●	●	●		●		●			●
SOUTHERN DISTRICT																						
Tigertail Beach	●		●	●	●	●	●		●			●	●	●	●		●		●			●
Point Marco Beach							●		●								●		●			●

Lely Barefoot Beach is 600 feet of developed barrier island beach just south of and adjacent to Lee County's Bonita Beach. Access is from S.R. 865 (Bonita Beach Rd.).

Wiggins Pass State Recreation Area has 166 acres of undeveloped beach separated from the mainland by mangrove swamp and tidal creeks. It lies south of Barefoot Beach and Wiggins Pass on the Gulf of Mexico. Access is on Gulf Shore Blvd. N. (U.S. 41).

Vanderbilt Beach provides walkway access to 2 miles of undeveloped beach along Gulf Shore Blvd. at Gulf Shore N., Seabreeze Ave., Channel Dr., Bayview Ave., and Gulf Shore S. There is also one road access at the west end of Vanderbilt Beach Rd. (S.R. 862).

Clam Pass/Pelican Bay South is a 3,200-foot undeveloped beach accessible only by boat or by walking north along the beach from Park Shore Beach about a half mile.

Park Shore Beach is an undeveloped beach with an asphalt walkway and boardwalk. Access is in Naples at the intersection of Gulf Shore Blvd. N. and Park Shore Dr.

Naples Municipal Beach provides street-end access to the Gulf of Mexico at Horizon Way, Vedado Way, Via Miramar, 7th Ave. N. through 21st Ave. S. off Gulf Shore Blvd., and 32d Ave. S., 33d Ave. S., and Sabre Cay off Gordon Dr. Fishing pier at 12th Ave. S.

Lowdermilk Park is a developed beach in the City of Naples. Access to 900 feet of beach frontage is at Bayan Blvd., west off Gulf Shore Blvd. N.

Tigertail Beach, on NW Marco Island, is accessible via North Collier Blvd. to Seaview Ct. W. This is a developed beach on the Gulf of Mexico. No alcohol allowed.

Point Marco Beach, on SW Marco Island, is accessible from Collier Blvd. It has less than an acre of undeveloped beach.

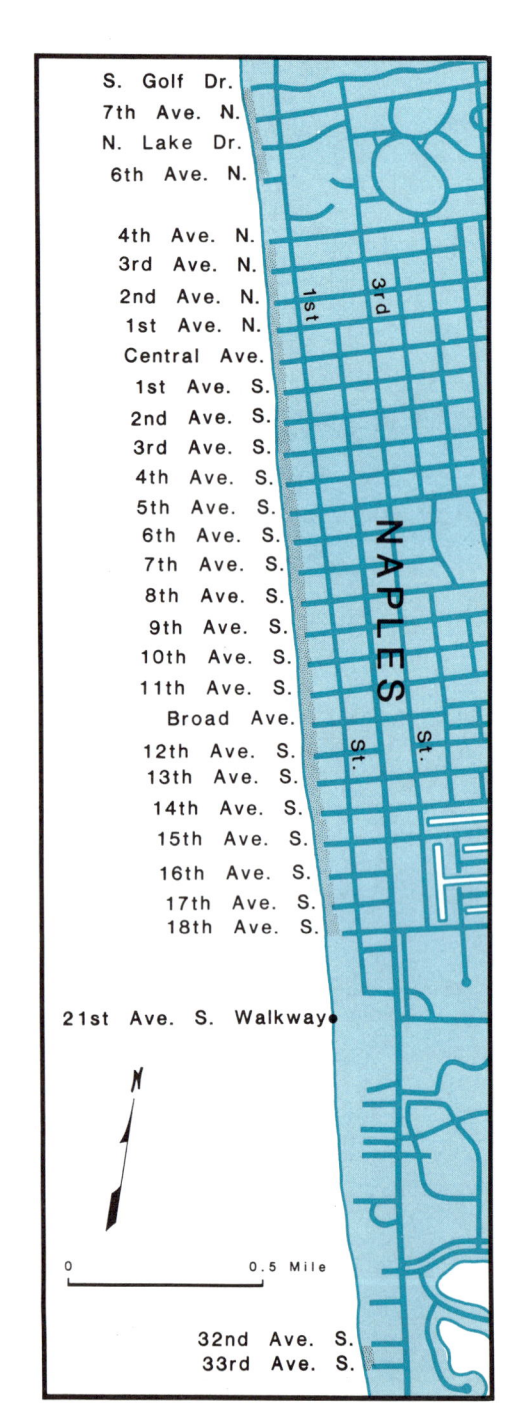

Naples (inset 1)   *Red mangrove*

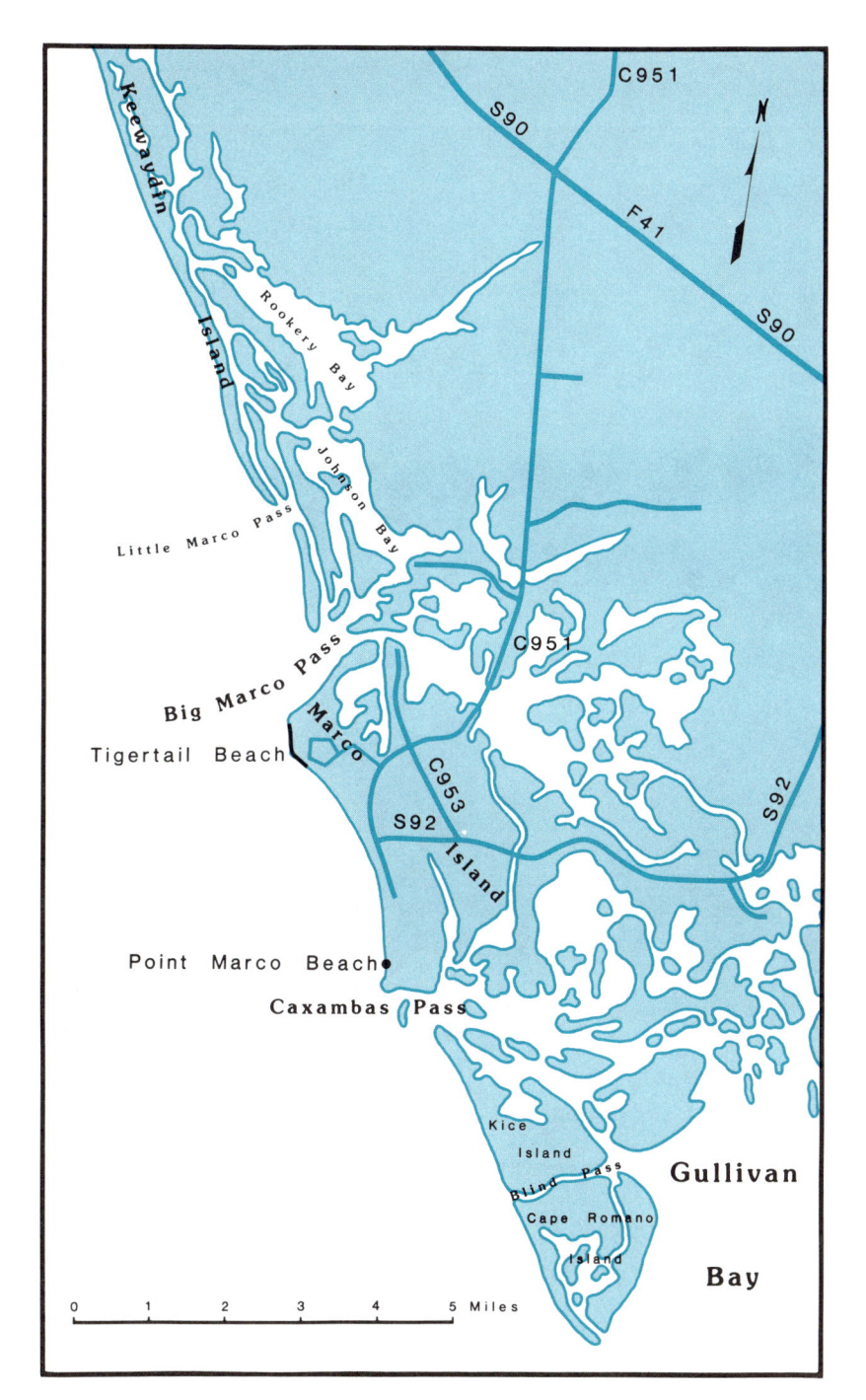

Great white heron

Central district

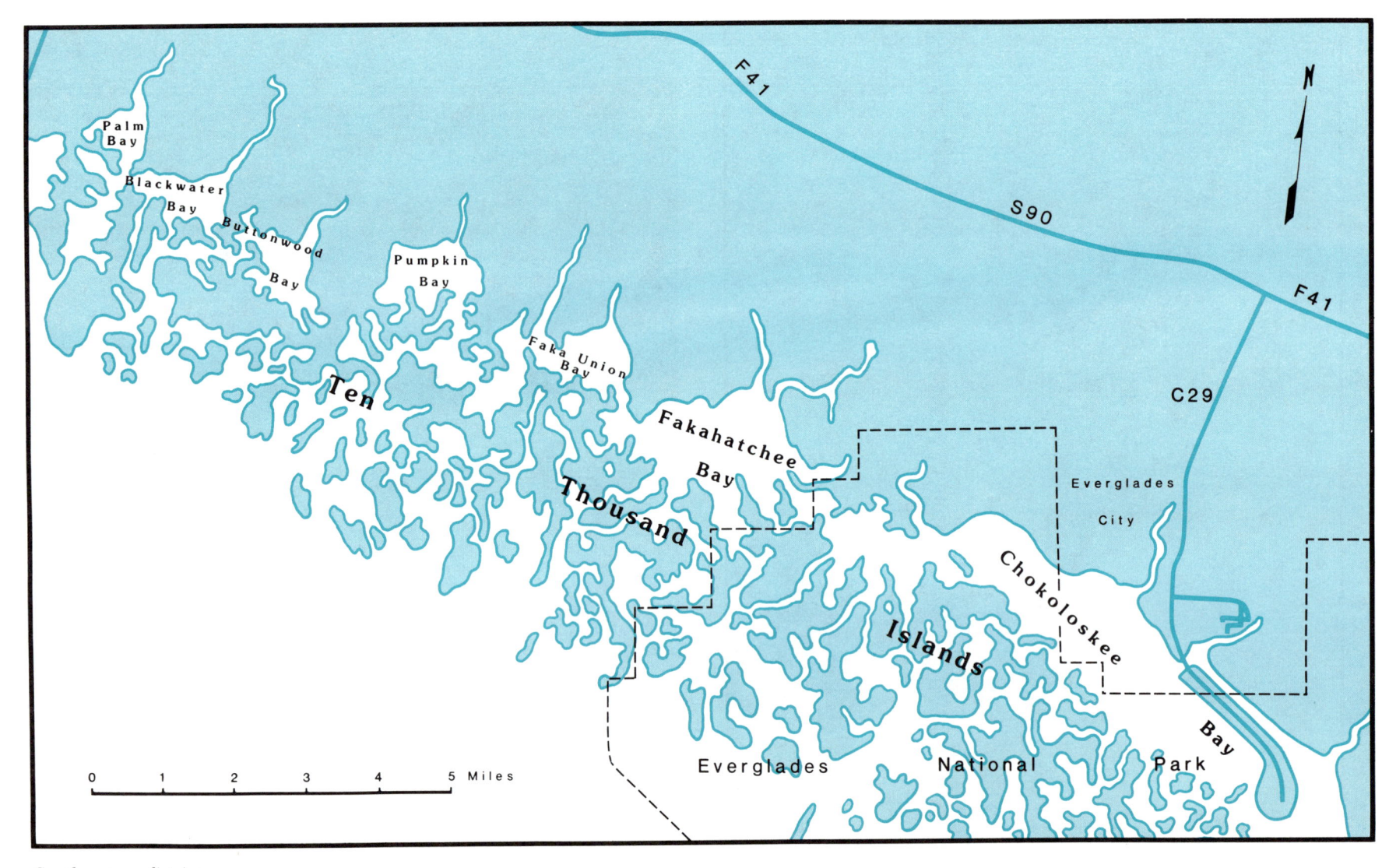

Southeastern district

Mangrove

Mangrove is a common name for certain shrubs and trees from three different families. They all grow in dense thickets or forests along coasts, tidal estuaries, and salt marshes. Distinguished by their masses of exposed, tangled roots, mangroves have adapted for survival in sheltered, saline habitats along much of Florida's peninsular coastline. The ability to procure fresh water from a saltwater environment permits them to flourish where other trees and shrubs cannot. By prospering here on the critical sea/land border, mangroves also assist in maintaining the quality of coastal waters.

In order to subsist in their saline setting, mangroves have evolved several unique characteristics. They are capable of maintaining a low salt content in their cells by secreting excess salt through specialized salt glands, or filtering salt out with their roots. Thick, wax-covered leaves help prevent water loss. To obtain oxygen, their roots have evolved distinct features. Either aerial roots, growing down to the sediment from branches, or pneumatophores (fingerlike projections), growing up from submerged roots, supplement the trees' need for oxygen.

Their roots also assist in shoreline stabilization and, by trapping debris and sediment, actually build the shoreline seaward. This accumulated matter serves as both a filter, removing pollutants from runoff, and a source of food for fish and invertebrates as it decomposes and releases nutrients into the surrounding water.

Three species of mangrove are found in Florida: red, black, and white. Red mangroves (*Rhizophora mangle*) are generally found seaward of the other species and as far north as Daytona Beach on the Atlantic and Cedar Key on the Gulf. Red mangroves are distinguished from the other two species by their prominent aerial roots and their long, pencil-shaped seeds.

The most cold-tolerant species, the black mangrove (*Avicennia germinans*), grows northward to Cedar Key on the Gulf and to St. Augustine on the east coast. Abundant pneumatophores, extending above the mud, distinguish them from other species. Their narrow, oblong leaves have a distinctive two-tone color, with dark green above and silver-green below. The fragrant white flowers attract honeybees.

White mangroves (*Laguncularia racemosa*) generally prefer higher ground than their cousins. Although lacking the characteristic root structures, a distinguishing feature of white mangrove is the presence of two small lumps at the base of each leaf, which serve as salt glands. The leaves are thicker and more rounded than red or black mangrove and often have a notch at the tip. Unable to tolerate cold temperatures, white mangroves are rarely found north of Cape Canaveral on the east coast or Tampa Bay on the west.

The large migration of new residents to Florida has unfortunately reduced the mangrove population of the state. However, as people have become aware of the trees' essential and unique function in the coastal ecosystem, steps have been taken to protect and expand their habitat. This effort has led to the currently popular trend of cultivating and transplanting mangroves on both private and public lands. The species are also protected by restrictive laws that govern their cutting and that set aside large state mangrove parks and wildlife areas.

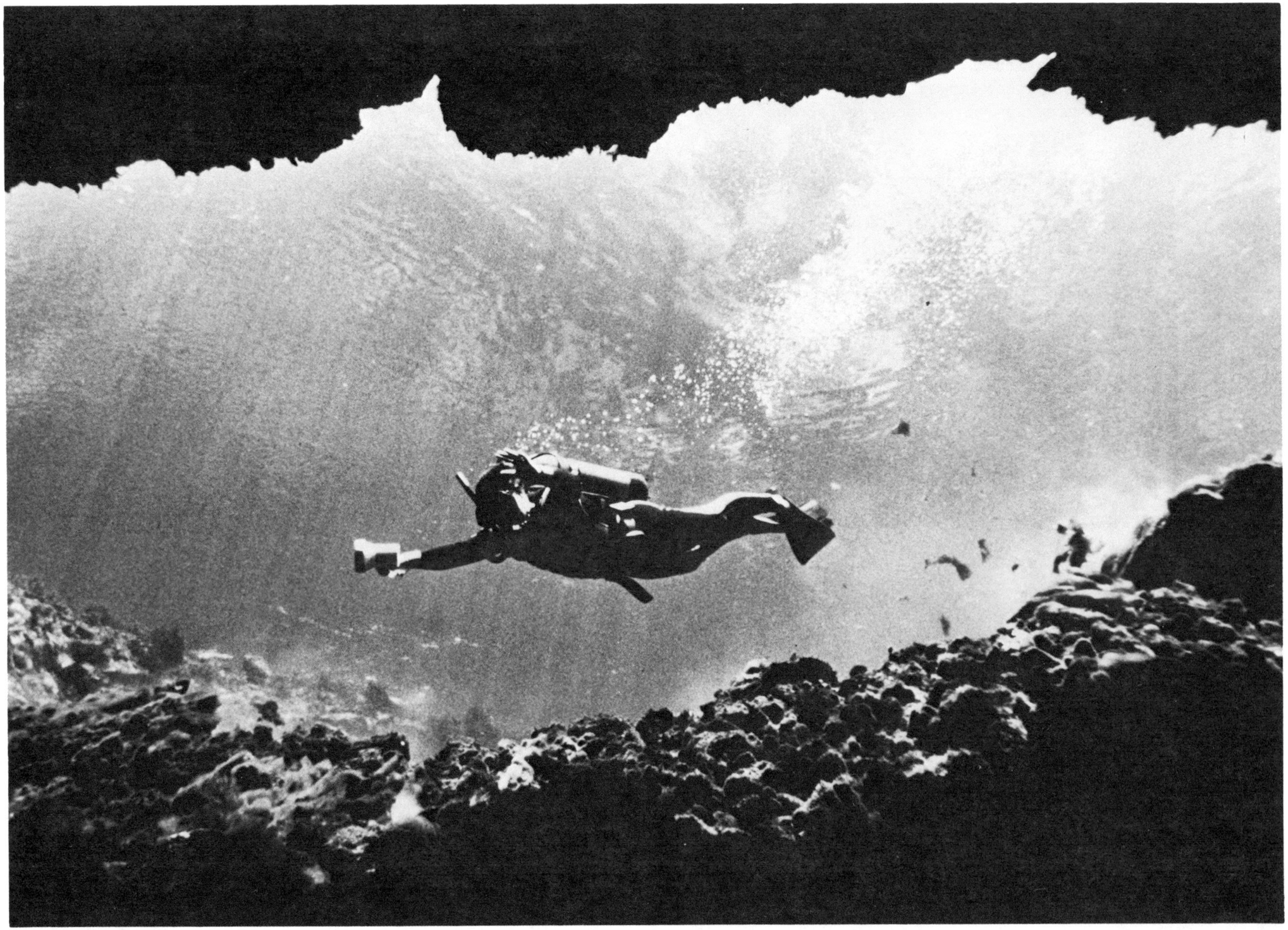

Northwest Coast

Hobie-cat regatta

Santa Rosa County

Santa Rosa County has the distinction of some of the newest yet most beautiful beaches of pure white sand in the state. The county does not directly face the Gulf of Mexico, but it has leased a section of beach from neighboring Escambia County on Santa Rosa Island at Navarre Beach. This facility is part of the "Miracle Strip" from Panama City to Pensacola (now called "Emerald Coast"). Santa Rosa County also has long shorelines on the Intracoastal Waterway and Escambia Bay available for water sports and boating. At Gulf Breeze fishermen have access to one of the world's longest fishing piers.

Blackwater River State Park, in the interior, is a popular destination for many visitors as well as local residents. Canoeing and tubing on the Blackwater River are enjoyed by people of all ages. Just to the east of Gulf Breeze on U.S. 98 is the Naval Live Oaks section of the Gulf Islands National Seashore. It is a historic remnant of the nation's first timber preserve. Here in 1828 the young American nation established the Santa Rosa Live Oak Timber Reservation. A tract of 30,000 acres was set aside to guarantee a source of timber for the nation's navy. Unfortunately, it was a short-lived project.

The earliest county inhabitants were Indians who thrived in the upland hardwood forests. Many came to the bays and Gulf for fish, which they dried in the sun. As more loggers and settlers arrived, the push to have their own county became stronger. Santa Rosa County was created in 1842.

The principal attractions in Santa Rosa County are swimming, sunbathing, and fishing. Freshwater fishing and hunting are popular in season.

Escambia County

Bordering Alabama, Escambia County is "the Western gate to the Sunshine State." It was here on Pensacola Bay in 1559 that Spain founded the first European settlement in the United States. Led by Don Tristan de Luna, 1,500 people established a colony. Unfortunately, a hurricane destroyed the expedition's ships and most of its supplies. Within two years the Spaniards left, shifting their interest to the Atlantic coast with the establishment of St. Augustine in 1565. Over the next three centuries the area saw a succession of Spanish, French, English, American, and Confederate governments.

In 1763, the British established the colonies of East and West Florida. During the Revolutionary War these were the only American colonies to remain loyal to Britain. The trading firm of Panton, Leslie and Company was formed and soon dominated trade with the Indians. It prospered with the active help of Alexander McGillivray, a Creek chief of most unusual background: his father was a Scot and long-time Indian trader and his mother was the daughter of a Creek mother and a French father. A miniature replica of one of the later Panton, Leslie trading posts is located at Main and Spring streets in Pensacola.

Andrew Jackson was named Florida's first governor. Escambia became one of its first two counties and Pensacola the first territorial capital. Shortly thereafter, the capital was moved to the new city of Tallahassee, halfway between St. Augustine and Pensacola.

In the second quarter of the nineteenth century a lighthouse and Fort Barrancas were constructed on land near the entrance to Pensacola Bay. These locations today are in the Pensacola Naval Air Station and are open to visitors. Fort Pickens, now part of Gulf Islands National Seashore, was built on Santa Rosa Island in the 1830s. It was held continuously throughout the Civil War by Union forces. In the late nineteenth century the famous Apache chief Geronimo was imprisoned there.

The Naval Air Station at Pensacola has additional points of interest for the visitor. It was the site where, in 1914, a small band of aviators were trained in naval aviation. The Naval Air Museum today honors the thousands of pilots who have earned their wings here and displays actual aircraft and spacecraft. Visitors are invited for tours aboard the aircraft carrier *Lexington*, which is maintained in active commission for actual flight deck training.

Downtown Pensacola has several museums, the Seville Square, a historic district, and exquisite wrought iron on numerous restored structures. All land south of Main Street was made from the rocks of ships' ballast brought from all over the world. In the late nineteenth century the habor was active as hundreds of sailing ships loaded lumber for export from sixteen wharves. Additional traffic was provided by numerous fishing boats loaded with red snapper caught for shipment throughout the nation. Most of the docks were destroyed by a 1926 hurricane.

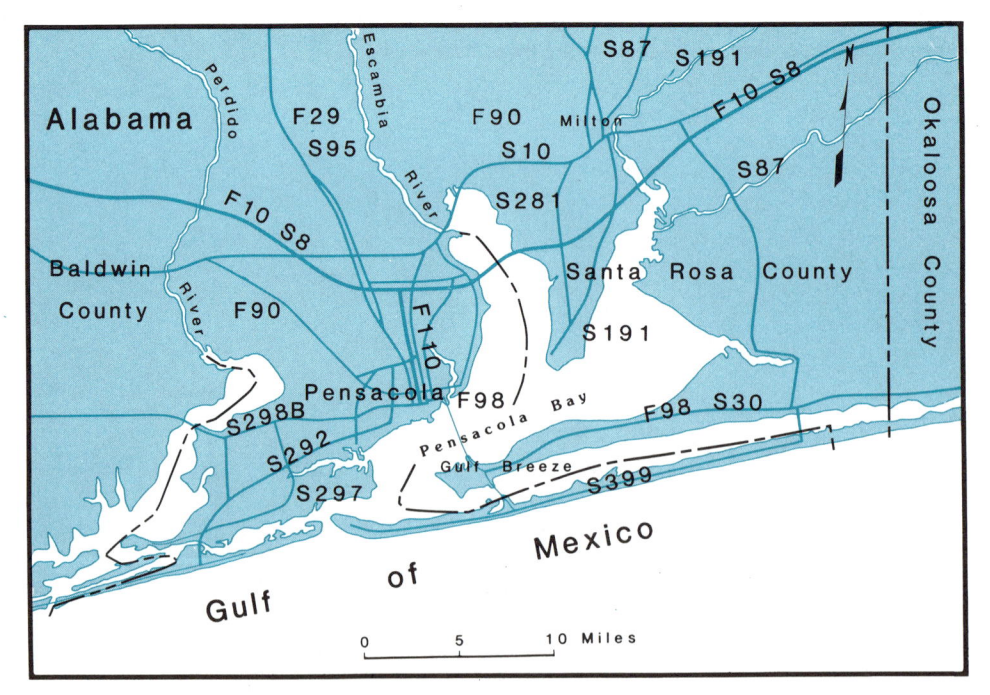

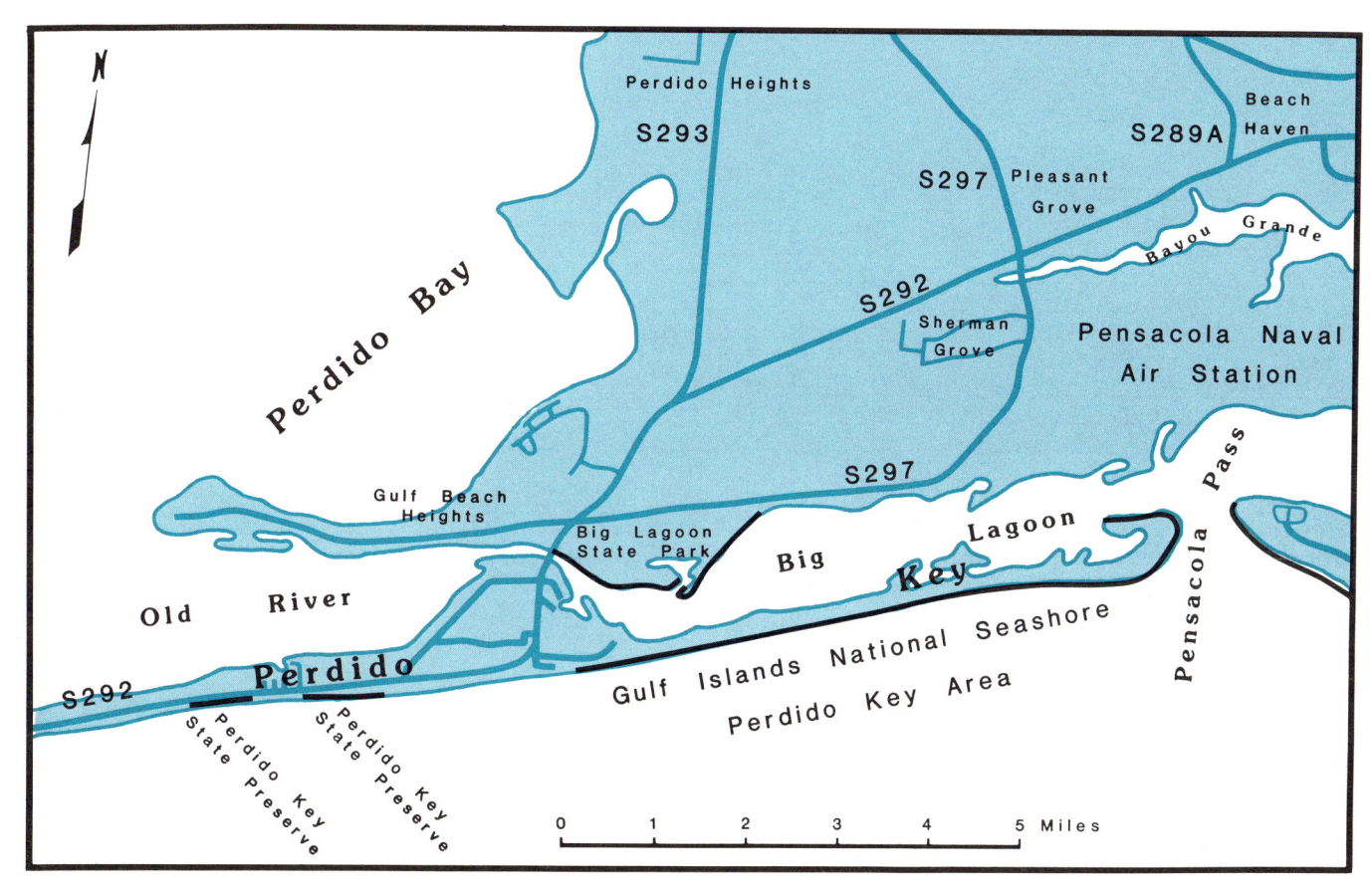

Western district

Perdido Key State Preserve: 250 acres of white sand, dunes covered with sea oats and rosemary, and pine flat woods located on S.R. 292 about 2 miles west of Big Lagoon State Recreation Area.

Big Lagoon State Recreation Area has 700 acres of coastal park on Old River with camping facilities. It is located about 13 miles west of Pensacola on S.R. 292 (Gulf Beach Highway).

Gulf Islands National Seashore/Perdido Key and Johnson Beach has 14 miles of natural sand and dune reserve reached from S.R. 292 southwest from Pensacola to Perdido Key.

Escambia/Santa Rosa Counties

NAME	Easy Access	Parking or Entrance Fee	Parking	Restrooms	Showers	Picnicking	Swimming	Lifeguards	Fishing	Boating Facilities	Shelters	Concession Stands	Handicapped Facilities	Public Transportation	Group Facilities	Nature Trails / Fitness	Atlantic / Gulf	Bay / Soundfront	Sandy Beach	Rocky Beach	Primitive Beach	Urban
WESTERN DISTRICT																						
Environmentally Endangered Lands																						
Perdido Key State Preserve # 1	●		●	●	●	●			●		●						●	●				
Perdido Key State Preserve # 2	●		●			●		●									●				●	
Gulf Islands National Seashore																						
Perdido Key and Johnson Beach		●	●	●	●	●	●	●	●		●		●		●	●	●	●	●			
Perdido Key State Preserve	●		●	●		●		●	●		●						●	●	●			
Big Lagoon State Recreation Area	●	●	●	●	●	●	●		●		●				●		●	●				
CENTRAL DISTRICT																						
Gulf Islands National Seashore																						
Fort Pickens and Langdon Beach	●	●	●	●	●	●	●	●	●			●	●	●	●	●	●	●	●			
Unnamed Beaches	●		●	●	●	●	●		●	●	●		●				●	●	●			
Quietwater Beach	●		●	●	●	●	●			●	●	●	●					●	●			
Casino Beach	●		●	●	●		●		●		●	●	●				●		●			
Pensacola Beach	●		●			●		●									●		●			●

Pensacola billfish tournament, July 4–6

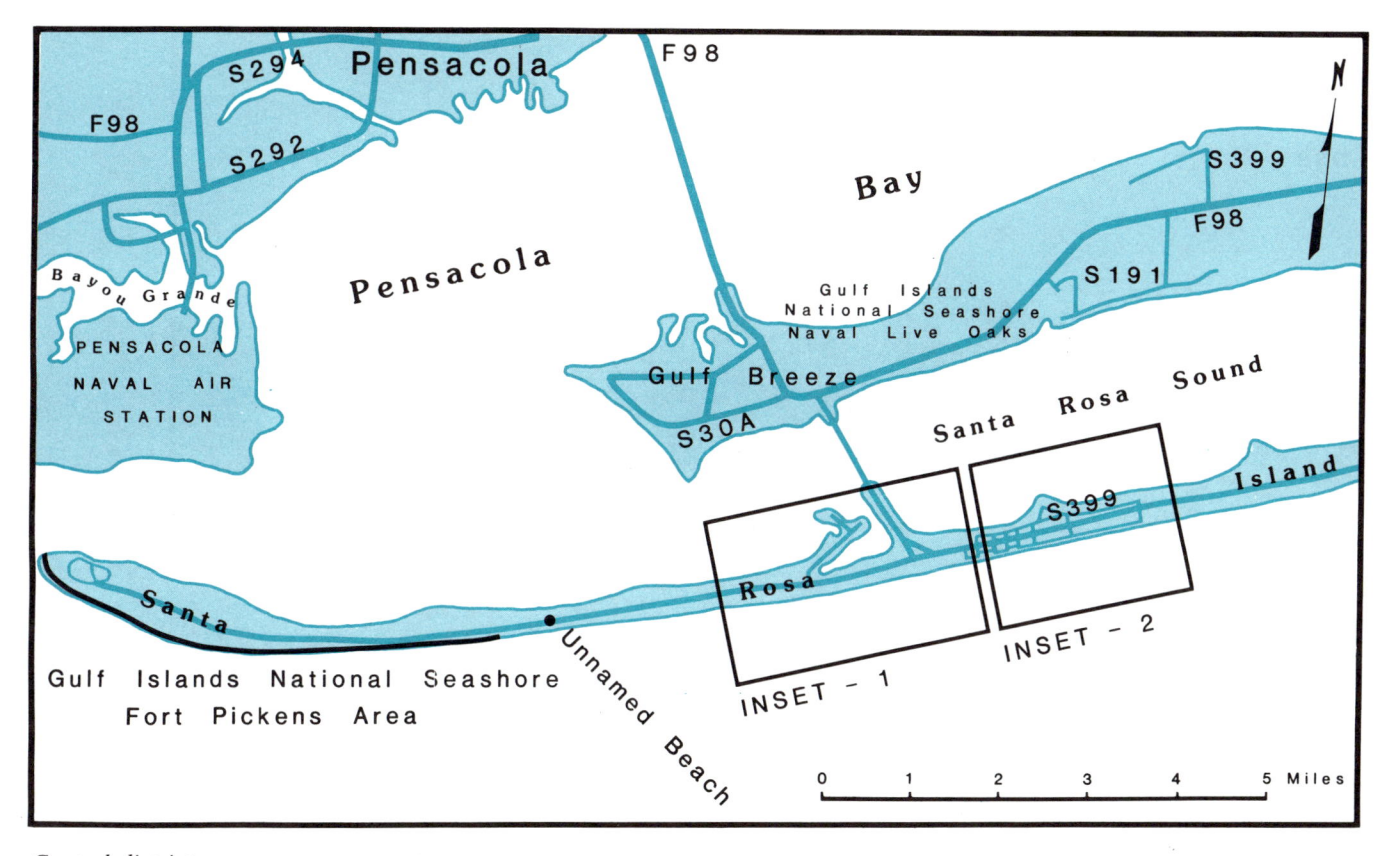

Central district

Gulf Islands National Seashore/Fort Pickens and Langdon Beach. Reached on S.R. 399, 9 miles west from Gulf Breeze to Pensacola Beach. Open daily throughout the year. Historic tours of fort facilities, 160-site campground, museums, and 17 miles of natural beach on the Gulf of Mexico. Scuba diving area and fishing pier on Santa Rosa Sound; Langdon Beach on the Gulf (lifeguards June–August).

Unnamed beaches have over 1,000 feet of Gulf and sound beach. They lie just outside of the Ft. Pickens entrance on both sides of Ft. Pickens Road, about 2.4 miles west of S.R. 399.

Quietwater Beach has 2,500 feet of area for swimming, boat launching, and picnicking on Santa Rosa Sound. Access is on Pensacola Ave. (S.R. 399) just after crossing the Bob Sikes Bridge.

Gulf Islands National Seashore/Naval Live Oaks lies east of Gulf Breeze on both sides of U.S. 98.

Casino Beach is the largest developed area for recreation on Santa Rosa Island. It is located at the intersection of Ft. Pickens Rd. and Via de Luna at the end of Pensacola Beach Blvd. (S.R. 399), about 0.8 mile south of the Bob Sikes Bridge.

Pensacola Beach public access. Avenida 10 through Avenida 23 off Ariola Drive.

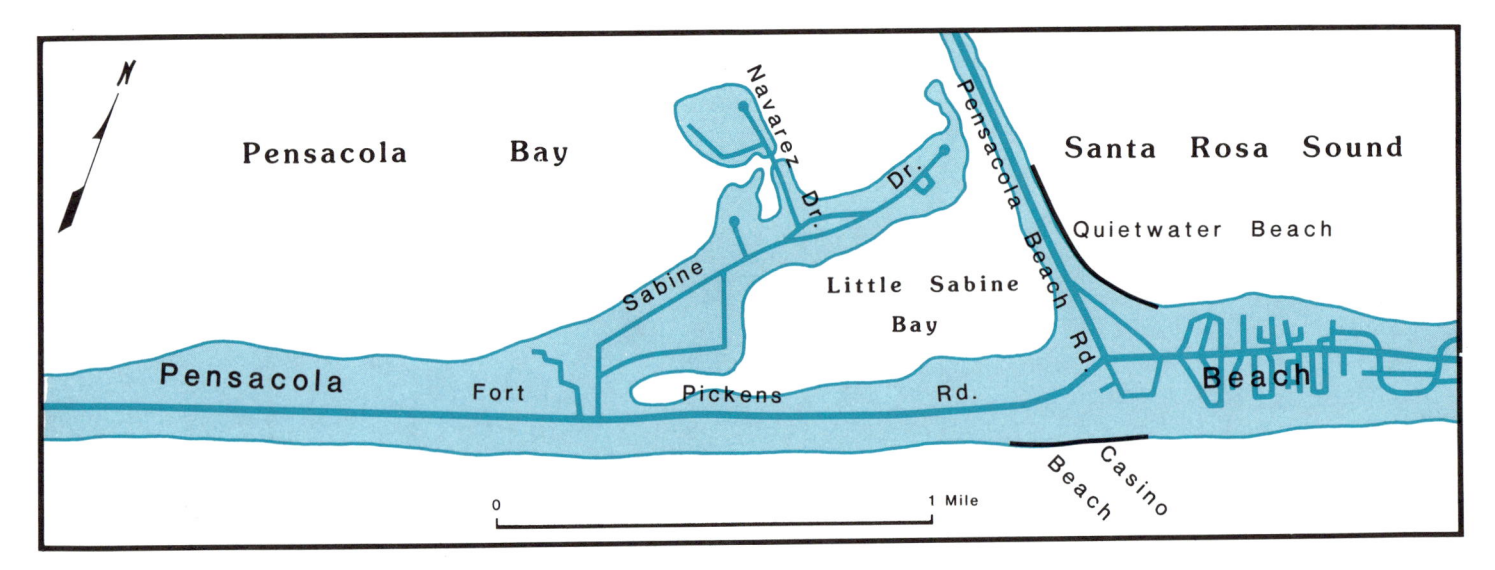

Pensacola Beach West (inset 1)

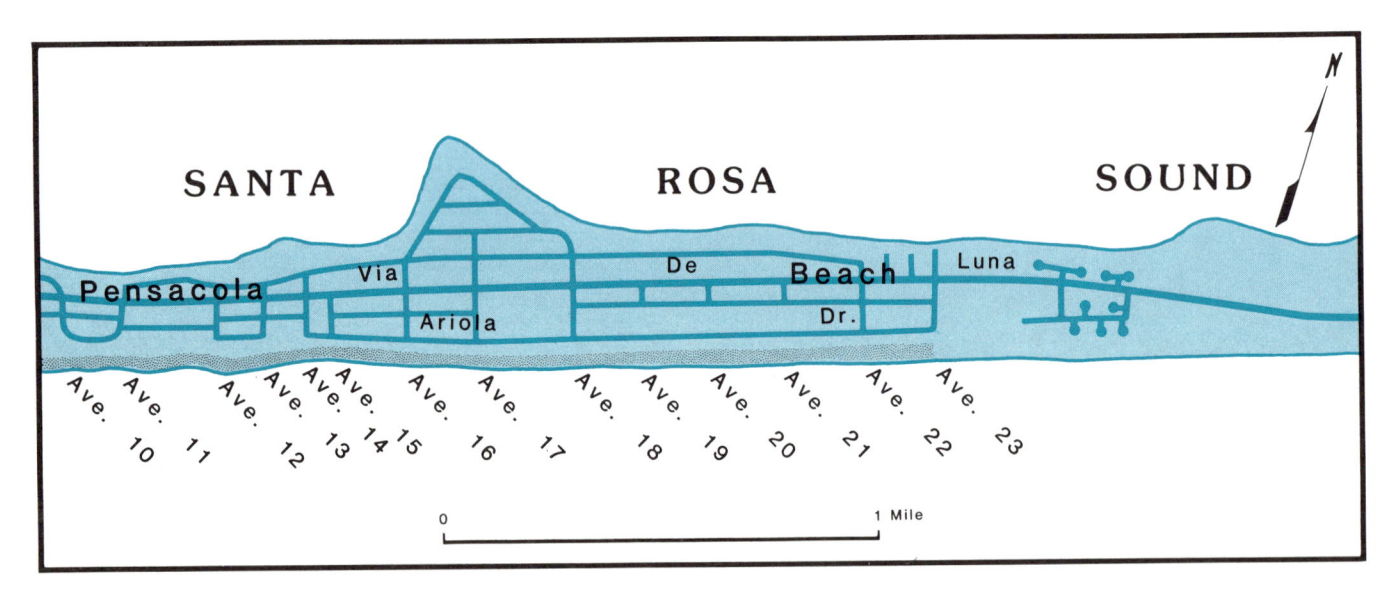

Pensacola Beach East (inset 2)

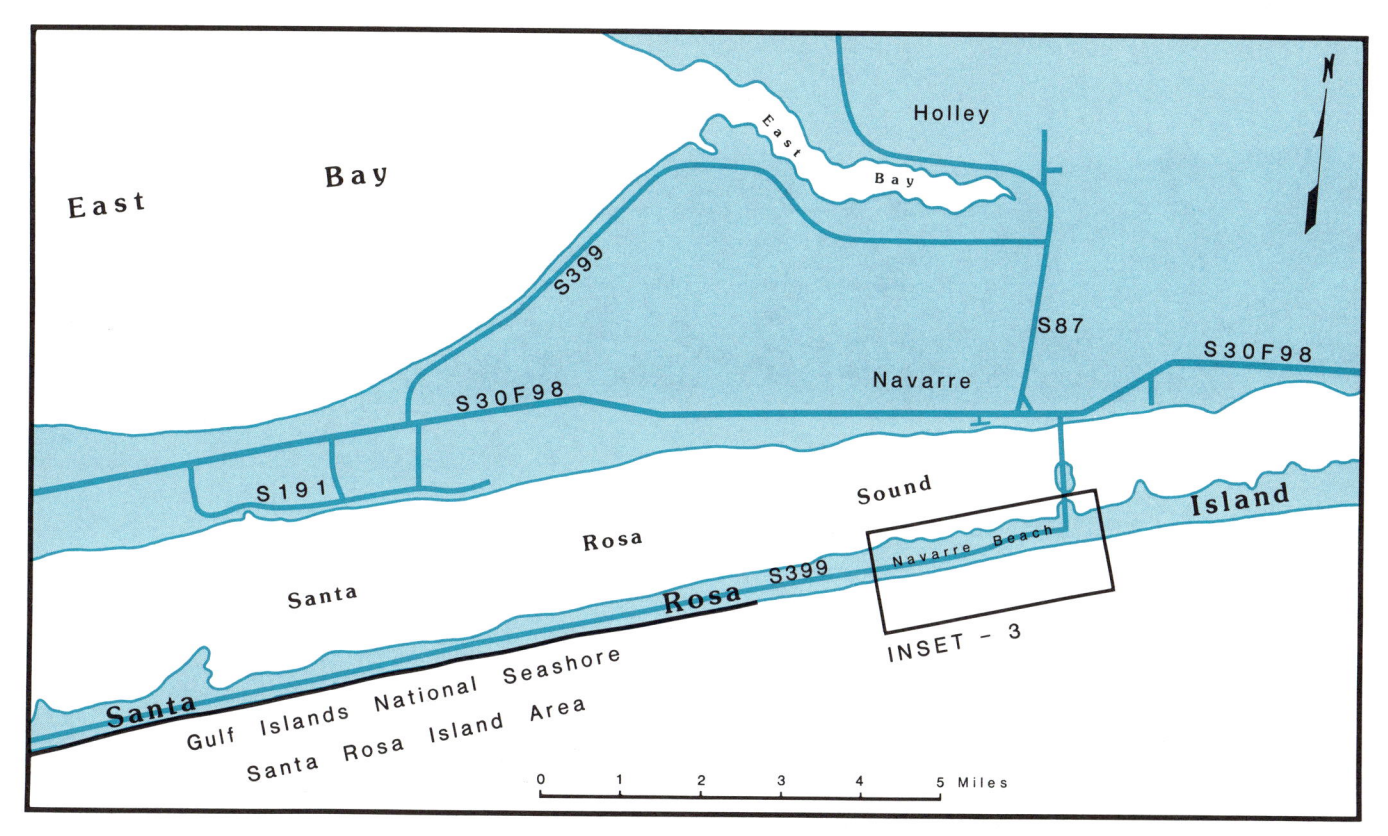

Eastern district

Gulf Islands National Seashore/Santa Rosa Recreational Facility. 106 acres of developed recreation beach on both Santa Rosa Sound and the Gulf of Mexico, located 10 miles east of Pensacola Beach on S.R. 399.

Navarre Beach public access: S.R. 399 West at Wisconsin St., South Carolina St., Ohio St., New Jersey St., Missouri St., Michigan St., east of Louisiana St., east of Indiana St., and west of California St.

Navarre Beach Fishing Pier: 8525 Gulf Blvd. off S.R. 399.

Shoreline Park, a small, community recreation area on Santa Rosa Sound, lies on both sides of the Navarre Beach Bridge.

Escambia/Santa Rosa Counties

FACILITIES / ENVIRONMENT

NAME	Easy Access	Parking or Entrance Fee	Parking	Restrooms	Showers	Picnicking	Swimming	Lifeguards	Fishing	Boating Facilities	Shelters	Concession Stands	Handicapped Facilities	Public Transportation	Group Facilities	Nature Trails / Fitness	Atlantic / Gulf	Bay / Soundfront	Sandy Beach	Rocky Beach	Primitive Beach	Urban
EASTERN DISTRICT																						
Gulf Islands National Seashore / Santa Rosa Recreational Facility	●		●	●	●	●	●	●	●		●	●	●		●		●	●	●			
Navarre Beach Public Access	●		●			●	●										●		●			●
Navarre Beach Fishing Pier	●	●	●	●			●		●			●					●		●			●
Shoreline Park	●		●	●	●	●			●	●					●	●		●				●

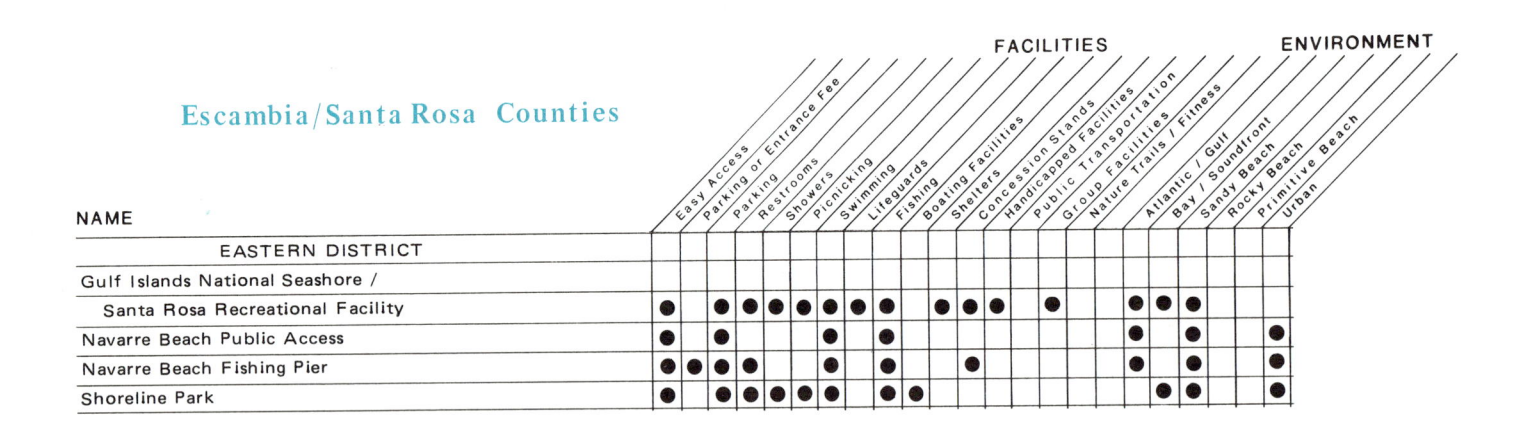

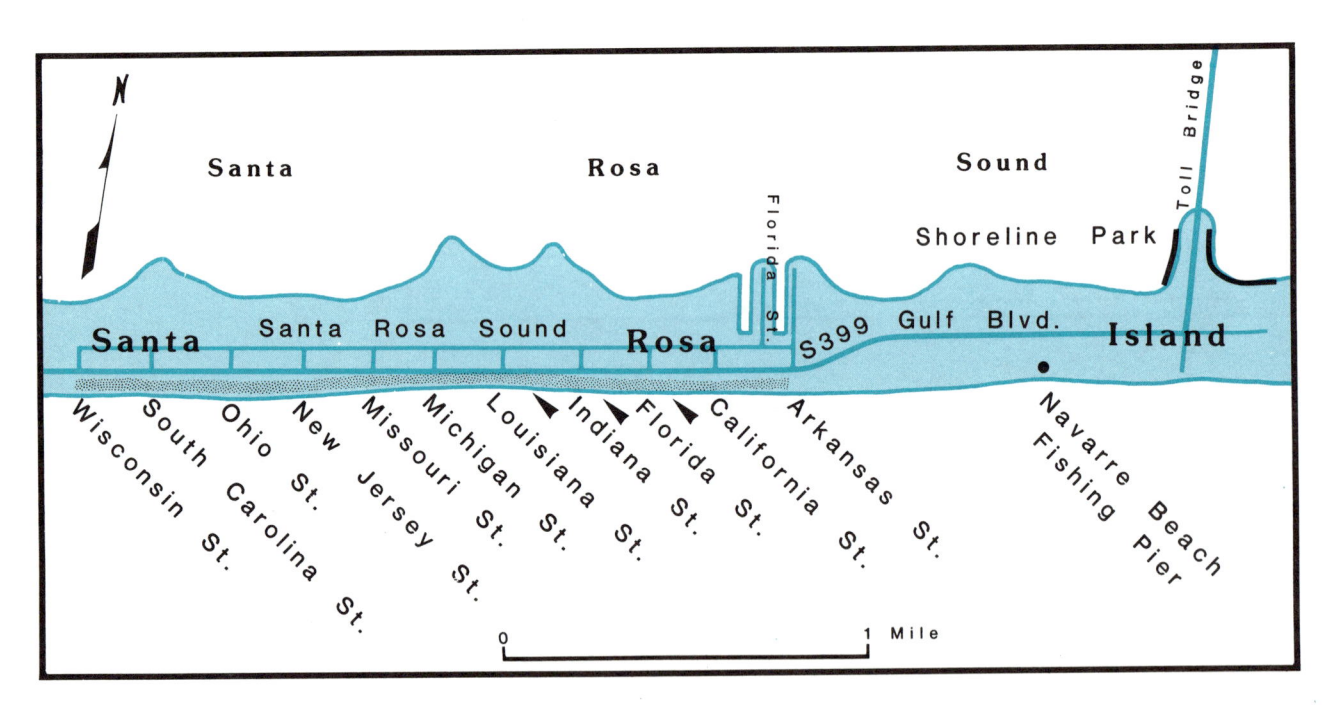

Navarre Beach (inset 3)

Okaloosa County

Okaloosa is derived from the Indian words for "pleasant place." Evidence of pre-Columbian occupation has been found throughout Okaloosa County; mounds, prehistoric encampments, and discarded shell middens reveal much of the Indian cultures that inhabited the area.

The Fort Walton Temple Mound and Museum show the daily life of the early inhabitants of the Gulf coast. Designated by the Department of the Interior as a national historic landmark, Temple Mound has been restored to its original structure of 500,000 baskets of dirt. The mound and museum, located in Fort Walton Beach on U.S. 98, are open Tuesday–Saturday 11–4 and Sunday 1–4.

The first beach acquisition under the governor's Save Our Coast program, designed to ensure adequate public beaches for the future, was the Burney Henderson Beach State Recreation Area. Highway 98 runs through the 1.2-mile stretch of vacant beach which exhibits white, sugary sand, pale green water, and dunes characteristic of most Okaloosa beaches.

The Okaloosa area of the Gulf Islands National Seashore is on Santa Rosa Island east of Fort Walton Beach on U.S. 98. The area, formerly known as Fort Walton Beach Park, has a picnic area and a boat launch to Choctawhatchee Bay.

Both the Gulf of Mexico and Choctawhatchee Bay waters are well known for hobie cat racing and wind surfing. Hobie Nationals and (in 1984) the Hobie World Championship Regatta took place on demanding courses, providing a colorful panorama of multihued sails.

A fishing pier extends into the Gulf of Mexico from which many of the well-known warm-water fish may be caught. There is an entrance fee.

Destin has a well-protected harbor, home to many commercial and recreational fishing boats. Named the "Luckiest Fishing Village in the World," it is surrounded by white beaches and East Pass and access to Choctawhatchee Bay and to the Intracoastal Waterway.

Some of the state's largest and prettiest dunes are found on Eglin Air Force Base property between Fort Walton Beach and Destin.

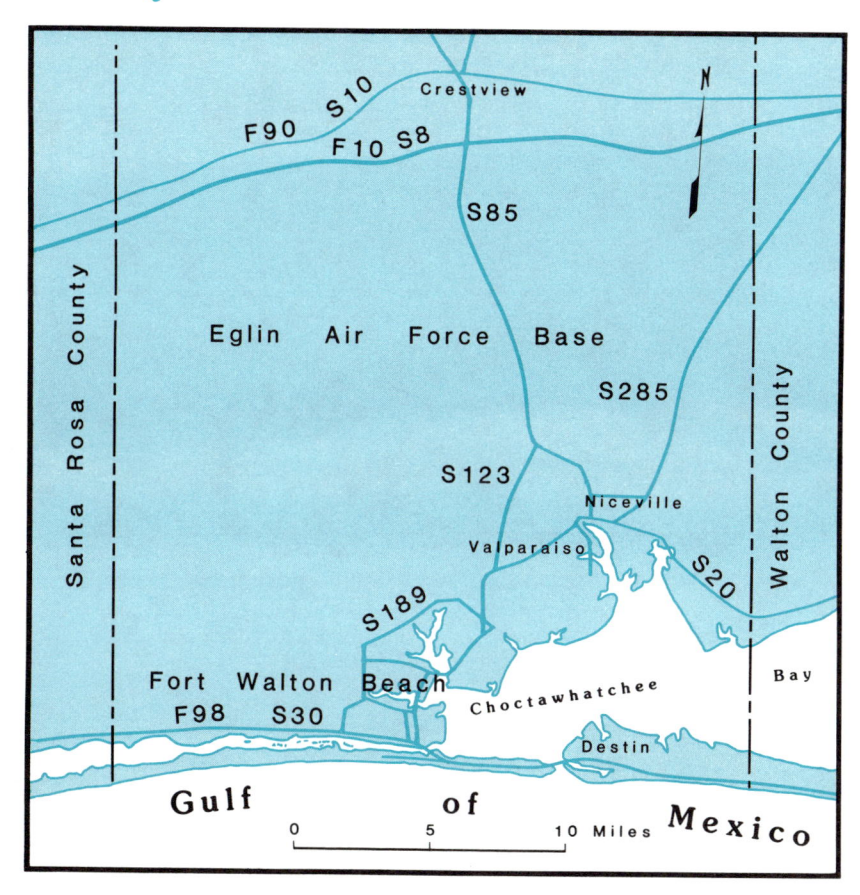

East Pass/Destin area

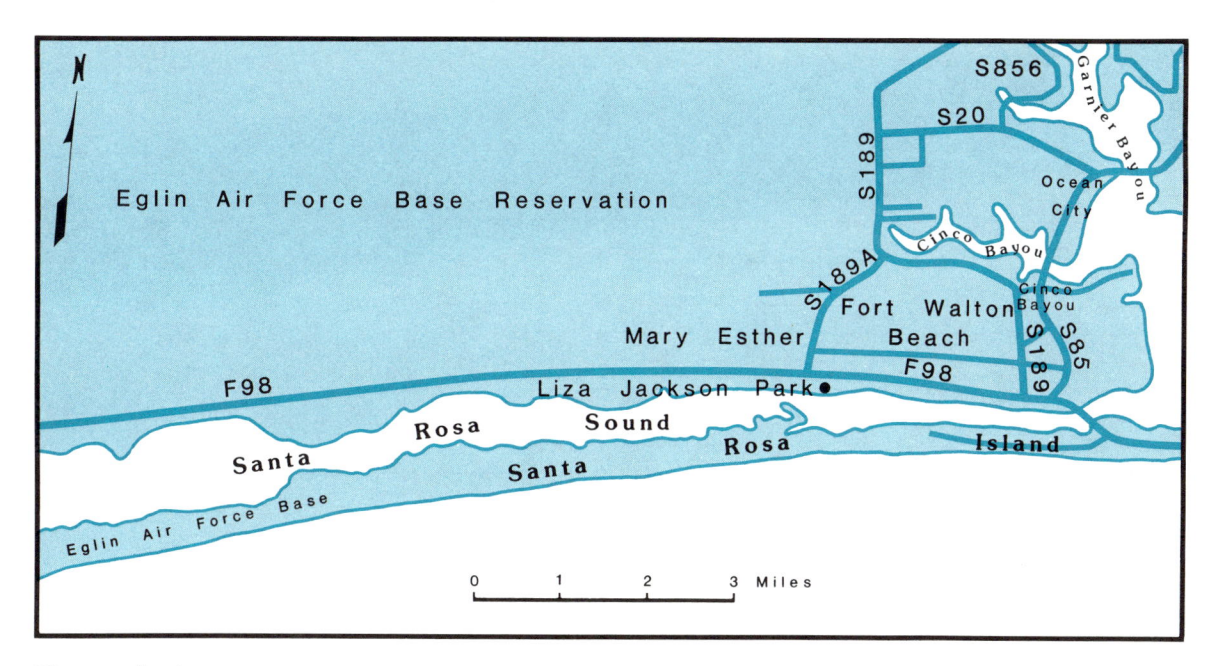

Western district

Regatta

Okaloosa County

NAME	Easy Access	Parking or Entrance Fee	Parking	Restrooms	Showers	Picnicking	Swimming	Lifeguards	Fishing	Boating Facilities	Shelters	Concession Stands	Handicapped Facilities	Public Transportation	Group Facilities	Nature Trails / Fitness	Atlantic / Gulf	Bay / Soundfront	Sandy Beach	Rocky Beach	Primitive Beach	Urban
WESTERN DISTRICT																						
Liza Jackson Park		●	●	●			●	●	●									●	●			
EASTERN DISTRICT																						
Garnier Beach	●		●	●	●	●	●	●	●	●								●	●			●
Santa Rosa Island/Okaloosa County Beach Access	●		●				●	●	●								●	●	●			●
Ross Marler Park	●		●	●	●	●	●		●	●	●		●					●	●			
Newman Brackin Wayside Park/ Okaloosa Island Pier	●		●	●	●	●	●			●		●		●			●		●			
John C. Beasley Park	●		●	●	●	●	●	●		●							●		●			
Gulf Islands National Seashore/ Okaloosa Area	●		●	●		●			●	●							●		●			
Burney Henderson Beach State Recreation Area	●					●	●										●		●	●		
Silver Beach Wayside Park	●		●	●		●	●	●	●		●		●		●		●		●			

Liza Jackson Park lies a half mile east of the Mary Esther cutoff (S.R. 189A) on U.S. 98. This 13-acre park fronts Santa Rosa Sound; it is open 7 A.M.–10 P.M. No alcohol allowed.

Garnier Beach, a small bayfront beach located off Beach View Dr. N.E. in the City of Fort Walton Beach.

Santa Rosa Island/County beach access (off Santa Rosa Blvd. west of U.S. 98): Seventh Beach Freeway (access to the Gulf of Mexico only); Sixth, Fifth, and Third Beach Freeways (boardwalk to Gulf and sound); Second and First Beach Freeways (access to sound and Gulf).

Ross Marler Park (formerly Jaycee Park), located on Santa Rosa Island off U.S. 98, about 1,000 feet east of Santa Rosa Sound. This developed community park has 55 acres on the Intracoastal Waterway facing Choctawhatchee Bay.

Newman Brackin Wayside Park/Okaloosa Island Pier, 40 acres with recreation facilities. U.S. 98 east on Santa Rosa Island.

John C. Beasley Park, a 22-acre community park off U.S. 98 east on Santa Rosa Island fronting the Gulf of Mexico.

Gulf Islands National Seashore/Okaloosa Area (formerly Fort Walton Beach Park), located on Santa Rosa Island east of Fort Walton Beach on U.S. 98. Small craft launch to Choctawhatchee Bay.

Burney Henderson Beach State Recreation Area, 208 acres recently purchased as "Save Our Coast" land acquisition. Still under development in 1985, this natural area lies 2 miles east of Destin on U.S. 98.

Silver Beach Wayside Park: U.S. 98 east, about one mile from Walton County.

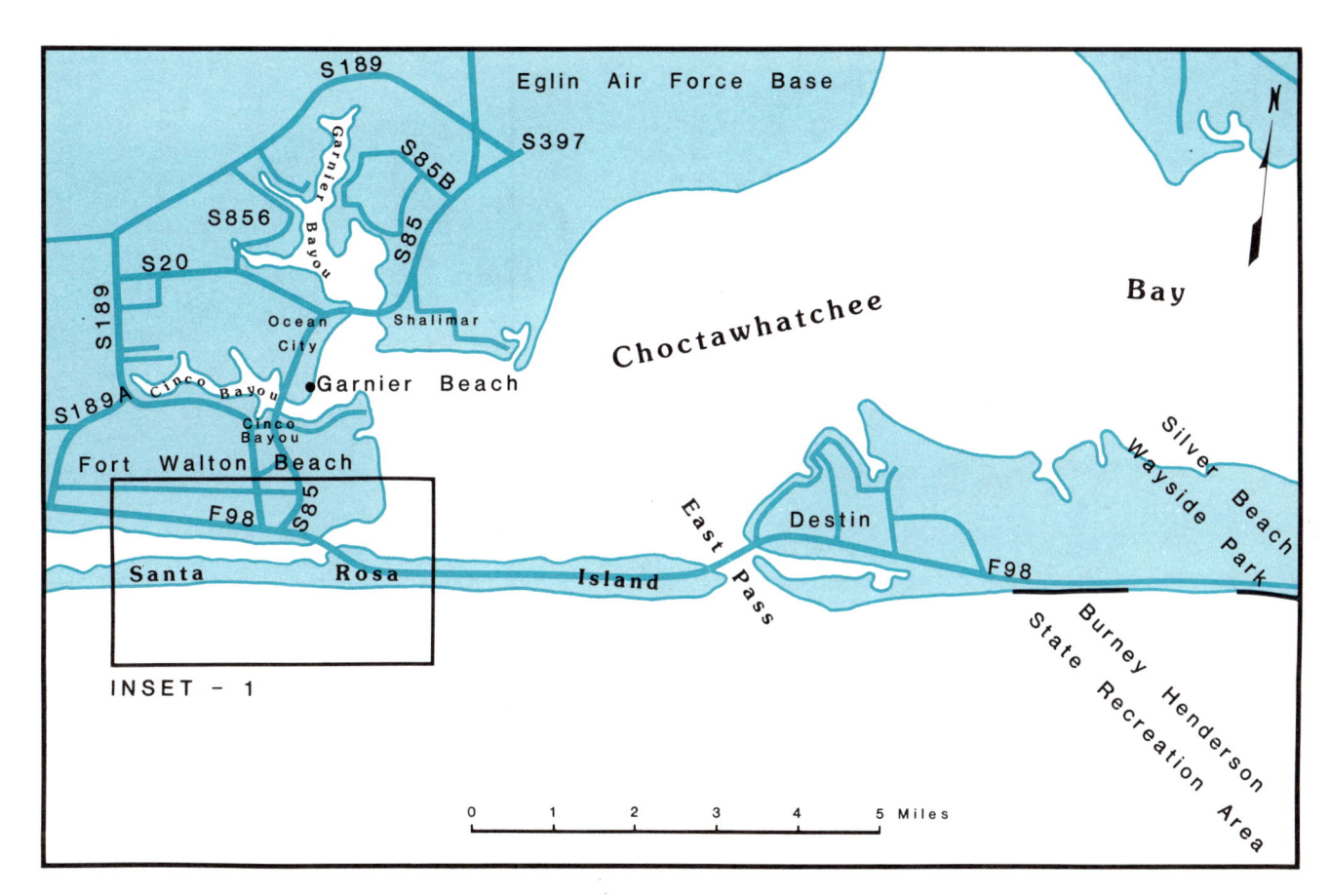

Eastern district

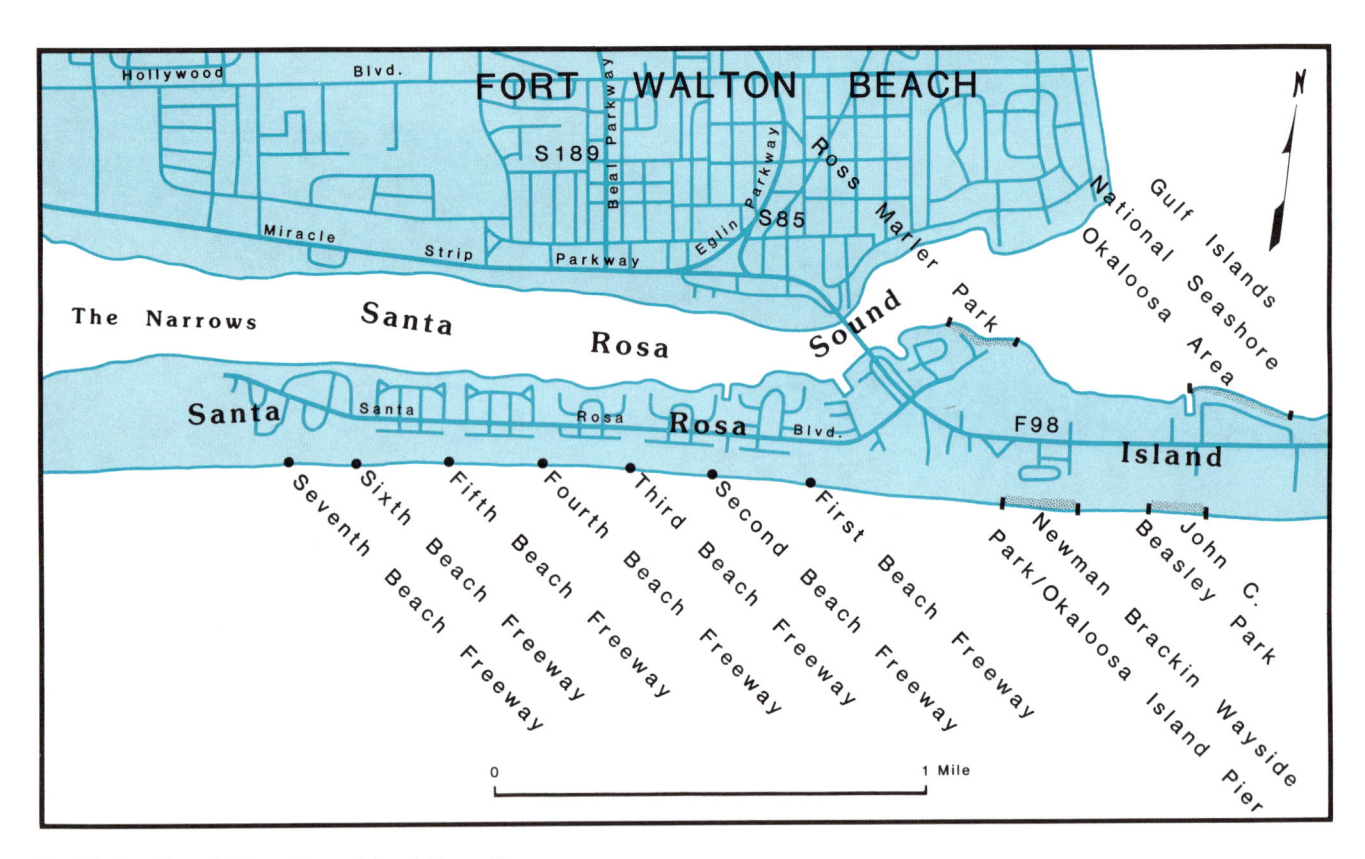

Ft. Walton Beach/Santa Rosa Island (inset 1)

Florida's Saltwater Laws

Since local regulations governing the taking of saltwater products may exist, contact the Marine Patrol District Office nearest the location where these activities will take place.

No license is required for saltwater sport fishing; however, any person who wishes to sell or barter any saltwater fish or other saltwater product must first obtain an appropriate license. Contact the nearest Marine Patrol District Office for further information concerning licensing.

Minimum Legal Lengths

All measurements are from the tip of nose to the rear center edge of tail.

Blue fish	10 inches
Pompano	9.5 inches
Flounder	11 inches
Mackerel	12 inches
Black mullet	11 inches
(West of Aucilla River)	9 inches
(Aucilla River to Citrus-Hernando county line)	10 inches
*Trout, spotted sea (weakfish)	12 inches
(no size limit in Gulf, Wakulla, or Franklin counties)	
*Red fish	12 inches
Snook	18 inches
(cannot possess more than two)	
Bonefish	15 inches
(cannot possess more than two)	
*Grouper	12 inches
(including red grouper, jewfish, Nassau grouper, black grouper, and gag)	
Striped bass	15 inches

*Possible increases in size limit in 1985–86. Contact the nearest Marine Patrol District Office for information.

Closed Seasons

Crawfish	April 1–July 25
Oysters	June 1–September 1
Stone crab claws	May 15–October 15
Snook	Jan.–Feb., June–July
Shad (sport)	No closed season
Shad (commercial)	Check nearest FMP office

Sports Fishermen's Crawfish Season

July 20 and 21 (2 days)

May not possess more than 6 crawfish on July 20 or more than 12 crawfish cumulatively for July 20 and 21.

Limits

Snook. Only 2 in possession. Unlawful to take by means other than pole and line. Can't buy or sell. Snatching prohibited.

Sailfish. Only 2 in possession; can't buy or sell.

Tarpon. Only 2 in possession; can't buy or sell.

Shad. Ten a day by hook and line.

Queen conch. Ten per day; only 20 in possession.

Striped bass. Only 2 in possession. Can't buy or sell except when raised artificially under permit.

Permit fish (of over 20 inches). Only 2 in possession. Illegal to sell, buy, or harvest by net; 20 inches or less not protected.

Stone crab claws. Forearm measurement must equal or exceed 2.75 inches. No trapping except under permit from the Department of Natural Resources. Legal claw or claws may be taken, but live crab must be released. Cannot possess intact stone crab.

Oysters. Three inches long.

Crawfish

Carapace must have a measurement of more than 3 inches or a tail measurement of 5.5 inches. Crawfish must remain in a whole condition at all times while being transferred on or below the waters of the state. The practice of wringing or separating the tail (segmented portion) from the body (carapace or head section) shall be prohibited on the waters of this state except by special permit issued by the Division of Law Enforcement. Any tail so separated under the provisions of a special permit shall measure not less than 5.5 inches measured lengthwise from the point of the separation along the center of the entire tail in a flat straight position with tip of the tail closed. No egg-bearing females. No spearing. It is unlawful to possess, have on board, or remove from the waters of the state, within any 24-hour period, more than 24 crawfish without first obtaining a crawfish license. No trapping except under license from the Department of Natural Resources. Traps may be worked during daylight hours only.

Blue Crabs

Illegal to use more than five (5) traps for the taking of blue crabs without a permit from the Department of Natural Resources. Egg-bearing females cannot be sold. Traps may be worked during daylight hours only. Possession for sale of blue crabs less than 5 inches from point to point across the carapace is prohibited. Unlawful to willfully molest any crawfish, stone crab, or blue crab trap, line, or buoy. Punishable as third-degree felony.

Manatee (Sea Cow)

No persons, firm, or corporation shall kill, injure, annoy, molest, or torture a manatee or sea cow. Boat speed is regulated in certain waters between November 15 and March 31 where the manatee is known to congregate. Manatees are endangered. Please take every precaution to avoid collision or contact with these large and ponderous mammals. Report any harassment observed to the nearest Marine Patrol District Office.

Porpoise

Unlawful to take or kill porpoises except under federal permit, or to molest, injure, or annoy them. All marine mammals or parts thereof are protected by federal law. If any marine mammal is found injured, beached, or dead, report it to the District Florida Marine Patrol Office.

Manta Ray

It is unlawful for any person, firm, or corporation intentionally to destroy a manta ray.

Marine Turtles

No person, firm, or corporation shall take, kill, disturb, mutilate, molest, harass, or destroy any marine turtle. Marine turtles accidentally caught will be returned to the water alive immediately.

No person may take, possess, disturb, mutilate, destroy, cause to be destroyed, sell, offer for sale, transfer, molest, or harass any marine turtle nest or eggs at any time.

Spearfishing

Illegal to possess in the water any spear, gig, or lance by a person swimming at or below the surface of the water in a prohibited area.

Illegal to spearfish in Pennekamp Coral Reef State Park, Collier County, that part of Monroe County from Long Key north to the Dade County line and the immediate area of the following: (1) all public bathing beaches; (2) commercial or public fishing piers; (3) bridge catwalks; (4) jetties.

Also illegal to spearfish in fresh water or for freshwater fish in brackish water except for rough fish in special areas designated by the Game and Fresh Water Fish Commission.

Coral

Unlawful to take, possess, or destroy sea fans, hard corals, or fire corals unless it can be shown by certified invoice that it was imported from a foreign country. Coral may not be taken or possessed in John Pennekamp Coral Reef State Park.

Drugs and Poisons

Illegal to place drugs or poisons in the marine waters unless a permit for such use has been obtained from the Department of Natural Resources.

Pelicans and Other Seabirds

The brown pelican, an endangered species, is one of the most common victims of fishhook and line injuries. To prevent such injuries:

> Always look around before casting to see whether a pelican or other seabird is flying above.
> Never leave fishing tackle out in the open, unattended—especially when it is baited.
> Do not leave hooks hanging from the end of an exposed fishing rod.
> Take care not to foul fishhooks or lines around piers or other structures. Seabirds frequently get caught in dangling fishlines.
> Do not throw waste fishline or other discards into the water.

Diver's Down Flag

All persons diving must display a "Diver's Down" flag. The flag is red with a white diagonal stripe. In addition to this flag, a rigid replica of the international code "alpha" flag must be displayed from a boat if anyone is diving from it. Use *extreme* caution when operating a boat in the area where these flags are displayed.

Explosives

The use of explosives or the discharge of firearms into the water for the purpose of killing food fish is prohibited. The landing ashore or possession on the water by any person of any food fish that has been damaged by explosives or the landing of headless jewfish or grouper if the grouper is taken for commercial use is prohibited.

Division of Law Enforcement, Florida Marine Patrol, Headquarters, 3900 Commonwealth Boulevard, Tallahassee, FL 32303

District Offices

Panama City, St. Andrew Marina (904) 763-3080
Carrabelle, South Marine Street (904) 697-3741
Homosassa Springs, U.S. Hwy. 19 (904) 628-6196
Tampa, 5110 Gandy Blvd. (813) 272-2516
Fort Myers, 1820 Jackson Street (813) 334-8963
Miami, 1275 N. E. 79th Street (305) 325-3346
Titusville, 402 Causeway (305) 267-4021
Jacksonville Beach, 2510 2d Ave. N. (904) 241-7107
Marathon, 2835 Overseas Highway (305) 743-6542
Jupiter, 19100 S.E. Federal Hwy. (305) 747-2033
Pensacola, 1101 East Gregory St. (904) 438-4903

From regulations, October 1983, Fla. Department of Natural Resources.

Fishing pier at sunset

Walton County

Located in the eastern part of Florida's famed Panhandle "Miracle Strip," Walton County welcomes visitors to 25 miles of pure, white, unspoiled sand beaches on the Gulf of Mexico. These sands, like those of the entire "Miracle Strip" from Panama City to Pensacola, are known worldwide for their beauty and unusual character. Most sands are composed of varying amounts of quartz, shell fragments, broken coral, and dark-colored minerals. Those of the "Miracle Strip" are almost pure quartz, with the result of sand that is both extremely fine-textured and glistening white in color. At times under Florida's bright sun, these broad beaches take on the appearance of snowfields. Walton County is also distinguished by impressive dunes, the largest in the state. Many additional miles of shoreline for recreational pleasure adjoin the sparkling waters of Choctawhatchee Bay.

For the visitor the open expanse of the Gulf of Mexico provides swimming, snorkeling, and deep-sea fishing for red snapper, grouper, king mackerel, and marlin. Shoreline surf fishing for pompano, flounder, and whiting is productive. Gathering crabs and other shellfish along the shore is also popular, as are sunbathing and carefree beachcombing.

Sightseers find several fascinating county attractions. DeFuniak Springs, the county seat, boasts the first memorial erected in the South in honor of Confederate dead. It was first raised in 1871 by the Walton County Female Memorial Association. Later it was moved to the courthouse square at DeFuniak Springs when that town became the county seat. Also in DeFuniak Springs is the little Walton-DeFuniak Public Library, believed to be the oldest public library building still operating in the state. Ponce de Leon Springs Recreation Area is a 370-acre park around two springs that flow at a rate of 14 million gallons a day and maintain a temperature of 68 degrees year-round. Along the Gulf, Grayton Beach State Recreation Area has been developed for camping, fishing, boating, and diving; nearby Eden Gardens has 11 acres of landscaped grounds encompassing a restored mansion with guided tours and picnicking and fishing facilities.

Today the county is developing its beaches for both visitors and new residents. Other activities in the county include a large poultry-raising and -processing industry; commercial farming of such crops as wheat, corn, and soybeans; and tree cutting from the county's extensive forests for forest products industries.

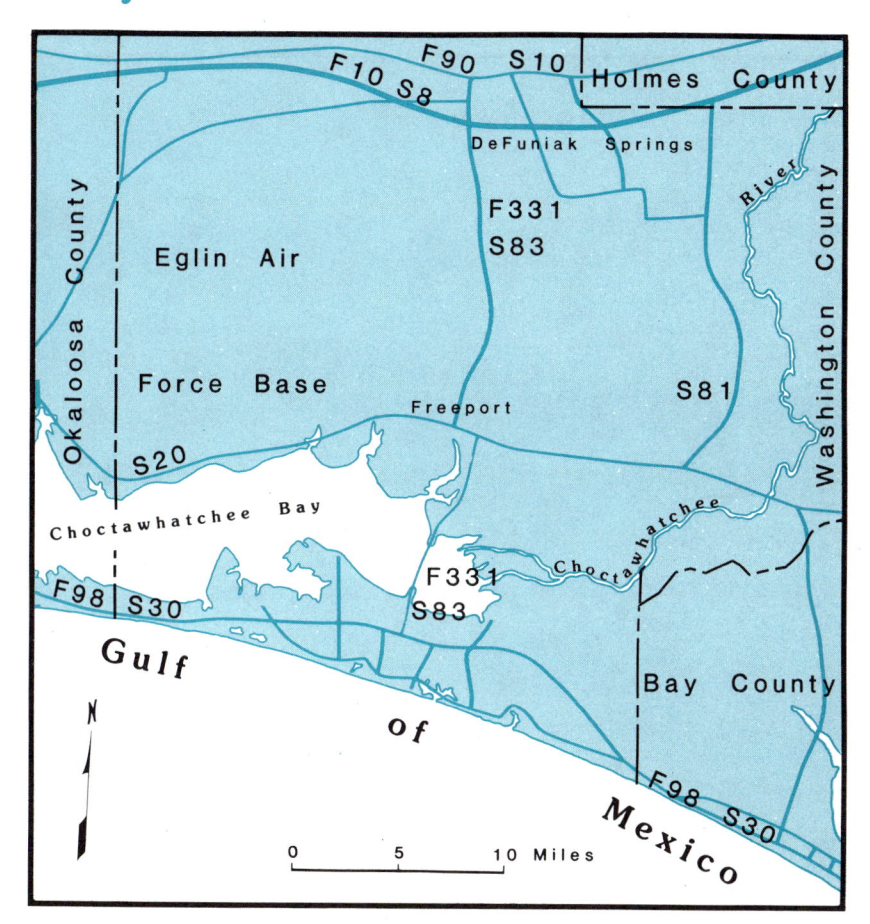

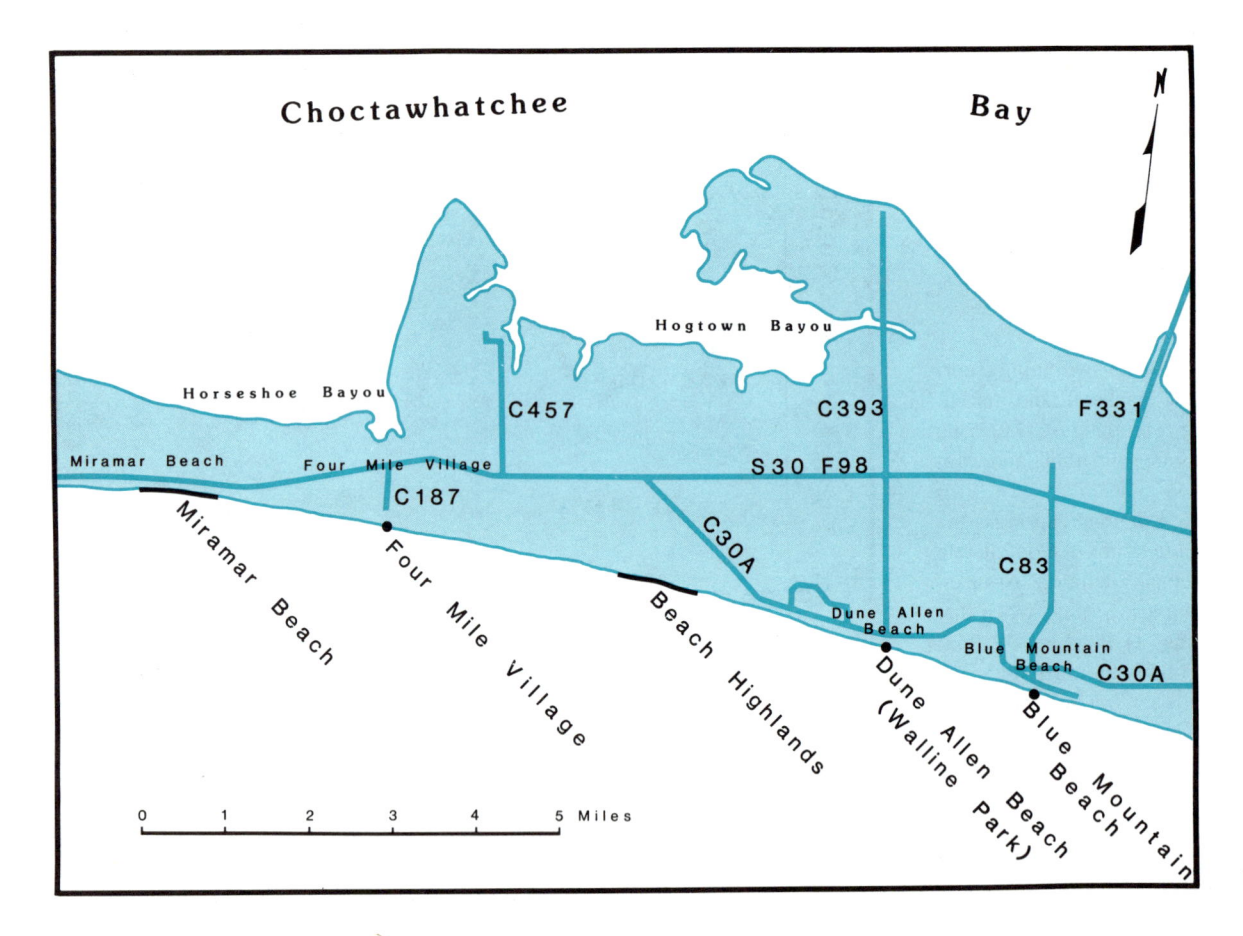

Western district

Miramar Beach has 5,500 feet of beach on the Gulf side of S.R. 30 about 2 miles east of Okaloosa County.

Four Mile Village lies 2.5 miles east of Miramar Beach at the end of S.R. 187 off U.S. 98/S.R. 30.

Beach Highlands is 6,000 feet of undeveloped beach off S.R. 30A about 2 miles south of S.R. 30.

Dune Allen Beach (Walline Park), near the intersection of state roads 393 and 30A, has 200 feet of undeveloped beach.

Blue Mountain Beach is 200 feet of undeveloped beach at the intersection of state roads 83 and 30A.

Walton County

NAME	Easy Access	Parking or Entrance Fee	Parking	Restrooms	Showers	Picnicking	Swimming	Lifeguards	Fishing	Boating Facilities	Shelters	Concession Stands	Handicapped Facilities	Public Transportation	Group Facilities	Nature Trails / Fitness	Atlantic / Gulf	Bay / Soundfront	Sandy Beach	Rocky Beach	Primitive Beach	Urban
WESTERN DISTRICT																						
Miramar Beach	●		●			●	●										●		●		●	
Four Mile Village			●			●	●										●		●		●	
Beach Highlands			●			●	●										●		●			
Dune Allen Beach	●		●			●	●	●									●		●			
Blue Mountain Beach	●		●			●	●	●									●		●			
EASTERN DISTRICT																						
Grayton Beach State Recreation Area	●	●	●	●	●	●	●	●	●	●	●	●		●	●		●		●			
Seagrove Beach	●					●	●	●	●								●		●	●		
Phillips Inlet Area	●		●				●	●									●		●			

Wind-sculpted dunes

Sea oats on the dunes

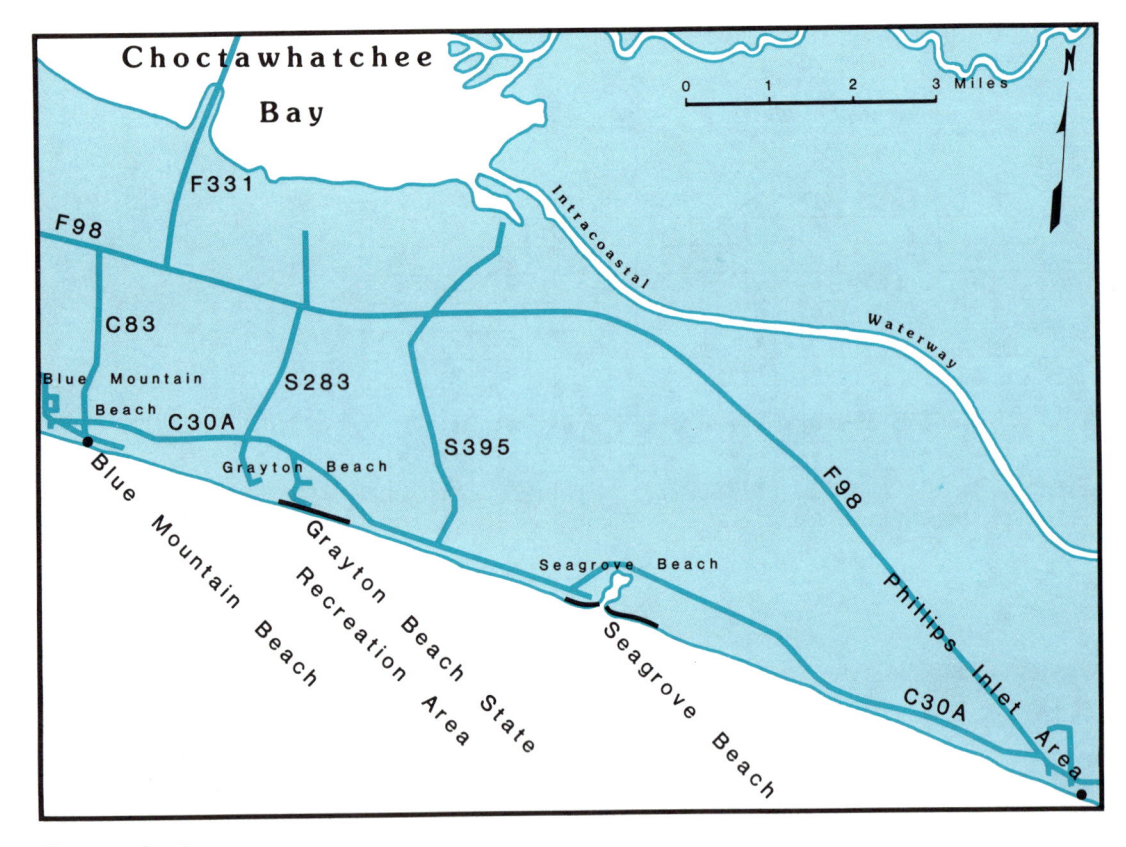

Eastern district

Grayton Beach State Recreation Area is 356 acres of pine woodlands, freshwater lakes, and white beaches situated between Western Lake and the Gulf of Mexico and reached at the southern end of S.R. 283 on S.R. 30A.

Seagrove Beach is 1.5 miles fronting the Gulf, reached by turning east from S.R. 395 onto S.R. 30A to a point near Deer Lake.

Phillips Inlet Area is 200 feet of undeveloped beach located 4,500 feet west of the Bay County line on S.R. 30.

Bay County

Located on the Gulf of Mexico at the middle of the Florida Panhandle, Bay County is noted for snowy sands and abundant fishing. Panama City's annual Miracle Strip Marlin Tournament opens the first blue marlin catch of the season and runs through November. Bay County was established in 1913 and took its name from St. Andrew's Bay, which it fronts.

Panama City, on the famed Miracle Strip, is one of the principal tourist centers in western Florida. Long known for their many vacation-related activities, hotels, motels, and restaurants, the beaches continue to lure visitors seeking the white sand and surf. During the peak tourist season (March–September) the coastal strip can overflow with out-of-state vacationers.

During World War II, the city was a major ship-building and war industry center, as well as a temporary home for thousands of war workers. Today, Tyndall Air Force Base and a naval base are important factors in the economy of Bay County. Panama City was so named by its original developer because it is on a direct line from Chicago to its namesake in the Canal Zone. Local shippers hoped to capitalize on the transshipment of products going and coming through the Panama Canal.

St. Andrews State Recreation Area is on more than 10,000 acres of land between St. Andrew's Bay and the Gulf of Mexico. Located west of Panama City, this is a popular recreation park. Near the park beach is an old "cracker" turpentine still, at one time a common sight in this part of the state. It has been restored for exhibition. The jetties, Gulf channel, bay, and lagoon are famous for fishing. Fishing piers are located on Grand Lagoon and the Gulf beach area.

Gulf beach

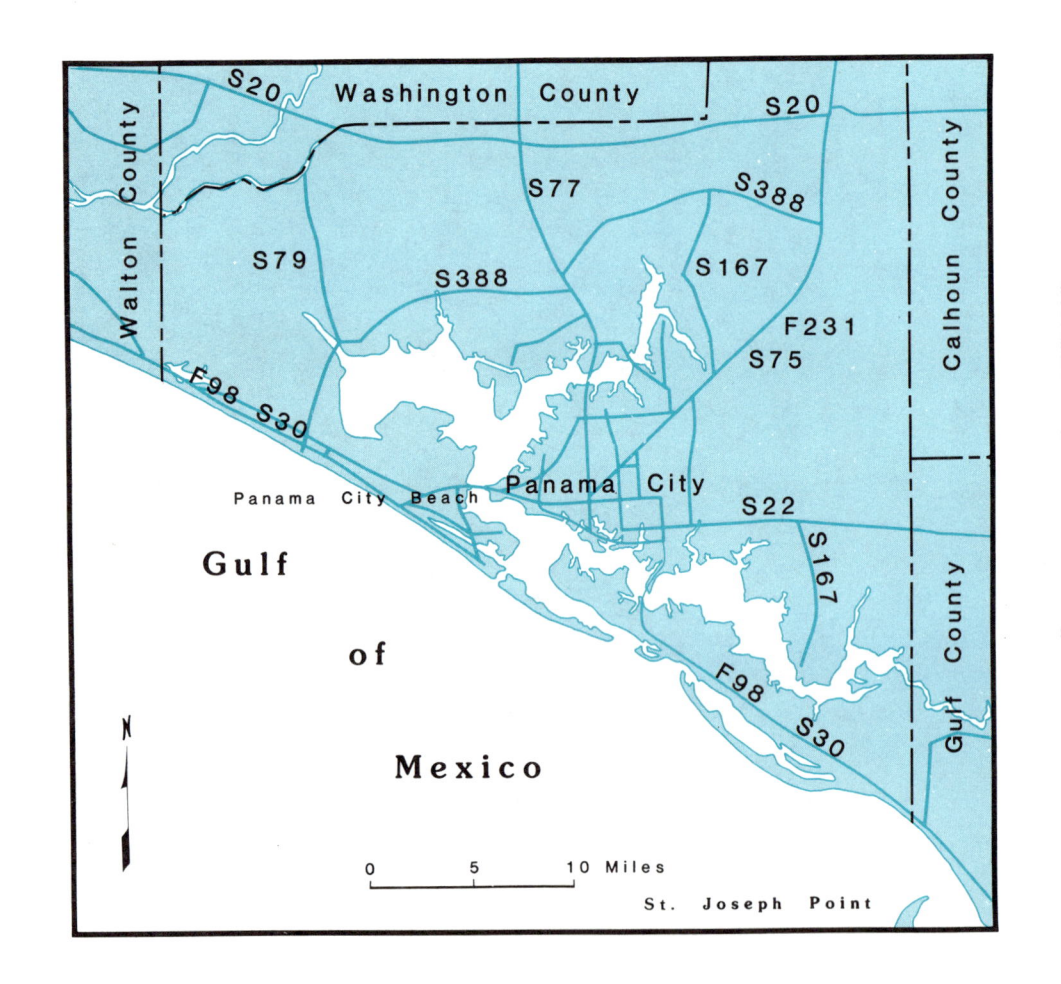

Sea oats

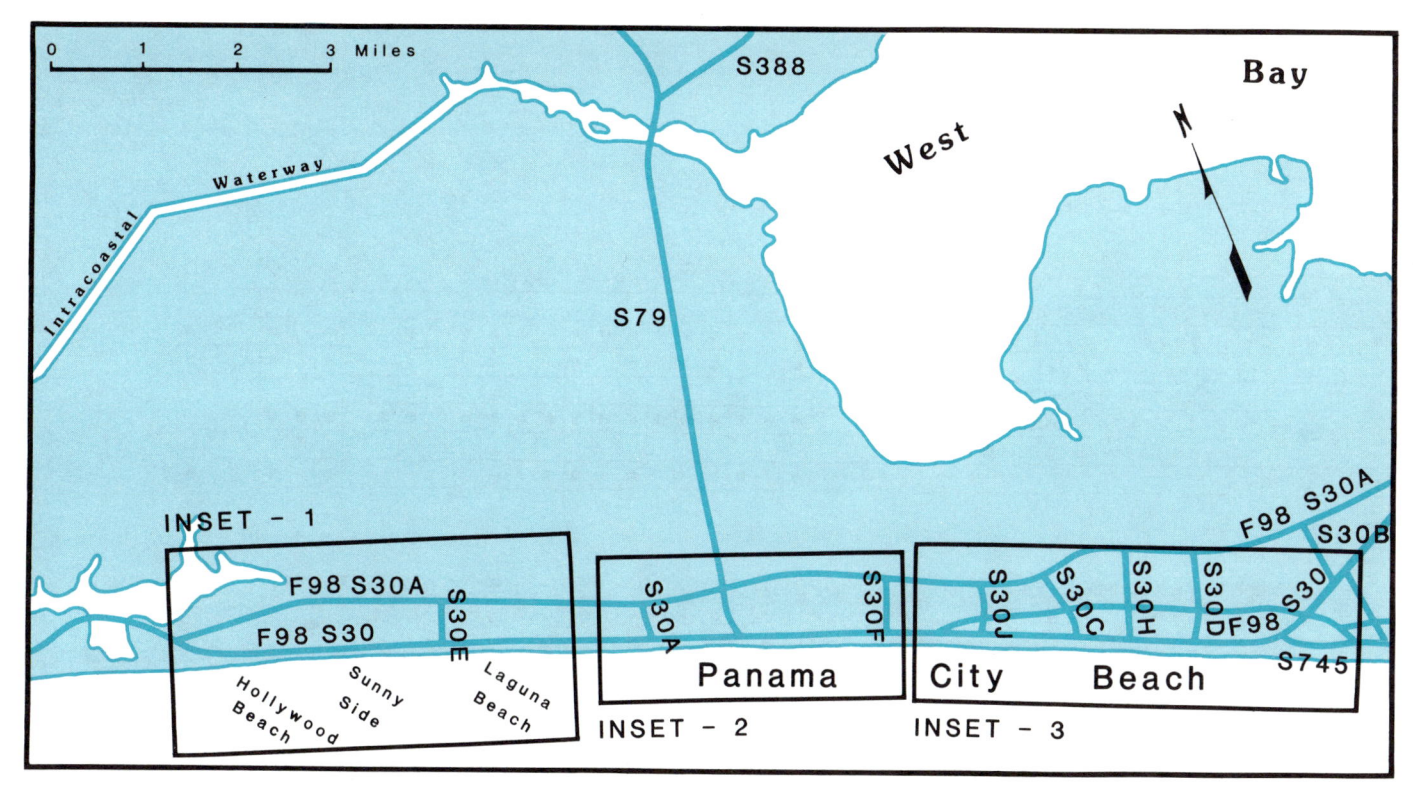

Western district

Dolphin with pup

193

Bay County

NAME	Easy Access	Parking or Entrance Fee	Parking	Restrooms	Showers	Picnicking	Swimming	Lifeguards	Fishing	Boating Facilities	Shelters	Concession Stands	Handicapped Facilities	Public Transportation	Group Facilities	Nature Trails / Fitness	Atlantic / Gulf	Bay / Soundfront	Sandy Beach	Rocky Beach	Primitive Beach	Urban
WESTERN DISTRICT																						
Panama City Beach and Pier	●		●	●	●	●	●	●	●		●	●	●		●		●		●			●
Bay County Pier and Park	●		●	●		●		●		●					●		●		●			●
CENTRAL DISTRICT																						
St. Andrews State Recreation Area	●	●	●	●	●	●	●	●	●	●	●	●			●	●	●	●	●			
EASTERN DISTRICT																						
Mexico Beach	●		●			●		●	●						●		●		●			●

194

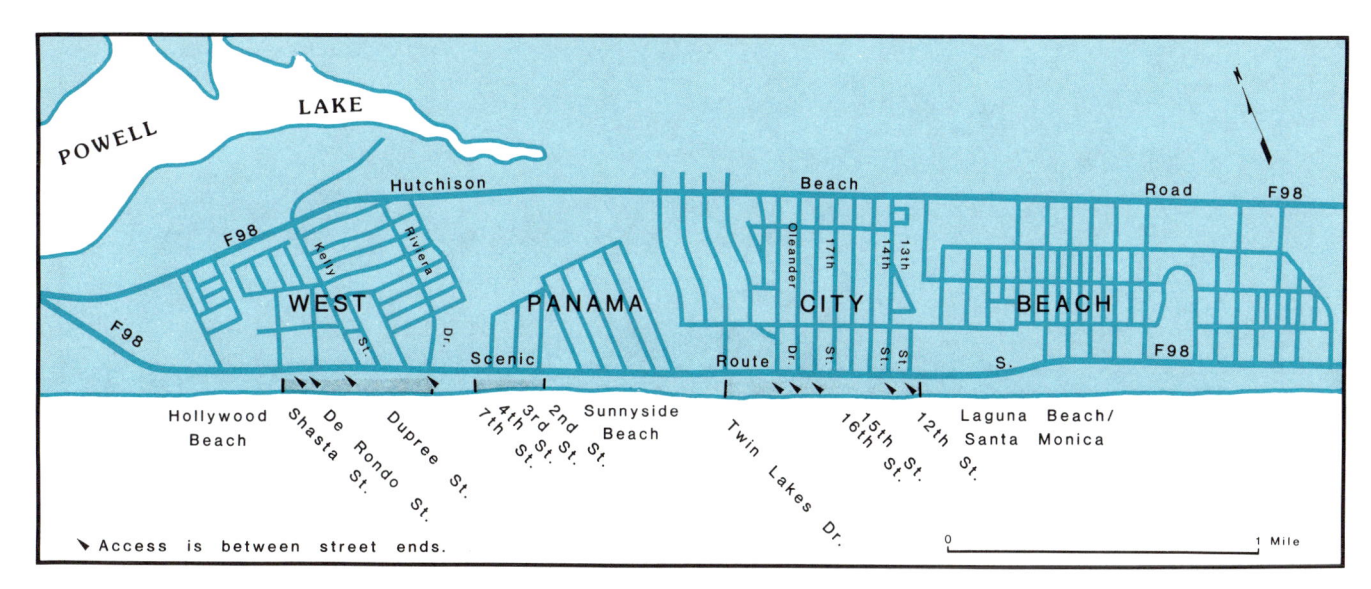

Hollywood Beach/Laguna Beach (inset 1)

Hollywood Beach accesses: Shasta St., east of Shasta St., between Derondo St. and Shasta St. (across from post office), Derondo St., between Kelly St. and Derondo St., Dupree St., and west of Riviera Dr.

Sunnyside Beach accesses: 7th, 4th, 3d, and 2d streets.

Laguna Beach/Santa Monica accesses: Twin Lakes Dr., west of Oleander Dr., between Rose Ln. and Oleander Dr., west of 17th St., 16th St., 15th St., east of 14th St., and 12th St.

Panama City Beach has about 1,400 feet of developed, urban beach facing the Gulf of Mexico about 0.3 mile west of Powell Adams Dr. Walkways and street-end easements to the 9 miles of beach maintained within the corporate city limits.

Bay County Pier and Park lie a half mile west of Beckrich Rd. on Miracle Strip Parkway. This popular area has 160 parking spaces. The pier is open 24 hours and the beach from 8 A.M. to midnight.

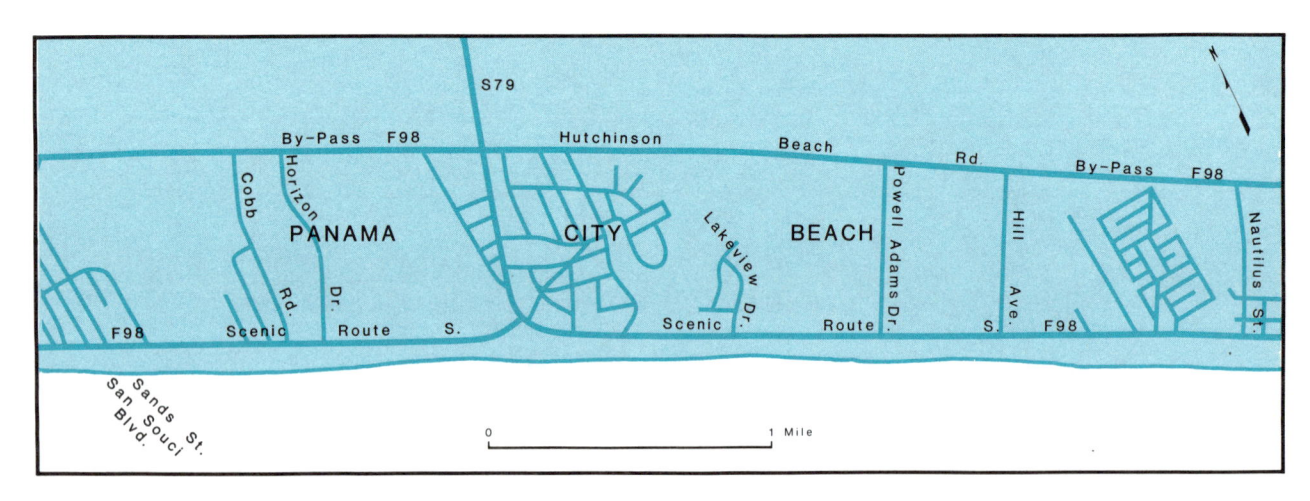

Panama City Beach (inset 2)

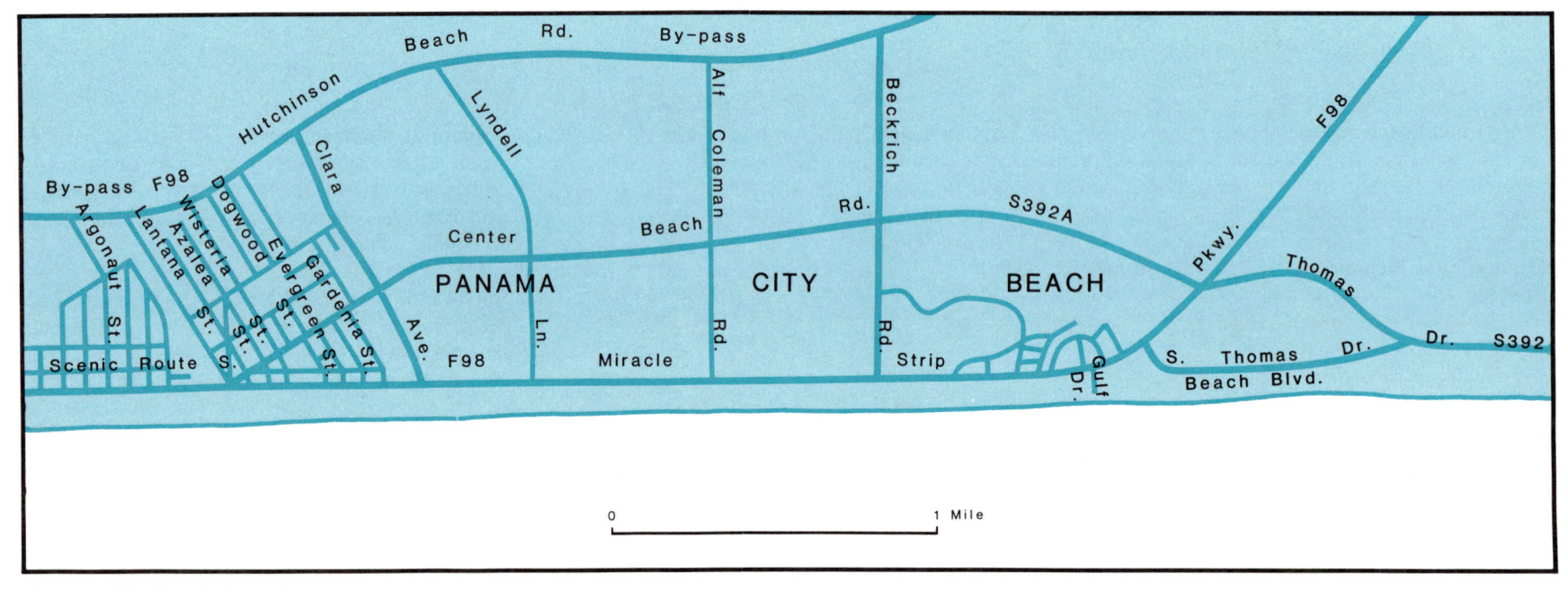

Panama City Beach (inset 3)

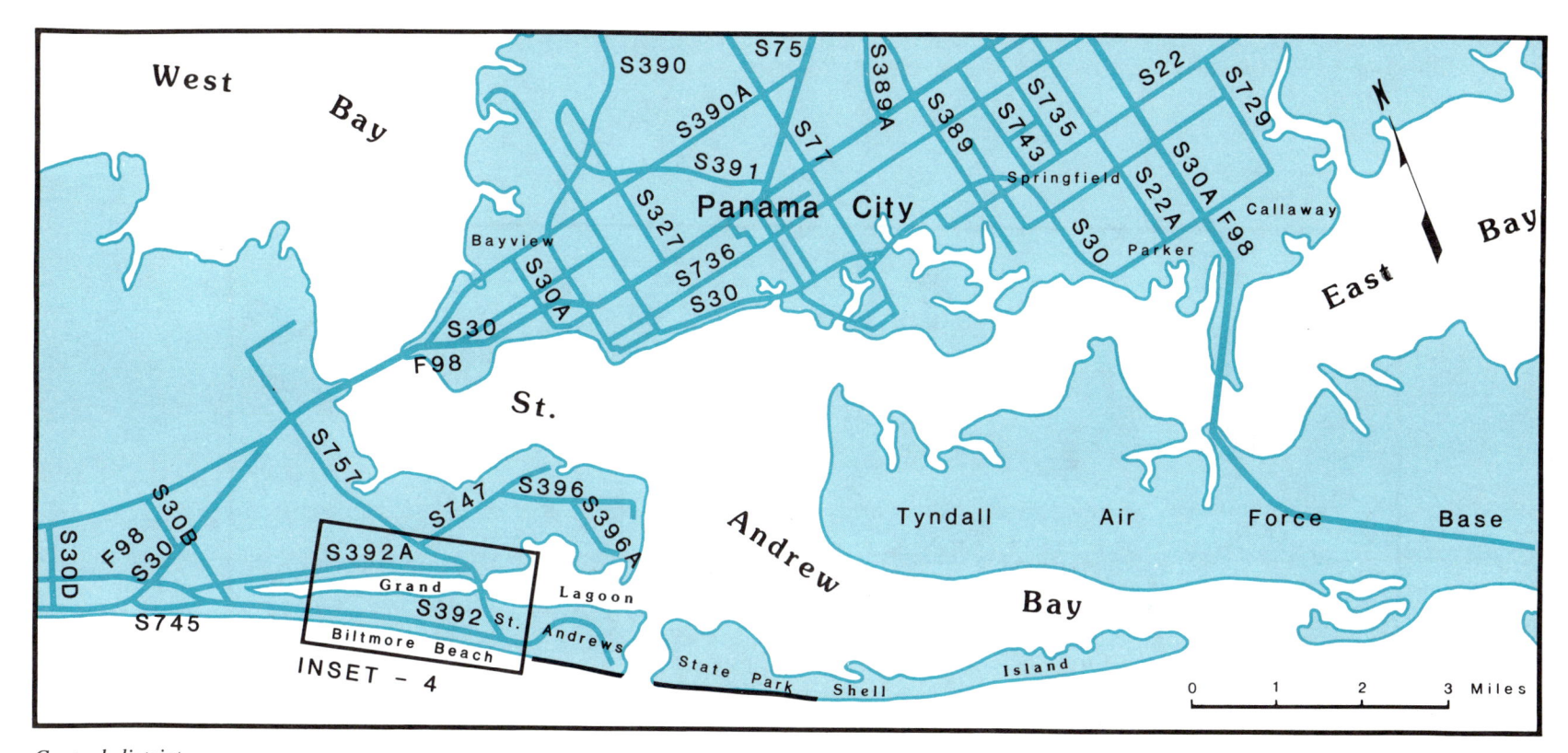

Central district

Surf Drive Area/Belair Beach accesses: west of Bonita St., Tarpon St., Marlin St., Ocean St., between Cobia St. and Ocean St., west of Cobia St., Appalachee St., and Choctaw St.

Gulf Drive Area/Biltmore Beach accesses: between Safari St. and Holiday Dr., Luff St., between Biltmore Dr. and Luff St., Biltmore Dr., Hurt St., between Irwin St. and Hurt St., Irwin St., Huff St., west end of Spy Glass Dr., Bristol St., and Lookout St.

St. Andrews State Recreation Area is located 3 miles east of Panama City Beach on S.R. 392. This 1,063-acre developed park also includes a primitive area on the western third of Shell Island, accessible only by boat.

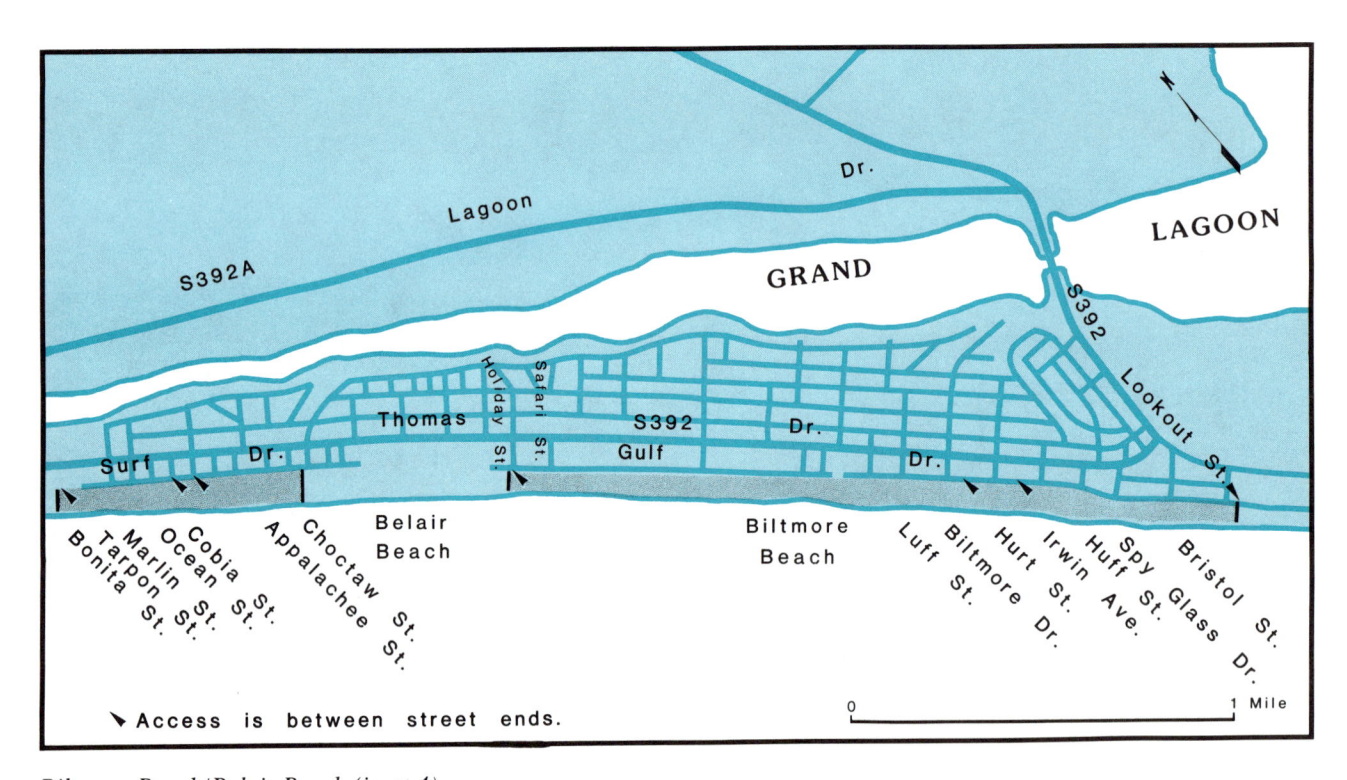

Biltmore Beach/Belair Beach (inset 4)

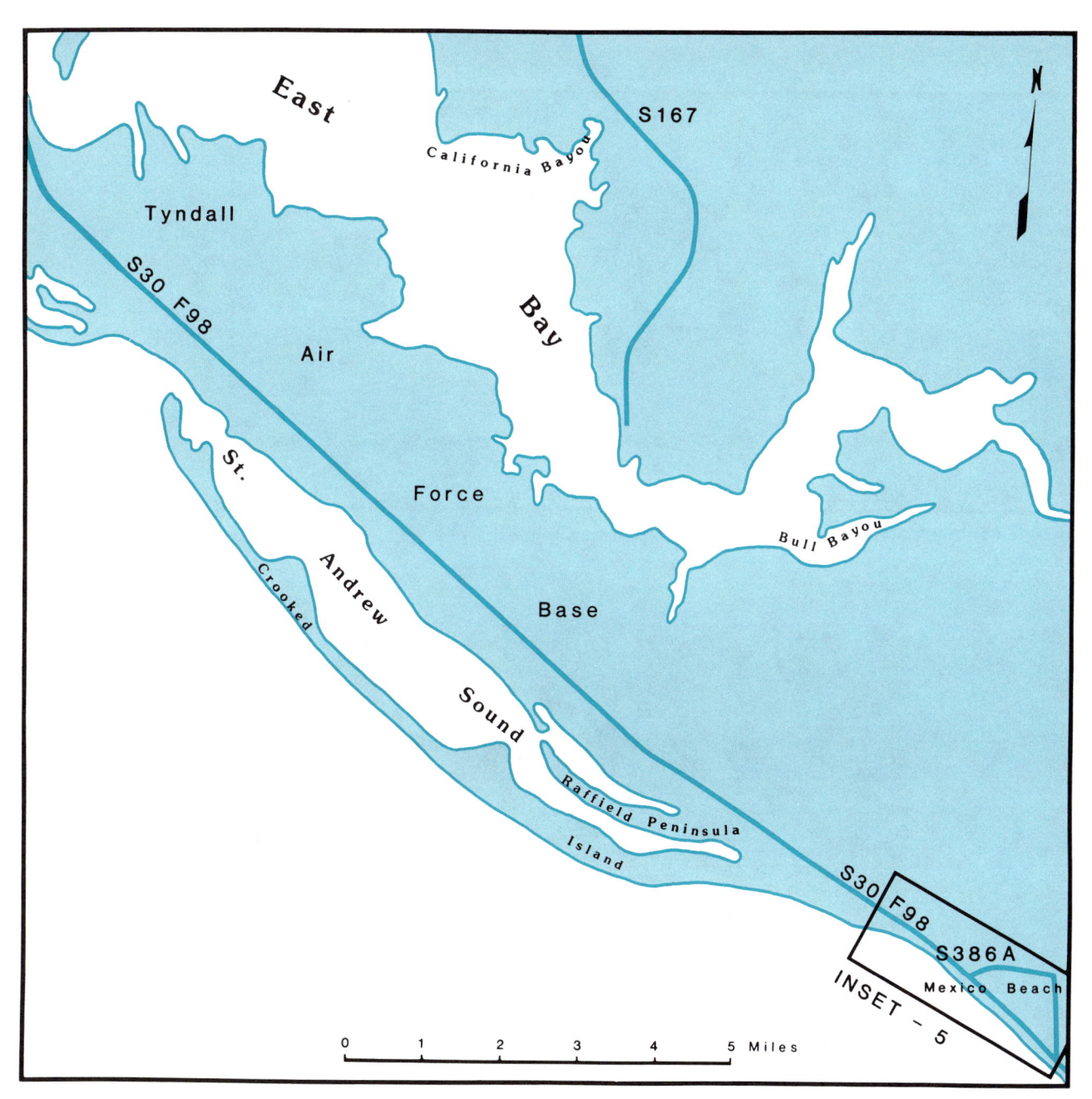

Eastern district

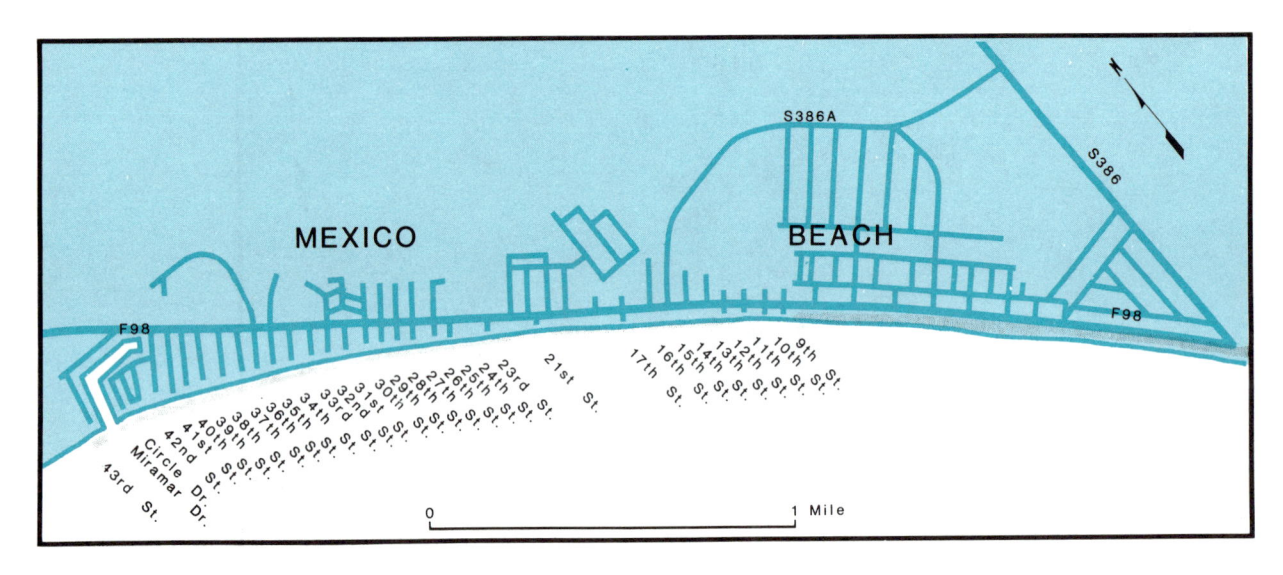

Mexico Beach (inset 5)

Mexico Beach reaches 3 miles west from the Gulf County line. The town has 26 points of access to the Gulf of Mexico: between S.R. 386A and 9th St., between 9th St. and 17th St., 21st St., 23d St., and between 24th St. and 43d St. A fishing pier is located 0.3 mile west of the landing field on 37th St. off S.R. 98.

Dune Grasses

Were it not for sand dunes, many inland areas would lie unprotected from the wind and waves generated by storms. The vulnerable dunes absorb much of the potentially destructive storm energy and help rejuvenate the torn beach with sand.

Formation of the protective dunes occurs under the stressful conditions of salt spray, wind, and heat. Sand, washed onshore, is transported by the wind. The dunes are formed naturally when this blowing sand is trapped by vegetation and gradually accumulates. The difficult process of dune formation is dependent on sand-trapping vegetation growing in this harsh environment.

There are certain conditions necessary before dune-building vegetation can become established. The area must be relatively protected from wind and wave motion, and there must be sufficient organic content in the sand to support vegetation. Once established, the dune-building plants can tolerate occasional water immersion and sand burial. As the sands accumulate around the grass, it grows upward, and an extensive, stabilizing root system develops throughout the dune.

This initial stage of dune formation usually begins in the "pioneer zone," which is closest to the open sea. In Florida there are three common dune-building grasses in this zone. Most common is the sea oat (*Uniola paniculata*), easily identified by the tall, graceful stems of 3 feet or more and seed heads or "oats." This perennial grass has an extensive underground root system and is often used in revegetation projects. Another common dune-building grass is marsh hay or salt meadow cord grass (*Spartina patens*). It is a tall, creeping grass with soft leaves that flower in early summer. Cord grass forms soft, lush meadow that tolerates a slow buildup of sand. The third is dune panic grass (*Panicum amarum*), a clumping grass that flourishes in wet sand and flowers in spring and summer.

After a time, these dune-building grasses provide enough organic material for other types of vegetation to become established and create the next stage of dune formation, the "scrub zone."

Warning

Sea oats are protected by Florida law. They are not to be picked, broken, or trampled.

St. Joe Beach

Gulf County

Florida's first constitution was written in Gulf County. The territorial convention met at the now-vanished city of St. Joseph on December 3, 1838, and remained until the completion of the constitutional draft on January 11, 1839. The State Constitutional Museum, a 13-acre park and museum open to the public 9:00 A.M. to 5:00 P.M. seven days a week, is located on S.R. 71 just east of U.S. 98 in Port St. Joe.

For over 125 years the lighthouse at Cape San Blas has endured storms, floods, and beach erosion. The first lighthouse was built in 1847 as a warning for those sailing near the treacherous shoals that extend for several miles south of the cape. The tower that stands today is a short distance inland from the original site.

Just north of the lighthouse on S.R. 30E is a 20-mile stretch of white sands with striking dune formations, the St. Joseph State Recreation Area. Prehistoric Indians inhabited this peninsula and the area between Port St. Joe and Apalachicola Bay. Here they harvested oysters, scallops, and fish in the salt marshes, lagoons, and deep-channel harbors. Today, these waters are ideal for cast-netting mullet, crabbing, shrimping, and surf-casting for trout and big red fish.

Wewahitchka, the county's earliest permanent settlement, is located in the northeast near the Dead Lakes bordering the Gaskin Wildlife Management Area. The nearby Dead Lakes Recreation Area covering 80 square miles provides excellent freshwater fishing. The park is named for the dead cypress, oak, and pine trees that were drowned by the natural overflow of the Chipola River. The white tupelo gum tree that grows only in the swampy waters of the Dead Lakes makes this area world famous. For over 100 years beekeepers in the river valleys have produced a honey that never granulates or becomes rancid, from the May blossom of the rare white tupelo. This swampy area between the Apalachicola and Chipola rivers annually produces over half a million pounds of honey, making Florida the third largest honey-producing state in the nation.

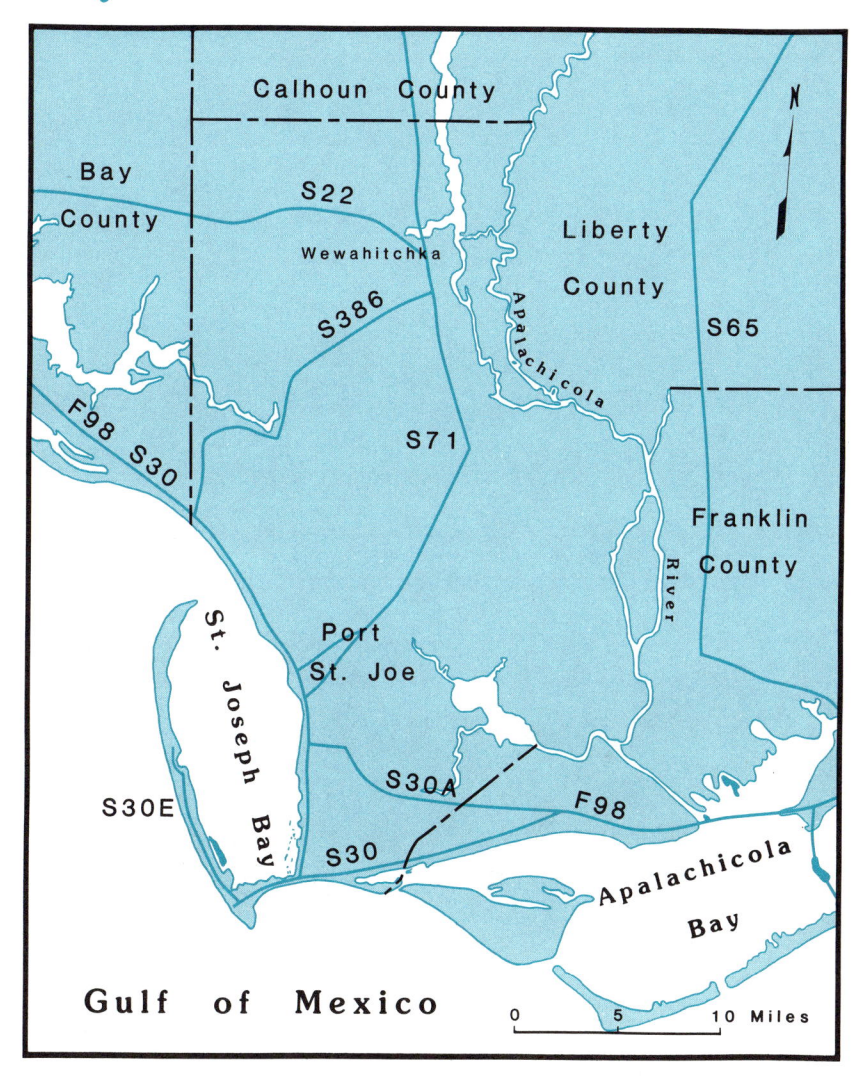

St. Joe peninsula

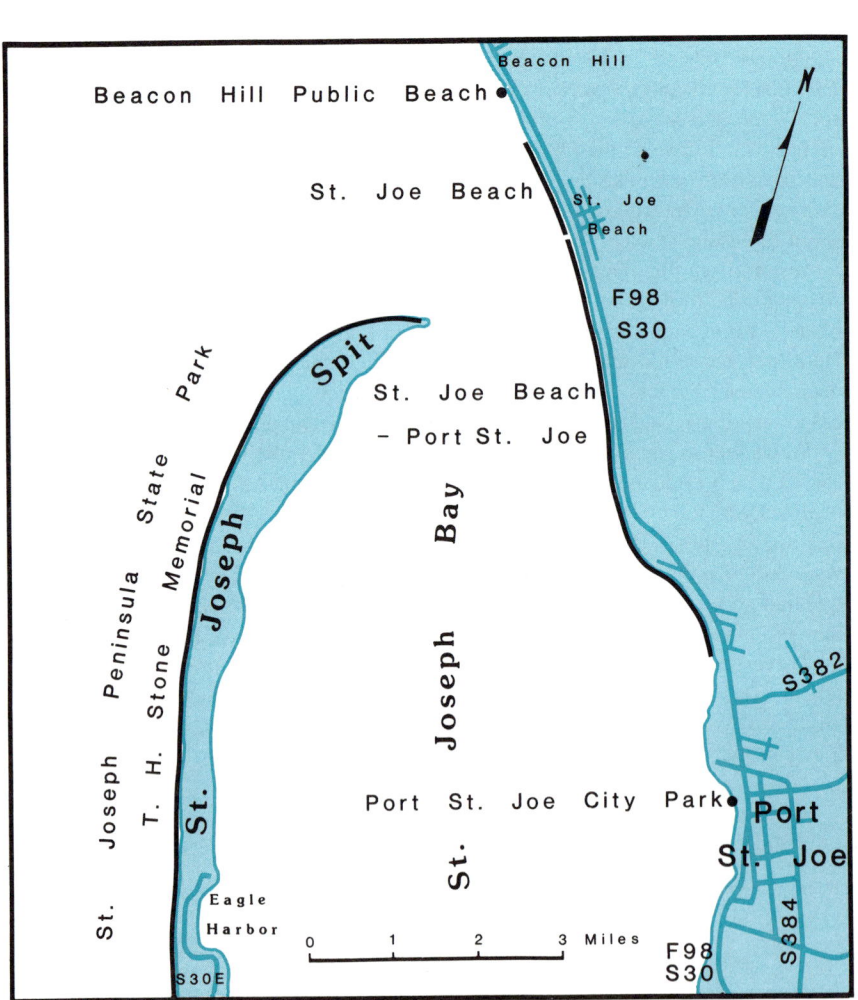

Northern district

Gulf County

NAME	Easy Access	Parking or Entrance Fee	Parking	Restrooms	Showers	Picnicking	Swimming	Lifeguards	Fishing	Boating Facilities	Shelters	Concession Stands	Handicapped Facilities	Public Transportation	Group Facilities	Nature Trails / Fitness	Atlantic / Gulf	Bay / Soundfront	Sandy Beach	Rocky Beach	Primitive Beach	Urban
WESTERN DISTRICT																						
Beacon Hill Public Beach	●		●			● ●		●			●							●	●	●		
St. Joe Beach	●		●			●	●											●	●			
St. Joe Beach–Port St. Joe	●		●			●	●											●	●	●		
Port St. Joe City Park	●		●			●		● ●										●	●			●
EASTERN DISTRICT																						
Indian Pass	●		●				● ●	● ●										●	● ●			
Cape San Blas	●		● ● ●	● ●		●		●		●				● ●				●	●			
St. Joseph Peninsula State Park / T.H. Stone Memorial	● ●	● ● ● ● ● ● ● ● ● ● ● ● ● ● ● ●																●	● ●			

Beacon Hill Public Beach lies 0.7 mile from the Bay County line on U.S. 98. This small, undeveloped beach has side-of-the-road parking only.

St. Joe Beach has 1.2 miles of undeveloped Gulf beach with off-road parking. It is located 1.7 miles southeast of Bay County on S.R. 30 (U.S. 98).

St. Joe Beach/Port St. Joe. This north county coastal area is open beach without facilities. Parking is along S.R. 30 (U.S. 98).

Port St. Joe City Park maintains a small boat-launching ramp on St. Joseph Bay at the end of 5th St.

Indian Pass is located at the eastern end of the Miracle Strip beaches on Indian Peninsula. Situated between St. Vincent's Island and St. Joseph Peninsula State Park, this area is the natural opening from Apalachicola Bay to the Gulf of Mexico.

Cape San Blas is located on the south end of St. Joseph Spit about 14 miles from Port St. Joe. Named for the lighthouse first built in 1847, the existing loran station (long-range aid to navigation) serves as a modern aid to boating.

St. Joseph Peninsula State Park/T. H. Stone Memorial at the north end of S.R. 30E has 20 miles of white-sand beaches, freshwater ponds, and salt marshes. The park is surrounded by water and is characterized by striking dune formations.

Southern district

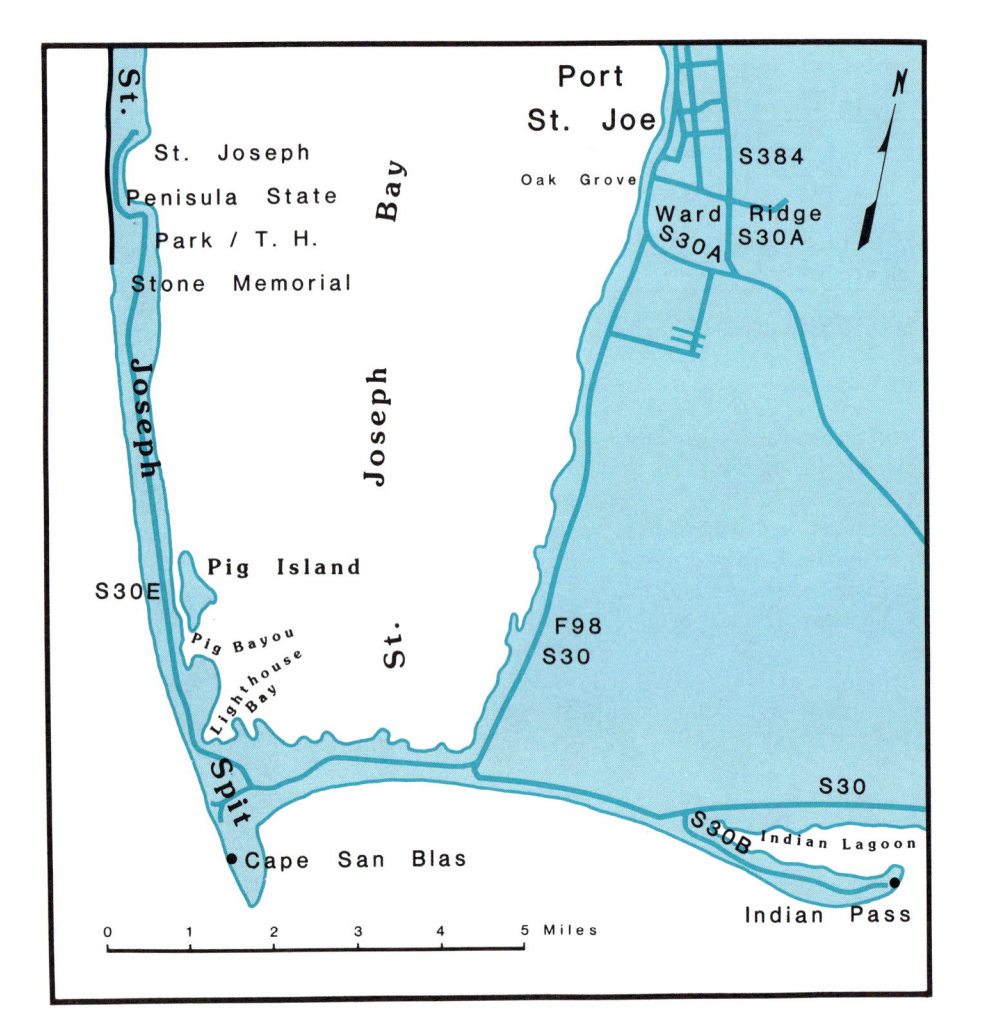

Surf casting

Save Our Coast

Florida has over 8,000 miles of tidal shoreline, more barrier islands than any other state, and one of the fastest growth rates in the nation. Three-fourths of the state's people reside in the 159 cities and 35 counties along the coasts; over 900 persons move to Florida each day. As development occurs, there is a resulting decrease in land available for recreation to satisfy the demand of Florida's residents and the 38 million annual visitors. In addition, the aesthetic features that draw people to the coast become degraded.

To preserve the ecologically sensitive coastal areas, Governor Bob Graham and the Florida Cabinet launched the "Save Our Coast" program in September 1981. The keystone of the program is a $200 million bond program to purchase coastal lands, primarily beaches. The lands selected for acquisition will be used for public recreation in harmony with conservation and preservation objectives. Nominations for parcel proposals have been submitted to the Department of Natural Resources (DNR), Division of Recreation and Parks, by interested citizens, private organizations, and government agencies. Criteria for selection are established by the joint Interagency Management Committee–Outdoor Recreation Advisory Committee and approved by the governor and cabinet.

The acquisition bond debt will be paid from revenues that accrue to the Land Acquisition Trust Fund, created in 1963 by the Outdoor Recreation Advisory Act. Revenue accrues to the fund primarily through a percentage of the Documentary Stamp Tax collected on real estate and other transactions in Florida. The DNR has developed a long-term plan for selling "Save Our Coast" bonds over the next three years. To date, $50 million worth of bonds have been sold.

The Henderson property in Okaloosa County became the state's first purchase on February 2, 1983. The 208-acre parcel extends for over a mile along the Gulf of Mexico. It was purchased for $13.1 million. The land, characterized by sugar-sand beaches and scenic dunes, has been designated as the Burney Henderson State Recreation Area.

Dune walkovers to protect the vegetation

Franklin County

Franklin County, in the Panhandle, is located on Apalachicola Bay and the Gulf of Mexico. The Apalachicola area was the second largest Confederate port during the Civil War. Several houses shelled by federal gunboats at that time still stand.

Like many Florida coastal towns, Apalachicola was plagued by yellow fever that took many lives in the 1830s. In an attempt to cool room temperatures to improve the condition of his patients, Dr. John Gorrie developed an ice-making machine in 1850. Although he died in obscurity in 1855, Dr. Gorrie is one of two Florida men honored in the Statuary Hall in Washington. The early history of Apalachicola and Dr. John Gorrie are shown in the Gorrie State Museum in Apalachicola.

The name Apalachicola is derived from a Creek Indian tribe, the Apalachee. Over 90 percent of the state's oysters are harvested in the area surrounding the Apalachicola harbor, and the name is often associated with these excellent oysters. Each year on the first Saturday in November a seafood festival is held in Apalachicola. Activities include art shows, a parade, entertainment, a grand ball, and a fireworks display. The Apalachicola Bay area is the largest national estuarine sanctuary in the United States.

St. Vincent National Wildlife Refuge, off St. Vincent Sound, is a 12,500-acre island federal reserve. Earlier owners brought exotic species to the island. Some zebra and eland were removed before the refuge was established, but the sunburn deer still remain. All travel to St. Vincent is over coastal water by boat.

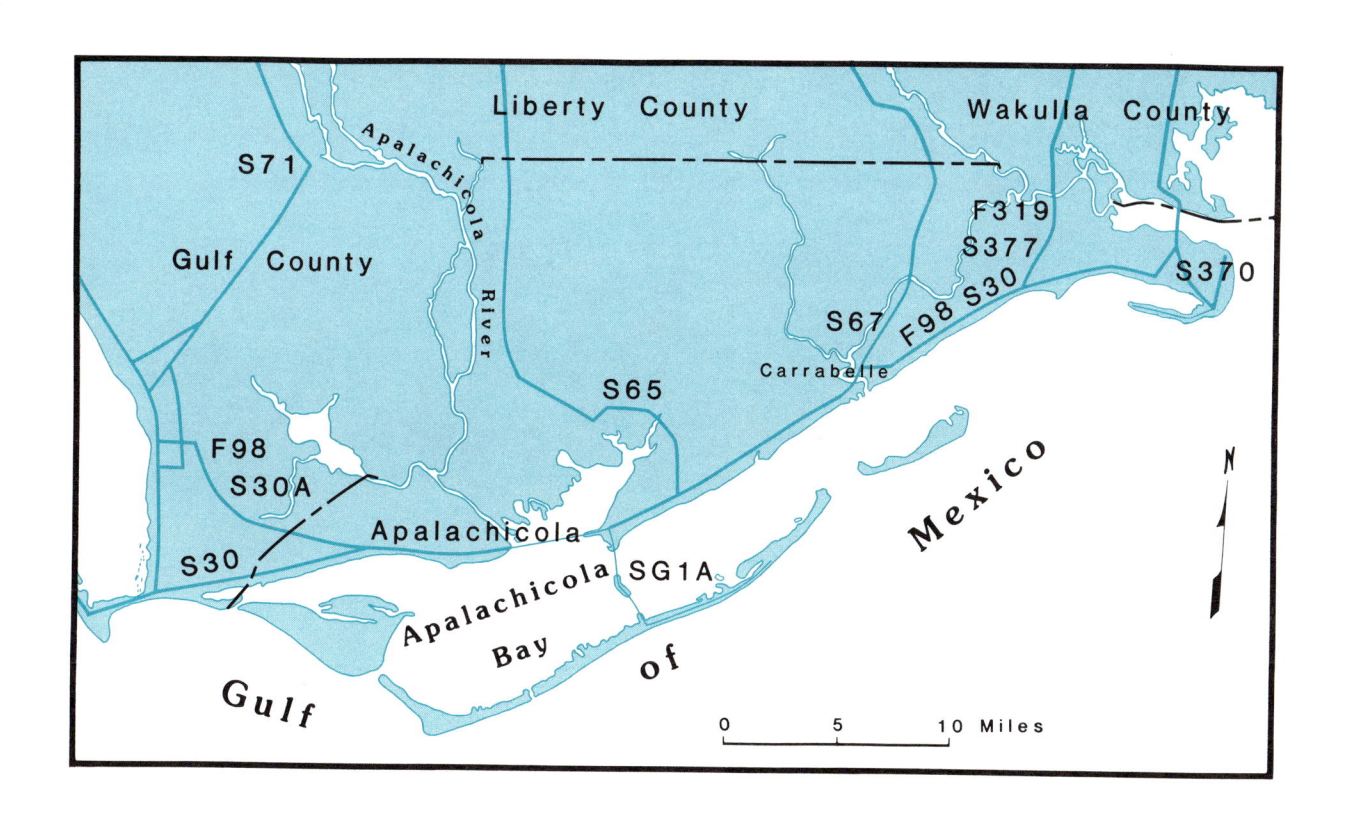

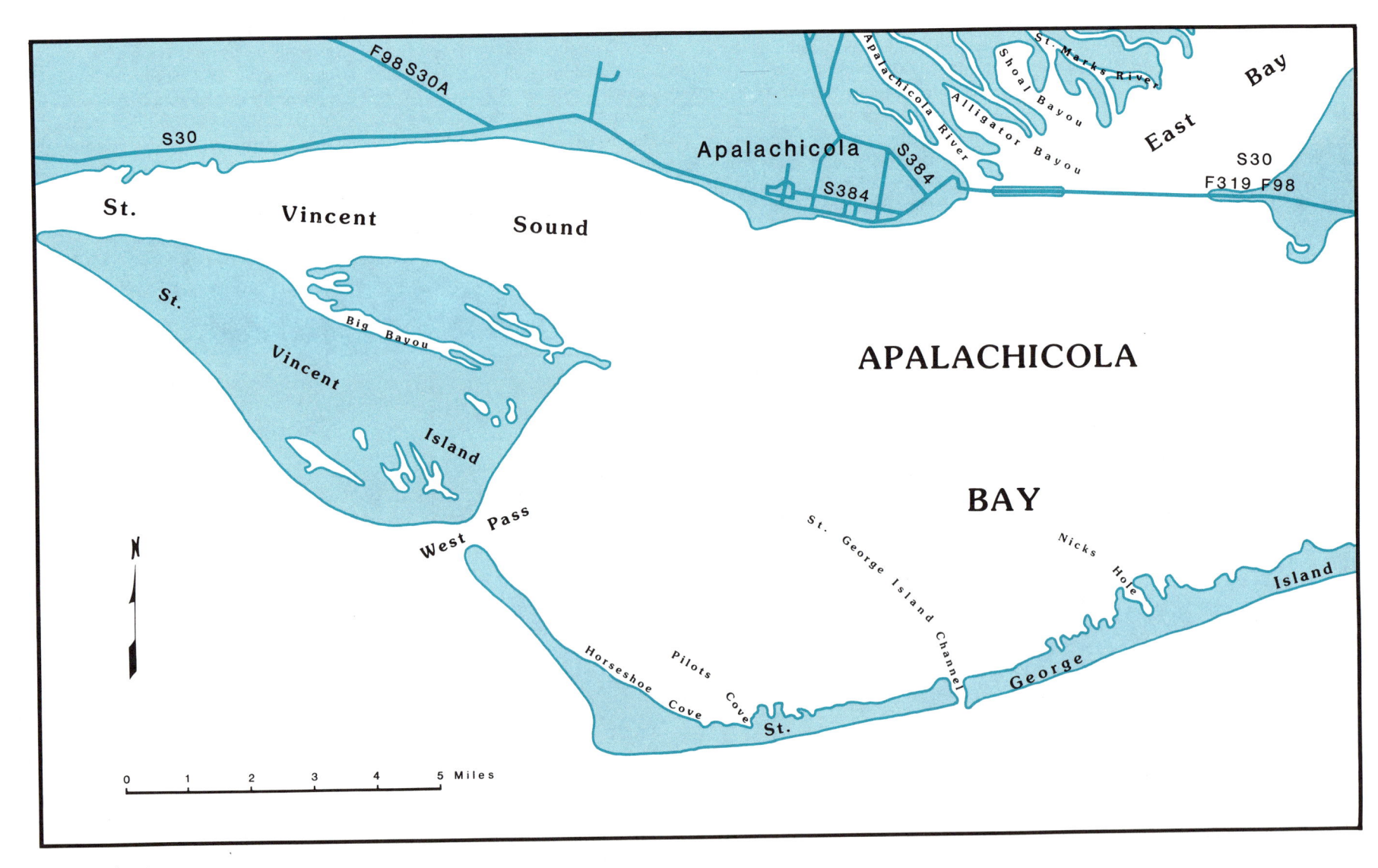

Western district

Franklin County

NAME	Easy Access	Parking or Entrance Fee	Parking	Restrooms	Showers	Picnicking	Swimming	Lifeguards	Fishing	Boating Facilities	Shelters	Concession Stands	Handicapped Facilities	Public Transportation	Group Facilities	Nature Trails / Fitness	Atlantic / Gulf	Bay / Soundfront	Sandy Beach	Rocky Beach	Primitive Beach	Urban
WESTERN DISTRICT																						
St. Vincent National Wildlife Refuge					●		●	●									●	●	●		●	
EASTERN DISTRICT																						
St. George Island			●			●		●									●	●	●			
St. George Island State Park/ Dr. Julian G. Bruce	●	●	●	●	●	●	●	●	●		●	●	●		●	●	●	●	●			
Carrabelle Beach	●		●	●	●		●	●	●			●					●		●			
Dog Island						●		●									●	●	●		●	

St. Vincent's National Wildlife Refuge is located on an unusual triangular-shaped barrier island, 9 miles long, 4 miles wide on the east end, and gradually narrowing to a point on the west end of Indian Pass. All travel to the island is over water; no public transportation is available. Boat launch sites are located at Apalachicola and Indian Pass, 21 miles west of Apalachicola on U.S. 98 (S.R. C 30). There are 14 miles of beaches along the south and east shores and 80 miles of inland trails.

St. George Island, located near Apalachicola and across the bay from East Point, has 29 miles of near-primitive beaches. Bordered by the Gulf on the south and Apalachicola Bay to the north, the white sandy beaches and towering dunes create rare beauty. Linked to the mainland by a bridge and causeway 4.2 miles long, the island is easily reached from U.S. 98/319.

St. George Island State Park/Dr. Julian G. Bruce occupies 1,883 acres at the eastern end of the long, narrow St. George Island. Access to 9 miles of undeveloped beaches and dunes is from U.S. 98/319 at East Point.

Carrabelle Beach, located one mile west of Carrabelle on U.S. 98/319, has a gradual sloping beach on St. George Sound.

Dog Island has isolated white-sand beaches just off the coast of Carrabelle.

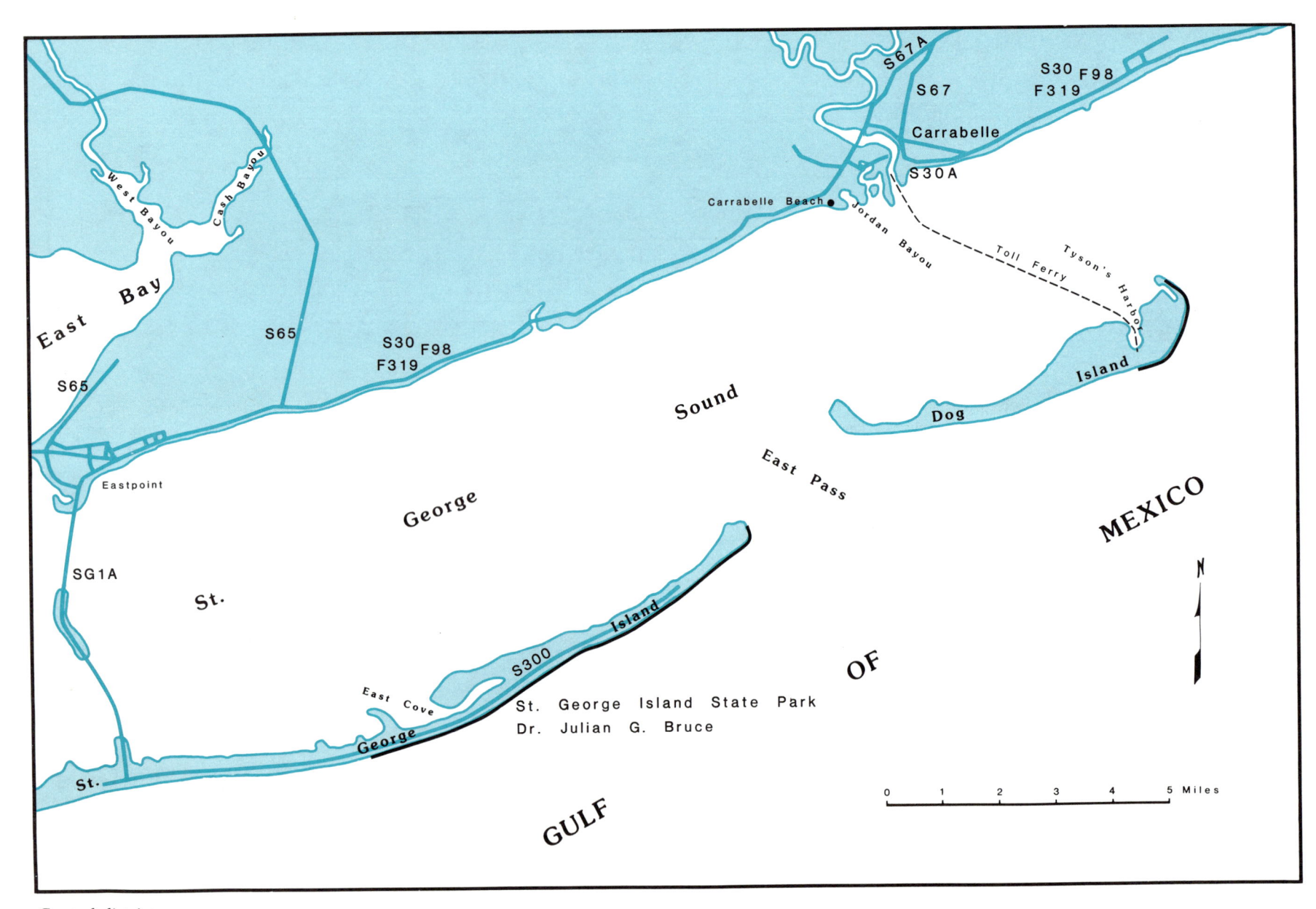

Central district

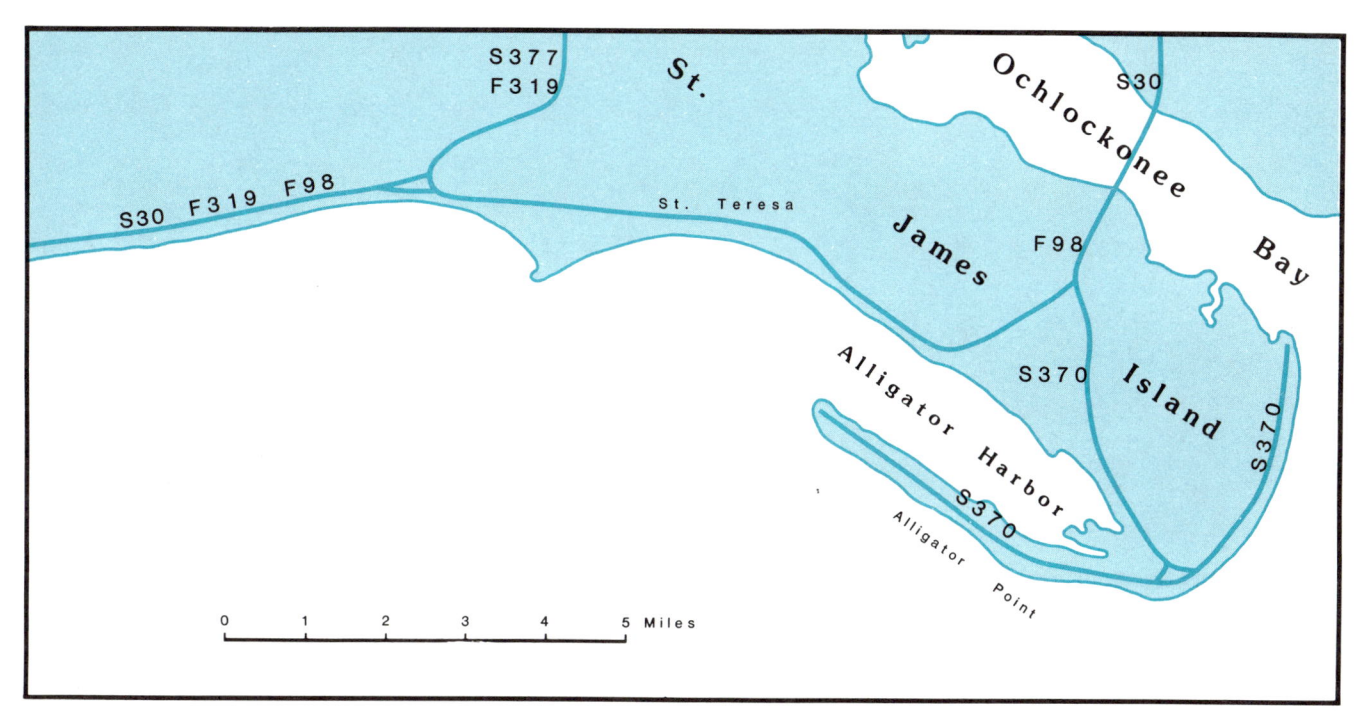

Eastern district

Dog Island

Oyster cluster

Sand dollar

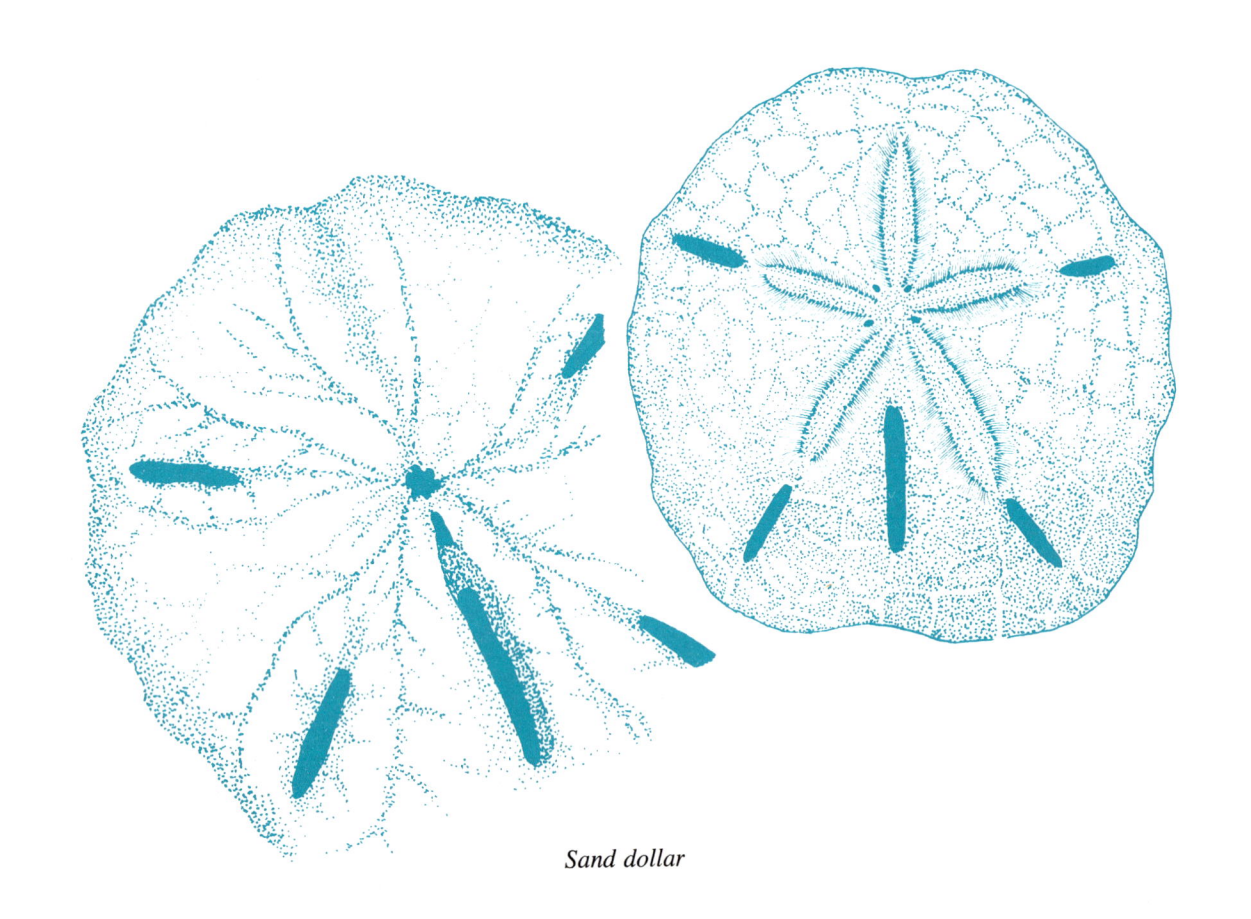

Index

A

Algiers Beach, 154
Ambersand Beach Park, 54
Amelia City Beach, 5
Amelia Island Beach, 3, 4, 5, 13
American Beach, 5
Anastasia State Recreation Area, xviii, xx, 17
Anclote Key, xviii, 117
Anna Maria Beach, 129, 130
Anna Maria Key, 127
Apalachicola, 209
Apollo Beach, xxi, 38
Archibald Memorial Beach, 121
Artificial reefs, 134
Atlantic Beach, 9, 11
Atlantic Dunes Park, 79
Atlantic Intracoastal Waterway, 6
Avalon Park access, 63
A. W. Young Park, 57

B

Bahia-Honda State Recreation Area, xviii, xx, 109
Bal Harbor/Surfside, 93
Banyan Rd. access, 63
Bath Tub Reef, 71
Bay County, xix, 191–200
Bay County Pier and Park, 195
Bayfront Park, 129
Beach erosion and restoration, 89
Beach Highlands, 188
Beacon Hill Public Beach, 205
Beer Can Island, 129
Belair Beach, 197, 198
Belleair Beach, 120
Belleair Shores Beach access, 119
Bicentennial Park, 49
Bicycle laws, 146
Big Cypress National Preserve, 159
Big Lagoon State Recreation Area, xviii, xx, 171
Big Pine Key Park, 109
Bill Baggs–Cape Florida State Recreation Area, xviii, 98
Biltmore Beach, 197, 198
Biscayne National Park, xxii, 91, 112
Black Island, 154
Blackwater River State Park, 169
Blind Creek access, 65
Blind Pass Beach, 141
Blowing Rocks Beach, 77
Blue crabs, 184
Blue Mountain Beach, 188
Boating, 29
Bob Graham Beach, 71
Boca Grande access ways, 150
Boca Grande Beach, 150
Bonita Beach, 154
Bowman's Beach, 154
Boynton Inlet Park, 78
Boynton Public Beach, 79
Bradenton Beach, 132, 133
Brevard County, xix, xxi, 43–51
Brohard Park Beach, 141
Broward County, xviii, xix, 83–88
Brown pelican, 142
Bryn Mawr access, 63
Bunch Beach, 154
Burney Henderson Beach State Recreation Area, 177, 180
Butler Beach, 17

C

Caladesi Island State Park, xviii, 115, 117
Canaveral National Seashore/Apollo Beach, xxi, 38
Canaveral National Seashore/Klondike Beach, xxi, 44
Canaveral National Seashore/Playalinda Beach, xxi, 44
Cape Canaveral access, 47
Cape Canaveral Jetty Park, 44
Cape Florida State Recreation Area, 91
Cape San Blas, 205
Captiva access ways, 150
Carlin Park, 77
Carl Johnson Park, 154
Carrabelle Beach, 211
Casey Key, 135
Casino Beach, 173
Caspersen Park Beach, 141
Cayo Costa Island Park, 149, 150
Cayo Hueso (Bone Key), 103
Charlotte County, 143–46
Charlotte Harbour, 143
Cherie Down Park, 49
Clam Pass/Pelican Bay South, 161
Clarence Higgs Memorial Beach, 107
Clearwater Beach, 118
Clearwater Beach Island, 117
Clearwater Beach Park, 117
Coastal Management Program, 58
Coastal state parks, xvii
Cocoa Beach, 47, 48
Coconut Drive Park, 63
Collier County, xix, 158–64
Collins Park, 95
Conn Beach, 57
Coquina Beach, 129
Coral, 185
Coral Cove Park (Blowing Rocks Beach), 77
Coral reef parks, 112
Corkscrew Sanctuary, 159
Cortez Beach, 129
Crabs, 184
Crandon Park, 91, 98
Crawfish, 184
Crescent Beach, 17

D

Dade County, xviii, 90–100
Dania Beach, 87
Daytona Beach, 13, 23, 31, 33, 35
Daytona Beach Shores, 37
Deerfield Beach, 84, 85
Deerfield Island Park, 84
DeFuniak Springs, 187
Delray Public Beach, 79
DeSoto National Memorial, xxi
Destin, 177
Diver's flag, 185
Dixie Beach, 154
Dog Island, 211
Dolphins, 20
Don Pedro Island complex, 144
Drive-on beaches, 13
Dr. Julian G. Bruce Memorial Park, 211
Dry Tortugas, 103
DuBois Park, 7
Dune Allen Beach (Walline Park), 188
Dunedin Beach, 117
Dune grasses, 201
Duval County, xix, 7–12

E

East Cape, 105
Eastern Keys, 106
Edgewater Beach, 38
Endangered Species Act of 1973, 81
Englewood Beach, 144
Escambia County, xviii, xix, xxi, 170–76
Estero Island, 156
Everglades National Park, xxii, 103, 105, 159

215

Exchange Park, 63
Explosives, ban on, 185

F
Fernandina Beach, xviii, 3, 5
Fishing, saltwater, 40, 183
Flagler Beach Municipal Beach, 25
Flagler Beach State Recreation Area, xviii, xx, 23, 25, 27
Flagler County, xviii, xix, 23–27
Flamingo Visitors' Center, 105
Florida Keys, 103
Flowing Well, 154
Forest zone, 68
Fort Clinch State Park, xviii, xx, 3, 5
Fort DeSoto County Park, 115
Fort Jefferson National Monument, 107, 110
Fort Lauderdale Beach, 83, 86
Fort Matanzas National Monument, xxi, 15
Fort Myers, 149
Fort Myers Beach, 154
Fort Pierce Beach, 61, 62, 63
Fort Pierce Inlet State Recreation Area, xviii, 63
Fort Walton Beach, 177, 182
Fort Walton Temple Mound, 177
Four Mile Village, 188
Frank B. Butler Park, 17
Franklin County, xix, 209–13
Frederick Douglass Memorial Park, 65

G
Garnier Beach, 180
Gasparilla Island, 149, 152
George Smathers Beach, 107
Golden Beach, 93
Government Cut Park, 95
Government Tracking Station, 57

Graveyard Creek, 105
Grayton Beach State Recreation Area, xviii, xx, 187, 190
Gulf County, xix, 202–6
Gulf Drive Area/Biltmore Beach access, 197
Gulf Islands National Seashore/Fort Pickens and Langdon Beach, xxi, 173
Gulf Island National Seashore/Naval Live Oaks, 173
Gulf Islands National Seashore/Okaloosa Area, 177, 180
Gulf Islands National Seashore/Perdido Key and Johnson Beach, xxi, 171
Gulf Islands National Seashore/Santa Rosa Recreational Facility, xxi, 175
Gulfside City Park/Algiers Beach, 154
Gulf Stream County Park, 79

H
Hallandale Beach, 87
Hannah Park, 9
Harry Harris County Park, 107
Haulover Beach, 93
Herman's Bay access, 65
Hobe Sound Beach, 71
Hobe Sound National Wildlife Refuge, 69, 71
Hollywood Beach (Bay Co.), 195
Hollywood Beach (Broward Co.), 87, 88
Holmes Beach, 129, 130, 133
Honeymoon Island State Park, xviii, 117
House of Refuge Beach, 71
Howard Beach (Diamond Head), 77

Howard County Park, 117
Hugh Taylor Birch State Park, xviii, 83, 86
Huguenot Memorial Park, 9
Humiston Beach Park, 57

I
Indialantic, 51
Indian Beach Park, 95
Indian Harbor Beach, 49
Indian Key, 105
Indian Mounds Park, 141
Indian Pass, 205
Indian River County, xix, 53–57
Indian River Shores Walkway, 57
Indian Rocks Beach, 119, 120
Indian Shores Beach access, 119

J
Jack Island State Preserve, 61, 63
Jacksonville Beach, 9, 12
Jaycee Park (Indian River Co.), 57
Jaycee Park (St. Lucie Co.), 63
Jensen Beach Park, 71
John C. Beasley Park, 180
John D. MacArthur State Recreation Area, 75, 77
John F. Kennedy Space Center, 43
John Pennekamp Coral Reef State Park, xviii, xx, 107, 112
Johns Pass Beach and Park, 121
John U. Lloyd Beach State Recreation Area, xix, 87
Jonathan Dickinson State Park, 69
Juno Beach Park, 77
Jupiter Beach Park, 77
Jupiter Island Park (Martin Co.), 71
Jupiter Island Park (Palm Beach Co.), 77

K
Key Biscayne, 98
Key Largo Coral Reef Marine Sanctuary, 112
Key West, 103, 110
Klondike Beach, xxi, 44

L
Laguna Beach/Santa Monica, 195
Lake Worth Beach, 75, 78
Lantana Park, 78
Lauderdale-by-the-Sea, 86
Laws: bicycle, 146; endangered species, 81; mammal protection, 133; saltwater fishing limits, 183; turtle protection, 39
Lee County, 148–56
Lely Barefoot Beach, 161
Lido Beach, 137
Lido Key, 135
Lighthouse Beach Park, 150
Lighthouse Park, 154
Little Duck Key County Park, 109
Little Talbot Island State Park, xix, xx, 9
Liza Jackson Park, 180
Loggerhead Park (Pegasus Park), 77
Loggia Beach, 93
Long Beach, 132
Longboat Key, 127, 135, 136
Longboat Key Beach access, 129
Longboat Key beaches, 137
Long Key State Recreation Area, xix, 107
Long Point Park, 50
Lori Wilson Park, 48
Lostmans Key, 105
Lover's Key Beach/Black Island/Carl Johnson Park, 154
Lowdermilk Park, 161
Lummus Park, 95

M

MacArthur State Recreation Area, 75, 77
McWilliams Park, 57
Madeira Beach, 122
Madeira Beach access, 121
Madeira Beach County Park, 121
Manapalan Beach, 78
Manasota Beach, 135, 141
Manatees, 126, 184
Manatee County, 127–33
Manatee County Beach, 129
Mandalay Park, 117
Mangrove, 165
Manta ray, 184
Marathon Recreation Complex, 109
Marineland Acres, 184
Marine life, dangerous, 28
Marine Mammal Protection Act of 1972, 133
Martin County, xix, 69–72
Martin County Park, 71
Matanzas Beach, 17
Matheson Hammock County Park, 98
Mayport, 7
Melbourne Beach, 51
Merritt Island, 38, 43
Mexico Beach, 200
Miami Beach, 91, 95, 96
Miami Beach Beachfront Park and Promenade, 101
Middle Cape, 105
Middle Cove access, 65
Miramar Beach, 188
Monroe County, xviii, xix, 102–11
Mullet, 67
Mullet Key/Ft. DeSoto Park, 125

N

Naples, 159, 162
Naples Municipal Beach, 161
Nassau County, xviii, 3–5
National beach areas, xxi
Navarre Beach, 176
Navarre Beach Fishing Pier, 175
Navarre Beach public access, 175
Neptune Beach, 9, 11
Newman Brackin Wayside Park/Okaloosa Island Pier, 180
New Smyrna Beach, 33, 37
Nokomis Beach, 141
Normandy Beach access, 65
North Beach, 17
North County Walkway, 54
North Jetty Park (St. Lucie Co.), 63
North Jetty Park (Sarasota Co.), 141
North Lido Beach, 137
North Patrick Beach, 48
North Redington Beach access, 121
North Sand Key, 105
North Shore Ocean Front Park, 95
North Shore Park, 95
Northwest Cape, 105

O

Ocean Beach, 95
Ocean Front Park, 95
Ocean Ridge Hammock Park, 78
Okaloosa County, xx, 177–82
Okaloosa Island Pier, 180
Ormond Beach, 13, 31, 33, 35
Ormand-by-the-Sea, 33, 34

P

Palma Sola Causeway, 129
Palm Beach County, 75–80
Palm Beach Municipal Beach, 78
Palm Beach Shores Park, 77
Palmer Point, 141
Panama City, 191
Panama City Beach, 195, 196
Paradise Beach Park, 46
Park Avenue Park, 57
Park Shore Beach, 161
Pass-a-Grille Beach Park, 125
Pelican Bay South, 161
Pelican Beach Park, 49
Pelicans, other seabirds, 185
Pensacola, 170
Pensacola Beach, 174
Pensacola Beach public access, 173
Pepper Beach State Park and Visitors' Center, xviii, 63
Perdido Key State Preserve, xix, 171
Phillips Inlet Area, 190
Phipps Ocean Park, 78
Pier Park, 95
Pinellas County, xviii, 114–25
Playalinda Beach, xxi, 44
Point Marco Beach, 161
Pompano Beach, 84
Ponce Inlet Park, 33
Ponte Vedra Beach, 13, 17
Porpoises, 184
Port Canaveral Jetty Park, 46
Port Charlotte, 143
Port Charlotte Beach Park, 144
Port St. Joe, 205
Port St. Joe City Park, 205
Punta Gorda, 143
Public access, xv–xvi

Q

Quiet Water Beach, 173

R

Rabbit key, 105
Redington Beach, 122
Redington Shores Beach, 121
Redington Shores County Park, 78, 121
Red Reef Park, 79
Richard G. Kreusler Park, 78
Riverfront Park, 57
Riviera Beach Municipal Park, 77
Rookery Bay National Estuarine Sanctuary, 159
Ross Marler Park, 180
Round Island Park, 57
Royal Palm Way access, 63

S

St. Andrews State Recreation Area, xix, xx, 191, 197
St. Augustine Beach, xxiii, 13, 17
St. George Island, 211
St. George Island State Park/Dr. Julian G. Bruce, xix, xx, 211
St. Joe Beach, 205
St. Joe Beach–Port St. Joe, 205
St. Johns County, xviii, xxii, 15–19
St. Joseph Peninsula State Park/T. H. Stone Memorial, xix, xx, 205
St. Lucie County, xviii, 61–65
St. Lucie Inlet State Recreational Area, xix
St. Petersburg Municipal Beach, 125
St. Vincent National Wildlife Refuge, 209, 211
Saltwater fishing piers, 40, 41
Saltwater fishing laws, 183
Sand Key County Park, 119
Sanibel, 154, 155
Sanibel Island, 149
Santa Monica, 195
Santa Rosa County, 169–76
Santa Rosa Island, 177, 182
Santa Rosa Island/County Beach access, xxi, 180

Sarasota County, xxi, 135–41
Satellite Beach, 49
Save Our Coast, 207
Scrub zone, 201
Seafood, 66
Sea turtles, 39
Seagrove Beach, 190
Seaway Drive access, 63
Sebastian Inlet State Recreation Area, xix, xx, 50, 53, 54
Seminole Blvd. access, 63
Sexton Plaza, 57
Sharks, 67
Shell collecting, southwest coast, 157
Sheppard Park, 48
Shoreline Park, 175
Shrimp, 66
Sidney Fisher Park, 48
Siesta Key, 135, 138
Siesta Key Beach accesses, 137
Siesta Key Beach Park, 137
Silver Beach Wayside Park, 180
Sixty-fifth Street Park, 95

South Beach, 107
South Beach Ocean Park, 63
South Beach Park (Palm Beach Co.), 79
South Beach Park (Indian River Co.), 57
South Cocoa Beach, 48
South County Walkway, 54
South Inlet Park, 79
South Jetty Park, 63
South Lido Beach, 137
South Patrick Beach, 49
South Ponte Vedra, 17
Spanish River Park, 79
Spearfishing, 184
Spessard Holland Park, 50
Stuart Beach, 69, 71
Sunny Isles Beach & Pier, 93
Sunnyside Beach, 195
Sunset Beach, 117
Surf Drive Area/Belair Beach access, 197
Surfside, 93

Surfside Park, 63
Swimming tips, 28
Switlick County Park, 109

T
T. H. Stone Memorial Park, 205
Tigertail Beach, 161
"The Rocks," 23, 25
Treasure Island, 123
Treasure Island Beach, 121
Treasure Island Beach access, 121
Turkey Key, 105
Turner Beach, 154
Turtle Beach, 137
Turtles: marine, 184; sea, 39

U
Underwater archaeology, 73
Unnamed beach (St. Johns Co.), 17
Unnamed beaches (Escambia Co.), 173
Upham Beach, 125
Upper Matecumbe County Park, 107
Usina's Beach, 17

V
Vanderbilt Beach, 161
Venice, 135
Venice Municipal Beach, 141
Vero Beach, 53, 56, 57
Vilano Beach, 17, 18
Virginia Beach, 98
Volusia County, xxi, 31–38

W
Wabasso Beach Park, 54
Walline Park, 188
Walton County, xviii, 187–90
Washington Oaks State Park, xix, 23, 25
Water safety, 28–29
West Coast Intracoastal Waterway, 147
Western Keys, 106
Wiggins Pass State Recreation Area, xix, 161
Wilbur-by-the-Sea, 33, 37